History of Vaccine Development

Stanley A. Plotkin
Editor

History of Vaccine Development

Springer

Editor
Dr. Stanley A. Plotkin
Emeritus Professor of Pediatrics
University of Pennsylvania
Philadelphia, PA
USA
stanley.plotkin@vaxconsult.com

ISBN 978-1-4419-1338-8 e-ISBN 978-1-4419-1339-5
DOI 10.1007/978-1-4419-1339-5
Springer New York Dordrecht Heidelberg London

Library of Congress Control Number: 2011928252

Printed on acid-free paper

Springer is part of Springer Science+Business Media (www.springer.com)

The photograph was taken at a meeting organized by the Pan American Health Organization in 1965 on the subject of vaccination. Among the vaccine experts seen on this photo are Hilary Koprowski (1), Sven Gard (2), Hunein Maasab (6), Geoffrey Edsall (7), Gene Buynak (20), Pierre Lepine (30), John Beale (34), Anton Schwarz (44), Howard Tint (45), Frank Perkins (46), Sam Katz (47), Harry Meyer (49), Ed Buescher (56), Robert Weibel (61), Maurice Hilleman (62), John Fox (70), Paul Parkman (73), Joseph Stokes, Jr. (79), Victor Cabasso (80), Anatol Smorodintsev (83), Saul Krugman (85), Joseph Melnick (to the right of 95), John Enders (96), Henry Kempe (103), Thomas Weller (under 118), Drago Ikic (119), Alastair Dudgeon (121), Leonard Hayflick (to the right of 124) and Werner Henle (134). The editor of this book is number 31.

Contents

Contributors

Robert Austrian[†]
Department of Molecular and Cellular Engineering, University of Pennsylvania School of Medicine, Philadelphia, PA 19104-6088, USA

Ian Bailey[†]
Jenner Educational Trust, The Jenner Museum, The Chantry, Gloucestershire, GL13 9BH, UK

Derrick Baxby
Department of Medical Microbiology, Liverpool University, PO Box 147, Liverpool L69 3BX, UK

Hervé Bazin
Emeritus Professor University of Louvain and 4 rue des Ecoles, 92330 Sceaux, France

A. John Beale[†]
Wellcome Research Laboratories

H. Fred Clark
Children's Hospital of Philadelphia, Philadelphia, PA 19104, USA

Philippe Desmettre
Rhône Mérieux, 254 rue Marcel-Mérieux, 69007 Lyon, France

Frank Fenner[†]
The John Curtin School of Medical Research, The Australian National University, Mills Road, Canberra ACT 2601, Australia

Anne Gershon
Department of Pediatrics, Columbia University, New York, NY, USA

Marina Gheorgiu
Laboratoire du BCG, Institut Pasteur, 25 rue du Dr Roux, 75724 Paris cedex 15, France

Marta Granström
Department of Clinical Microbiology, Karolinska Hospital,
171 76 Stockholm, Sweden

Maurice R. Hilleman†
Merck Institute for Therapeutic Research, Merck Research Laboratories,
West Point, PA 19486, USA

Florian Horaud†
Institut Pasteur, 25, rue du Dr Roux75724, Paris cedex 15, France

Constant Huygelen†
SmithKline Beecham Biologicals, Rixensart, Belgium

Albert Z. Kapikian
Laboratory of Infectious Diseases, National Institute of Allergy and Infectious Diseases, National Institutes of Health, DHHS, Bethesda, MD, USA

Samuel L. Katz
Duke University School of Medicine, Box 2925, Durham, NC 27710, USA

Edwin D. Kilbourne†
Department of Microbiology and Immunology, New York Medical College,
Basic Science Building, Room 315, Valhalla, NY 10595, USA

Hilary Koprowski
Jefferson Cancer Institute, Thomas Jefferson University, 1020 Locus Street,
Philadelphia, PA 19107-6799, USA

Angela Ki Che Leung
Sun Yat-sen Institute for Social Sciences and Philosophy, Academia Sinica,
Taipei, Taiwan

Douglas R. Lowy
National Institutes of Health, Building 37, 9000 Rockville Pike, Bethesda,
MD 20892, USA

Joseph Melnick†
Department of Molecular Virology, Baylor College of Medicine,
One Baylor Plaza, Suite 735, Houston, TX, USA

Thomas P. Monath
Kleiner Perkins Caufield & Byers, Cambridge, MA 02139, USA

Paul A. Offit
Children's Hospital of Philadelphia, Philadelphia, PA 19104, USA

Stanley A. Plotkin
University of Pennsylvania and Vaxconsult,
4650 Wismer Rd., Doylestown, PA 18902, USA

Vince Pozsgay
National Institute of Child Health and Human Development,
National Institutes of Health, Bethesda, MD 20892-2720, USA

Edgar H. Relyveld
6 rue du, Sergent Maginot, 75016 Paris, France

John B. Robbins
National Institute of Child Health and Human Development,
National Institutes of Health, Bethesda, MD 20892-2720, USA

Frederick C. Robbins†
Case Western Reserve University School of Medicine, 10900 Euclid Avenue,
Cleveland, OH 44016, USA

Philippe Sansonetti
Unité de Pathogénie Microbienne Moléculaire, INSERM U389, Institut Pasteur,
28 rue du Docteur-Roux, 75724 Paris, France

John T. Schiller
Laboratory of Cellular Oncology, Center for Cancer Research,
National Cancer Institute, Bethesda, MD, USA

Rachel Schneerson
National Institute of Child Health and Human Development,
National Institutes of Health, Bethesda, MD 20892-2720, USA

Shouson C. Szu
National Institute of Child Health and Human Development,
National Institutes of Health, Bethesda, MD 20892-2720, USA

Catherine M. Wilfert
Duke University School of Medicine, Box 2925, Durham, NC 27710, USA

Introduction

Stanley A. Plotkin

Vaccine development has now entered its fourth century. It is therefore, time to look back and consider the history of the field, which is now long and illustrious. In 1995, I organized a meeting in Paris that brought together a number of people who were intimately familiar with particular vaccines, and who in some cases were the actual developers of those vaccines, to talk about how those vaccines were developed. Their talks were transcribed and edited but never received wide dissemination. In this book, I have brought together those talks. Regrettably, many of the writers who recounted their personal achievements are now dead, but that makes these documents even more important.

S.A. Plotkin (✉)
University of Pennsylvania and Vaxconsult,
4650 Wismer Rd., Doylestown, PA 18902, USA
e-mail: stanley.plotkin@vaxconsult.com

S.A. Plotkin (ed.), *History of Vaccine Development*,
DOI 10.1007/978-1-4419-1339-5_1, © Springer Science+Business Media, LLC 2011

Table 1 gives the history of vaccine development, divided into live and attenuated vaccines, although the distinction between the two is beginning to blur with the development of vectored vaccines.

Active immunization began in China, or in India with the practice of variolation, in which the smallpox virus itself, given artificially, prevented people from developing scarring from natural smallpox, although inevitably some variolated individuals died from the inoculation itself (see Chap. 2). The dawn of vaccinology came with the observations of Edward Jenner as to the efficacy of cowpox (or some virus related to cowpox, the identity of which is still uncertain) in preventing subsequent smallpox (Chaps. 3–5). This was the beginning of live, attenuated vaccines. More than 80 years later, Louis Pasteur found means of attenuating organisms in the laboratory (Chap. 6). Apparently, the first organism attenuated simply by ageing on the laboratory bench, was the agent of fowl cholera, now called *Pasteurella multocida*. Pasteur and his colleagues then studied heat, desiccation, exposure to oxygen, and passage in atypical host species as means to attenuate anthrax bacilli and rabies virus.

The next signal advance in vaccine development occurred later in the nineteenth century in the USA and in Pasteur's Institute, and that was the chemical inactivation of whole bacteria. Daniel Salmon and Theobald Smith described the principle by inactivating a *Salmonella* (later named after Salmon) that caused disease in pigs. This work and that of the French group led eventually to vaccines against typhoid, plague, and cholera, and subsequently pertussis, all based on inactivated whole bacilli (see Chaps. 9 and 10).

Progress continued in the first half of the twentieth century, based first on the Pastorian method. Two powerful vaccines developed at that time were Bacille Calmette Guérin for tuberculosis, which was a Mycobacterium bovis passaged in artificial culture medium by Albert Calmette and Camille Guérin; and yellow fever, which was a virus adapted to growth first in mouse brain and then in the chorioallantois of chicken eggs by Max Theiler (Chaps. 7 and 14). Later in the century Herald Cox used the embryonated egg to develop a vaccine against the rickettsial disease, typhus. In addition, the discoveries of Emil Behring, Emile Roux, and Shibasuro Kitasato relative to toxin production by the diphtheria and tetanus bacilli permitted Gaston Ramon to inactivate the toxins with formalin to produce what are now called toxoids. The toxoid vaccines rapidly controlled diphtheria and tetanus (Chap. 8).

Just after World War II the technology of cells grown in vitro for virus cultivation was demonstrated by Enders, Weller, and Robbins and then built upon by many other researchers (Chaps. 16 and 17). Virus culture permitted the development of numerous vaccines, including inactivated polio, live polio, measles, mumps, rubella, adenovirus, varicella, and later on rotavirus and zoster (Chaps. 18–26). It also permitted a switch from crude rabies vaccines grown in animal brain or embryonated eggs to a more refined and potent cell culture vaccine (Chap. 13). Japanese encephalitis and tick-borne encephalitis vaccines were also first developed in animal tissue and then switched more recently to cell culture. Hepatitis A is an example of a whole inactivated virus vaccine, similar to the inactivated polio vaccine (Chap. 25).

Table 1 Outline of the development of human vaccines

	Live attenuated	Killed whole organisms	Purified proteins of organisms or polysaccharides	Reassortants	Genetically engineered
Eighteenth century	Smallpox				
Nineteenth century	Rabies	Typhoid			
		Cholera			
		Plague			
Early 20th century	BCG (Tuberculosis)	Pertussis (whole cell)			
	Yellow fever	Influenza	Diphtheria		
		Rickettsia	Tetanus		
Late 20th century	Polio (OPV)	Influenza	Pneumococcus	Influenza (killed + live)	Hepatitis B recombinant
					Cholera toxin
					Pertussis toxin
	Measles	Polio (IPV)	Meningococcus		
	Mumps	Rabies (new)	Hepatitis B (plasma derived)		
	Rubella (ca)	Anthrax			
	Adenovirus	Japanese encephalitis	*H. influenzae*		
	Typhoid Ty12a	Hepatitis A	Typhoid (Vi)		
	Varicella	Tick-borne encephalitis	Pertussis (acellular)		
	Cholera CVD103	*E.coli*(+ CTB)	*H. influenzae*, meningococcus (protein conjugated)		
Twenty-first century	Zoster		Pneumococcal conjugate	Rotavirus	Human papillomavirus

ca=cold adapted
CTB=cholera toxin B

On the bacterial side advances were made through two means: identification of capsular polysaccharides or other components that could immunize without the remainder of the bacilli, and the discovery that conjugation with protein could greatly increase the immunogenicity of polysaccharides (Chaps. 11 and 12). Thus, powerful vaccines have been developed against the three major causes of meningitis in children and invasive infections in adults: *Haemophilus influenzae* type b, meningococci and pneumococci. For each of the three pathogens vaccines were developed both as polysaccharides and as protein-conjugated polysaccharides. For *H. influenzae* type b the conjugate vaccine has completely replaced the polysaccharide because infants do not respond to the latter; for meningococcal infections both types of vaccines are in use; and for pneumococcus the polysaccharide vaccine is given to the elderly whereas the conjugate vaccine is given to infants. However, in the pneumococcal case the serotypes contained in the two vaccines are different.

A capsular polysaccharide from the typhoid bacillus is used to vaccinate against that disease (Chap. 9).

In the way of vaccines made from purified proteins, aside from the diphtheria and tetanus toxoids, there are three made from naturally produced substances: the original hepatitis B surface antigen vaccine made from the plasma of infected donors (Chap. 25), the anthrax vaccine made from secreted protective antigen, and acellular pertussis vaccines containing 1–5 components of the organism (Chap. 10). In addition, most influenza vaccines depend on the viral hemagglutinogen protein, and thus consist of more or less purified proteins (Chap. 15).

Genetic engineering, first pioneered by Stanley Cohen and Herbert Boyer in 1973, has become the preferred way to produce vaccine antigens. At the moment, four vaccines contain proteins produced by genetic engineering: the current hepatitis B surface antigen vaccine (Chap. 25), the human papillomavirus L1 virus-like particle (Chap. 27), the Lyme OspA subunit vaccine, and the recombinant B component of toxin present in an oral cholera vaccine. Although not dependent on recombination, one of the two vaccines developed against human rotaviruses depends on reassortment of RNA segments coding for viral proteins from animal and human viruses in order to develop attenuated strains for vaccination (Chaps. 28 and 29).

Finally, although this book concerns mainly the development of vaccines for humans, it should not be forgotten that the history of veterinary vaccine development is rich and also worth telling (Chap. 30).

Thus, as we enter the fourth century of vaccination we have more tools than ever to develop vaccines, but still their successful development depends on the vision of scientists and physicians, many of whom recount their own stories in this volume.

"Variolation" and Vaccination in Late Imperial China, Ca 1570–1911

Angela Ki Che Leung

The first reliable record of smallpox in China can be dated to the fifteenth century[1] [1]. By early Song period, around the tenth century, smallpox had become essentially a childhood disease, but it remained one of the most fatal childhood diseases until the nineteenth century.

Variolation using human pox against smallpox in China was one of the ancient popular inoculation practices existing in different parts of the world before Jennerian vaccination [2]. This chapter deals with its historical development and its importance in the introduction of Jennerian vaccination in the country during the early nineteenth century.

The Development of Variolation in China

The practice of variolation in China can be documented to the seventeenth century, and be traced back to the sixteenth century.[2] Joseph Needham's claim that the method could be dated to the tenth century [3] cannot be substantiated by any reliable sources. In the sixteenth and early seventeenth century, variolation made its way rather slowly. The first extant and available written record which actually described variolation was in a 1695 medical book by Zhang Lu (1617–?), a famous doctor. He noticed that variolation, a technique "bestowed by a Taoist immortal," was first practiced in Jiangxi (right bank of the lower Yangzi River), and spread all over the country during his time [4].

[1]There are different hypotheses as to the time when smallpox was first recorded in China. The most authoritative argument remains that of Fan Xingzhun, who dates the first specific record of smallpox to the fifth century.

[2]Several variolation practitioners of the seventeenth century claimed that the technique was invented by a sixteenth century doctor. It is very likely that variolation was practiced in the sixteenth century before it was written down in the following century.

A.K.C. Leung (✉)
Sun Yat-sen Institute for Social Sciences and Philosophy, Academia Sinica, Taipei, Taiwan
e-mail: kcleung@arts.cuhk.hk

S.A. Plotkin (ed.), *History of Vaccine Development*,
DOI 10.1007/978-1-4419-1339-5_2, © Springer Science+Business Media, LLC 2011

He described three methods of variolation: putting a piece of cotton imbued with pox pus into the nostril of the healthy child, using squama the same way when a fresh pustule was not available, and making the healthy child wear clothes that had been worn by a child who had contracted the disease. After the child was thus variolated, he would have fever in about 7 days, with a slight and benign case of smallpox [4]. This would prevent the child from getting smallpox again.

The techniques were increasingly refined in the eighteenth century during which a greater number of medical books on smallpox with descriptions of variolation were published. A 1713 work described a fourth method using powdered squama blown into the nostril through a thin silver tube. This was said to be convenient for bringing the techniques to remote areas [5]. These four standard methods were later described in great details in the 1742 medical compendium endorsed by the Imperial court [6]. To a great extent, this compendium "legitimized" variolation's position in orthodox medicine which had until then snubbed the technique as being peculiar.

By the end of the eighteenth century, variolation was even divided into two schools, the Huzhou school (Zhejiang) which preferred the use of fresh pus, claiming that it was more effective and the other school was the Songjiang school (Jiangsu) which preferred to use older, medically-treated squama: "cooked pox," claiming it was safer [7].

The main reason for inhalation as the dominant variolation method was the belief that through the respiratory system, the effect of variolation could, starting with the *fei* (pulmonic orb), successively reach the five viscera (*zang*)[3] and circulate within them. The affected five *zang*, considered to be impregnated with innate toxic matters would, in about 7 days, release a "toxin" and produce external signs (fever, pox, thirst, etc.). The signs would gradually diminish as the poison was duly liberated by the variolated matter, in about 20 days. The elimination of such poison, it was believed, would prevent the person from getting smallpox again in his life.

Upper classes of the society seemed receptive to variolation before the imperial "recognition" in the mid-eighteenth century, though the progress was slow and gradual. This headway was revealed by both literary and medical sources [3, 8, 9]. Some late seventeenth century and eighteenth-century variolation experts wrote that they acquired the technique from their fathers or grandfathers who had inoculated thousands of children in their lifetime. The *Pére d'Entrecolles,* a Jesuit living in Peking in the early eighteenth century thought, probably after being told by *des médecins du palais*, that it had been in practice in China for about a century [7, 10, 11].

The rapid spread of variolation in the eighteenth century was likely to be a result of its effectiveness. One variolator of the time, Zhang Yan, boasted that out of the 8–9,000 persons he had inoculated, merely 20–30 died. Zhu Chungu, the expert who began to inoculate the Machu imperial court in the late seventeenth century even said that the technique had never failed [12–16].

[3]The five viscera are: *xin* (orb of the heart), *gan* (hepatic orb). *pi* (splenic/pancreatic orb), *fei* (pulmonic orb), *shen* (renal orb).

For this reason perhaps, in the mid eighteenth century, at about the same time variolation was "legitimized" by the imperial court: many literati, especially those of the Lower Yangzi region,[4] strongly recommended the technique in their writings, using the experience of their family as an illustration. A contemporary Japanese doctor was told by his Chinese colleagues that 80–90% of China's well-off families had their children inoculated [17]. Though such figures cannot be taken at their face value, they certainly reflected popularity of the technique, at least among the upper classes.

Variolation finally began to reach the poorer classes only in the beginning of the nineteenth century, just before Jennerian vaccination was popularized. At least one charitable institution in southern China began to provide the service free of charge around 1807 [18].

Variolation Practiced by the Manchu Imperial Family

The Manchus, like the Mongolians and Tibetans, were more vulnerable to smallpox than the Chinese, especially as they left their sparsely populated original habitat and entered densely populated Chinese cities like Peking. Manchu troops died of smallpox in great numbers during and after the wars of conquest in the early half of the seventeenth century. Various draconian quarantine strategies were thus taken to protect the imperial family from contracting the disease in the seventeenth century, though they did not prevent the first Manchu Emperor Shunzhi of dying from it in 1662[5] [16, 19].

Given such background, it is not difficult to understand why the second Emperor, Kangxi (1662–1722), was so intent in fighting the disease. In 1681, he summoned two famous Jiangxi smallpox experts (one of whom, the above mentioned Zhu Chungu) to the court to variolate the royal family and banner troops stationed in Manchuria and Mongolia. We know that Kangxi's policy was maintained long after his death as lists of variolated children of banner troops can be found later in the eighteenth century.[6] Some scholars even think that variolation may have something

[4] An eighteenth century variolator regretted that not many northerners were inoculated, so thousands of children died during epidemics. Such tragedies were much less frequent in the south [9].

[5] These strategies include creating sites for seclusion (*biduosuo*) during smallpox epidemics, setting-up "smallpox secretariat" to handle the banishment of all smallpox patients thirteen miles from the city wall with their families, forbidding those members of the imperial family who had never had smallpox to enter the capital. When the first Manchu emperor died at 23 of smallpox in 1662, the Kangxi Emperor was chosen to be the successor and not his elder brother precisely because he had had smallpox as a child and had a better chance to have a longer reign.

[6] One very interesting list is found in box 4717 of the "Imperial Pharmacy" section of the Qing Archives in the No 1 Archives in Beijing. This box contains documents dated 1744, 1749, and 1755. Though this list is not dated it should be of the mid-eighteenth century. It contained 73 names of inoculated children of the red and white banner troops stationed in Chahar in Manchuria, the oldest of whom was 18 *sui*, the youngest 3 *sui*, implying that these Manchu children were inoculated at a much older age than Chinese children.

to do with the long-term decline in infant mortality of the Manchu nobility in the eighteenth century[7] [20].

Kangxi's choice for variolation to protect the nobility promoted the position of the technique in the medical orthodoxy. However, despite the court's interest, there was never a national policy to apply variolation against smallpox.

Introduction of Jennerian Vaccination

In the Spring of 1805, the vaccine was carried on live subjects from Manila to Macau by a Portuguese merchant, Hewit. From Macau, Jennerian vaccination was introduced to China in Canton [21]. In this same place, a cowpox vaccination bureau was immediately established – under auspices of Alexander Pearson, surgeon of the East-India Company's factory at Canton – by some Cantonese merchants and medical experts who began to study and apply the technique. Pearson also wrote a tract on the subject, which was translated into Chinese as *Zhongdou qishu (Wonder book on inoculation)*, and published to popularize vaccination[8] [22]. Pearson noticed that this technique had been met with "fewer obstacles from prejudice than could be anticipated, especially in a Chinese community." In the first 12 months of the introduction of vaccination in Canton, thousands were inoculated and Chinese doctors or merchants who were associated with the Company soon became vaccinators [23].

One of the technical difficulties in the practicing antismallpox vaccination was the preservation of the vaccine. By 1816, it had already been twice extinct in Canton, and the "hope that … the vaccine might be found upon the cows in some of the remoter province proved fallacious," as Pearson observed [24]. Despite the obstacle, due to the effort of a number of Chinese enthusiasts, vaccination rapidly spread. By the early 1820s it was popular in other provinces through merchant guilds, concerned officials, and private individuals[9] [25].

[7]Lee, Wang, and Campbell claim that child mortality of the Machu nobility fell from 400 hundred per 1,000 during the early eighteenth century, to 100 and below by the late eighteenth century, at the same time, life expectancy at birth doubled from the low twenties to the high forties. They suspect that variolation could have contributed to this change.

[8]One of the earliest account of the activities of the Canton establishment is by Rev. William Milne, in his *Life in China,* London, 1859

[9]A report by Pearson on 19 March 1821 stated that vaccination had, by that time, spread to the provinces on Jiangxi (Kiangsi), Fujian (Fukien), Jiangsu (Kiangsu), and reached Beijing. The French surgeon was sent by Vannier, Minister of Cochinchina. *Ibid*, p 40.

The Chinese historian Chen Yuan (1880–1971) found out from Chinese sources that vaccination was spread chronologically in the 1820s from Canton to Hunan, Peking, Fujian, Jiangsu, Jiangxi, Sichuan. Some vaccinators paid wet nurses with infants to travel from one spot to another to transmit the vaccine arm to arm.

Unlike variolations, Jennerian vaccination was first tried on the poorer classes[10] [26]. Chinese indigenous charitable institutions, especially foundling homes, soon provided free vaccination, sometimes alongside with traditional variolation, as a service for the community (In the 1840s, many foundling homes vaccinated children of the district) [27].

One of the first charitable vaccination organizations was established in Nanking in 1834–1835. This "Vaccination Bureau" (*niudou ju*) was officially set up during a smallpox epidemic of the winter. Another early bureau was established by a scholar-official of the Weixian district of Shandong who sent some dozen children with their parents to Perking around 1833 to transmit the vaccine back home from arm to arm [28]. The vogue was only temporary halted by the Taiping upheaval in the 1850s. As soon as peace was restored in the 1860s, the spread of charitable vaccination bureaus quickly regained momentum. At least 43 vaccination bureaus were set up from the 1860s till the end of the Qing in 1911 all over the country.[11]

By the mid-nineteenth century, general rules about vaccination were already well-known and observed by charitable bureaus: vaccinators were urged to make the preservation of the vaccine a priority; children with skin diseases were not to be vaccinated; special attention was to be paid to avoid patients with leprosy; 4–5 days after vaccination, the child was to be inspected by the vaccinator, the healthy pustule of about 8 or 9 days was to be transmitted to other infants as vaccine; poor families were sometimes paid to have their infants vaccinated so that the vaccine would not become extinct; expenses were paid by merchant guilds, donations by local officials, notables, shops, and sometime by miscellaneous taxes [28, 29].

The acceptance of Jennerian vaccination by the Chinese society was relatively quick (less than 50 years) when we compare it with the slow progress of variolation (over a century). The reason is that the cultural and psychological block hindering the initial spread of variolation had already broken down when vaccination was introduced. For many, the two techniques were similar. In fact, for reason of convenience or technical difficulties, some of the late nineteenth century institutions provided both variolation and vaccination to fight smallpox (a foundling home in the Shanghai stated in 1883 that children would be variolated or vaccinated in the Spring) [30].

[10]Pearson reported in 1816 on the first vaccination in 1805–1806, "it was from the beginning conducted…by inoculation at stated periods among the native, and of them, necessarily, the poorest classes, who dwelt crowded together in boats or otherwise … "

[11]On the number of bureaus: See Angela Leung, "Charitable institutions of the Ming and Qing". Unpublished research report, National Science Council, Taipei, Taiwan, 1991. I have used more than 2,000 local gazetteers to count different types of charitable institutions in this project. For vaccination bureaus the preliminary count in 1991 was 34, but after recent checking, I found out at least nine bureaus had been erroneously left out. This is certainly still an underestimation.

Acceptance of Jennerain Vaccination

However, the Chinese did not accept Jennerian vaccination exactly as it was understood in the West. There was a construction of the Chinese interpretation of vaccination, deciphering it effectiveness in terms conforming to Chinese orthodox medical thought. Basically, the classic notion o *taidu* (foetal toxin), to which the principle of variolation and vaccination was accommodated, persisted. According to this concept, toxic matters from the father and the mother – a result of physical desire, emotional instability, or unbalanced nutritional habits – were inevitably passed onto the fetus the moment it was conceived. The toxin would express itself at one moment or another during the lifetime of the child. Smallpox, measles, chickenpox, all sorts of skin eruptions, boils or ulcerations, were different manifestations of *taidu*. Vaccination, like variolation, was a way of controlled release and elimination of the *taidu* before any occurrence of smallpox epidemic.

The principle for traditional variolation by inhalation is explained above. Chinese vaccinators justified the incision method of vaccination by borrowing from principles of meridian points in acupuncture. The two spots on each arm where the vaccine was to be injected were controlling the "five viscera and six bowels" (*wuzang liufu*).[12] Some vaccinators prescribed ways to measure the whereabouts of the spots (e.g., the first spot was at the length of the middle finger of the child up to the elbow, the other was at the palm's length from the first spot up the arm) [31]. In other words, the vaccine injected into the "correct" reflexive points would most effectively liberate *taidu* deep inside the body [32].

Moreover, Chinese vaccinators preserved certain traditional rituals: boys were to be vaccinated on their left arm first, and girls on their right arm. As for variolation, spring and winter were sometimes recommended as better seasons for vaccination. Postvaccination care, including the taking and application of medicine for the release of "remaining toxin," was also very similar to postvariolation care. Some early vaccinators even recommended the squama for the preservation of the vaccine as in variolation [33]. One of them, Deng Liu (1774–1842), suggested that powdered squama mixed with milk could be used as vaccine [34]. Jennerian vaccination was thus conceived as an improved version of variolation, perfectly understandable in Chinese medical terms. Vaccination therefore reinforced rather than changed the Chinese etiology of smallpox: it remained a disease caused by innate factors.

The sinicized vocabulary of vaccination, the familiar explanation of the way it worked made it easier for Chinese social elites and the general public to accept the western technique. Very rarely was it seen as an instrument of Western imperialism. When compared to opium, another importation from the West, many admitted that

---

[12] *Sanjiao*, the biggest of the six bowels (*liufu*), consisted of the three portions of the body cavity, commanding the circulation of fluid and air (*fi*). Some western medical doctors believe that the reflexive points commanding the *sanjiao* actually are controlling glandular excretion. The two reflexive points corresponding to *sanjiao* are called the *xiaoshuo* and the *finglengyuan*.

while one was detrimental to health, the other was unquestionable beneficial [35]. However, despite the initial ease of its introduction in early nineteenth century, vaccination had then only achieved its first step into China. It still had a few hurdles to clear: the licensing of vaccinators necessitating the institutional recognition of the technique (as late as 1909, scholars observed that "doctors of our country do not know how to vaccinate, and vaccinators are not doctors, how strange it is") [36], the uninterrupted supply of the vaccine (several vaccination bureaus noted that parents still needed constant persuasion to have their infant vaccinated; winter was a particularly poor season as parents were hesitant to bare the arms of the child. Thus, it was usually during the winter that the vaccine became extinct) [29, 37–39], just to name a few. These difficulties could only be solved much later in the twentieth century when China imported not only western technique, but also the medical thought and institutions that came with it.

References

1. Xingzhun. *Zhonggou yufang yixue sixiang shi* (The history of medical thought on prevention in China). Shanghai 1953;106–10
2. Moulin AM. *Le dernier langage de la medicine. Histoire de l'immunologie de Pasteur au sida.* Paris: 1991;21–2
3. Needham J. *China and the Origins of Immunology*. Hong Kong: 1980;6
4. Zhang Lu. *Zhangshi yitong* (the comprehensive book of medicine by Zhang Lu), 1695. Shanghai, 1990 reprint, 697
5. Zhu Chungu (1637-?). *Douzen dinglun* (Decisive discussion on smallpox) 1767 ed (first edition 1713), chap 2, 25a–b, 28a
6. *Yizon jinjian* (Golden mirror of the medical tradition), 1742. Beijing, 1990 (1963) reprint, chap 60
7. Zhu Yiliang. *Zhongdou xinfa* (Precious methods of variolation) 1808;2a–b
8. *Zengzi tong* (Comprehensive dictionary) compiled 1627, published 1671, chap *wu*, 9b
9. Leung AK. *Ming-Qing yufang tianhua zhi coushi* (Preventative measures against smallpox in the Ming and the Qing periods), *Guoshi shilunI.* Taipei: 1987;244
10. Zhu Yiliang. *Zhongdou xinfa* (Precious methods of variolation) 1808;244
11. Le Pére d'Entrecolles. La petite vérole, letter au reverend Père du Halde, *Pékin* 1726. In: *Lettres édifiantes et curieses de Chine par des missionaires jésuites 1702–1776.* Paris 1979;330–341
12. Zhu Chungu (1637-?). *Douzen dinglun* (Decisive discussion on smallpox) 1767 ed (first edition 1713), chap 2, 27b
13. Zhu Chungu (1637-?). *Douzen dinglun* (Decisive discussion on smallpox) 1767 ed (first edition 1713), chap 3, 3a
14. Leung AK. *Ming-Qing yufang tianhua zhi coushi* (Preventative measures against smallpox in the Ming and the Qing periods), *Guoshi shilunI.* Taipei: 1987;245
15. Zhang Yan, *Zhongdou xinshu* (New book on variolation) (1760 ed with a 1741 preface), chap 3, 3a
16. Chang Chia-feng. Strategies of dealing with smallpox in the early Qing Imperial family. Jami & Skar, eds. *East Asian Science: Tradition and Beyond*, Osaka, Hashimoto: 1995;199–205
17. Leung AK. *Ming-Qing yufang tianhua zhi coushi* (Preventative measures against smallpox in the Ming and the Qing periods), *Guoshi shilunI.* Taipei: 1987;245
18. Leung AK. *Ming-Qing yufang tianhua zhi coushi* (Preventative measures against smallpox in the Ming and the Qing periods), *Guoshi shilunI.* Taipei: 1987;246

19. Leung AK. *Ming-Qing yufang tianhua zhi coushi* (Preventative measures against smallpox in the Ming and the Qing periods), *Guoshi shilunI.* Taipei: 1987;247
20. Lee, Wang Campbell. Infant and child mortality among the Qing nobility: implications of two type of positive check. *Population Studies* 48/3. 1994;398:401–2
21. Pearson A. Vaccination. 1816. *Chinese Repository,* vol 2, May 1833;36
22. Milne W. *Life in China.* London: 1859;56–7
23. Milne W. *Life in China.* London: 1859;37–8
24. Milne W. *Life in China.* London: 1859;37
25. Chen Yuan. Niudou ru Zhongguo kao (On the introduction of Jennerian vaccination in China), *Yixue weisheng bao.* 6–7, Dec 1908 – Jan 1909, reprinted in *Chen Yuan zaonian wenji* (Early works of Chen Yuan). Taipei 1992;221
26. Pearson A. Vaccination. 1816. *Chinese Repository,* vol 2, May 1833;37
27. Leung AK. *Ming-Qing yufang tianhua zhi coushi* (Preventative measures against smallpox in the Ming and the Qing periods), *Guoshi shilunI.* Taipei: 1987;250
28. *Niudou ju changcheng* (Regulations of the vaccination bureau) *Jiangningfu chongjian puyutang zhi* (Monograph on the charitable institution in Nanking), 1871, chap 5, 13a; 2. *Weixian zhi* (Gazetteer of Weixian) 1941, chap 29, 24a; 3. Xu Dong, *Muling shu* (Book for the magistrate), 1848, chap 15, 29a–b
29. Xu Dong, *op cit*, chap 15,29a; *Baoshan xian xuzhi* (Sequal of the Gazetteer of Baoshan) 1921, chap 11, 2b-3a; *Jiangsu shengli* (Provincial rules of the Jiangsu province) 1876, vol 1, 9a; *Songjiang fu xuzhi* (Sequal of the Gazetteer of Songjiang) 1883, chap 9, 14a; *Deyi lu* (A record of philanthropic deeds) 1896, chap 3, 6b; *Juangning fu chongjian puyutana, op cit,* chap 4, 14a
30. *Songjiang fu xuchi* (Gazetteer of the Songjiang prefecture) 1883, chap 9, 14a. The same for the institute in Hongjiang, Hunan, see *Hongjiang yuying xiaoshi* (Account of the foundling institution in Hongjiang) 1888, chap 2 on vaccination, 4a
31. *Zhongxi douke hebi* (Combination of the Chinese and Western ways to treat smallpox), Shanghai 1929, 62
32. *Songjiang fu xuchi* (Gazetteer of the Songjiang prefecture) 1883, chap 9, 14a. The same for the institute in Hongjiang, Hunan, see *Hongjiang yuying xiaoshi* (Account of the foundling institution in Hongjiang) 1888, chap 2 on vaccination, 1b
33. *Niudou ju changcheng* (Regulations of the vaccination bureau) *Jiangningfu chongjian puyutang zhi* (Monograph on the charitable institution in Nanking), 1871, chap 5, 13a; 2. *Weixian zhi* (Gazetteer of Weixian) 1941, chap 29, 24a; 3. Xu Dong, *Muling shu* (Book for the magistrate), 1848, chap 3, 29a
34. Yang Jiamao. Deng Liu he niudou jiezhong fa (Deng Liu and vaccination). In: *Zhonghua yishi zazhi,* 1986; 16/4, 221
35. Chen Yuan. *Ba Ruan Yuan yindou shi* (Postscriptum to Ruan Yuan's poem on vaccination), 1908. Reprinted in Chen Yuan, *op cit,* 216
36. Chen Yuan. Yisheng canpo doushi zhuce (Registration of doctors, midwives and vaccinators). In: *Yixue weisheng pao.* 1909, reprinted in Chen Yuan 1992, *op cit,* 249–50
37. Xu Dong, *op cit*, chap 15,29a; *Baoshan xian xuzhi* (Sequal of the Gazetteer of Baoshan) 1921, chap 11, 2b-3a; *Jiangsu shengli* (Provincial rules of the Jiangsu province) 1876, vol 1, 9a; *Songjiang fu xuzhi* (Sequal of the Gazetteer of Songjiang) 1883, chap 9, 14a; *Deyi lu* (A record of philanthropic deeds) 1896, chap 3, 6b; *Juangning fu chongjian puyutana, op cit*, chap 5, 13a–14a
38. *Songjiang fu xuchi* (Gazetteer of the Songjiang prefecture) 1883, chap 9, 14a. The same for the institute in Hongjiang, Hunan, see *Hongjiang yuying xiaoshi* (Account of the foundling institution in Hongjiang) 1888, chap 2 on vaccination, 4a
39. *Hongjiang yuying xiaoshi. Op cit*; chap 2 (On vaccination), 9a; *Jiangning fu chongjian puyutang zhi. Op cit*, chap 5, 14a

Edward Jenner's Role in the Introduction of Smallpox Vaccine

Derrick Baxby

Introduction

Although Jenner's name is universally linked with smallpox vaccination, there had always been controversy about his role in its introduction, with over-enthusiastic supporters provoking those who would minimise his influence. Critics include those who advance the claims of earlier "discoveries" of vaccination, those who believe Jenner's vaccine was merely attenuated smallpox virus and that he simply employed a safe type of variolation (smallpox inoculation), and those who maintain that others did the bulk of the work, which established vaccination [1, 2]. Assessments of Jenner have varied from simple country doctor to genius by supporters and from misguided fool to deliberate deceiver by critics. The centenary of his first vaccination was marked by particularly partisan views [3], and the bicentenary provided an opportunity to attempt a more objective assessment. The original literature is not cited; it has been analysed elsewhere in some detail [4, 5]. Jenner performed his first vaccination on May 14, 1796, 3 days before his 47th birthday. As an experienced physician and surgeon trained in London, by John Hunter, he had a general practice in the Vale of Berkeley and a consultant practice in Cheltenham, to which his London friends referred patients who visited the fashionable spa. He had been elected to the Fellowship of the Royal Society in 1789 for his observations on the habits of the newly-hatched cuckoo. Thus, although he preferred country life, he was a well-trained and respected doctor-scientist with a wide circle of interests and friends.

D. Baxby (✉)
Department of Medical Microbiology, Liverpool University,
PO Box 147, Liverpool L69 3BX, UK

S.A. Plotkin (ed.), *History of Vaccine Development*,
DOI 10.1007/978-1-4419-1339-5_3, © Springer Science+Business Media, LLC 2011

Early Observations

The idea that cowpox, a mild localised disease, conferred immunity to smallpox was widespread in rural areas. Jenner, an experienced country doctor and variolator, would inevitably find this of interest and he collected information on it, particularly from the early 1780s [4]. Some cases of cowpox had occurred years before and were not seen by Jenner, but he did see other cases from 1782 onwards. Information about immunity to smallpox was gradually collected during routine variolations; many done by Jenner during 1792–1797, others by his nephew and assistant Henry Jenner. In all, Jenner collected data on 28 individuals, which provided the epidemiological evidence on which he based his hypothesis. They represented the cases where cowpox had occurred recently or many years before, where it had been acquired directly from the cow or a horse, and where immunity to smallpox was detected by natural exposure and/or variolation. Jenner appreciated the value of making his observations in an area where smallpox was uncommon, and of ensuring as best he could that any immunity was due to cowpox and not previous smallpox or variolation. Although circumstantial, these epidemiological dates clearly demonstrate Jenner's ability to formulate and modify hypotheses in the light of accumulated evidence [4, 5]. However, they needed testing by experiment.

Jenner's Vaccinations

It is unlikely that Jenner was the first person to inoculate cowpox. Claims were later made on behalf of those who were said to have priority. Particular attention was paid to Benjamin Jesty, a Dorsetshire farmer, who is reliably believe to have inoculated his wife and family with cowpox in 1774 [2]. However, all these early claims came to light only after Jenner published in *Inquiry*, and so had no influence on the theory or practice of medicine.

On May 14, 1796, Jenner took material from a lesion on the hand of Sarah Nelmes (Fig. 1), who had become infected from her master's cows, and inoculated two sites on the arm of 8-year-old James Phipps. The lesions developed in about a week, Phipps had some slight indisposition, but recovered uneventfully. Jenner variolated him on July 1, but "no disease followed." At this stage, Jenner submitted his work to the Royal Society for the publication. They declined and suggested that he need more information [5]. Jenner collected more epidemiological basis of his claims. He published his results in a monograph at his own expense. Generally referred to as the *Inquiry*, its full title (*An Inquiry into the Causes and Effects of the Variola Vaccinae*) gave no indication of what particular "effects" were described. He invented a Latin name for cowpox, *variolae vaccinae* (smallpox of the cow), and illustrated the monograph with four hand-coloured engravings which showed the hand of Sarah Nelmes and vaccinated lesions on the arms of three patients. At first, the terms "cowpox inoculation" and "vaccine inoculation" were to be used to describe the process, but these soon gave way to *vaccination*.

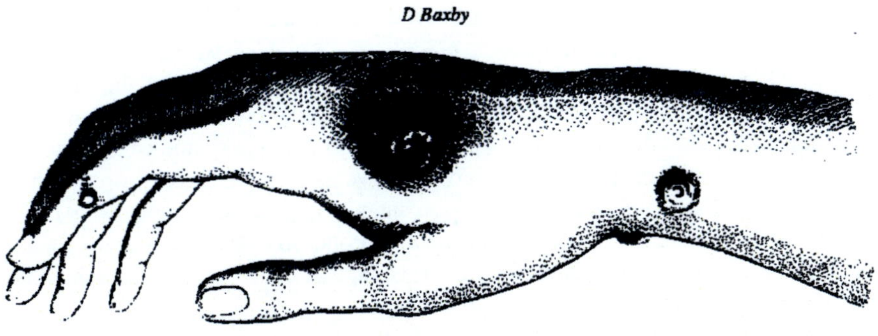

Fig. 1 The hand of Sarah Nelmes from Jenner's *Inquiry*

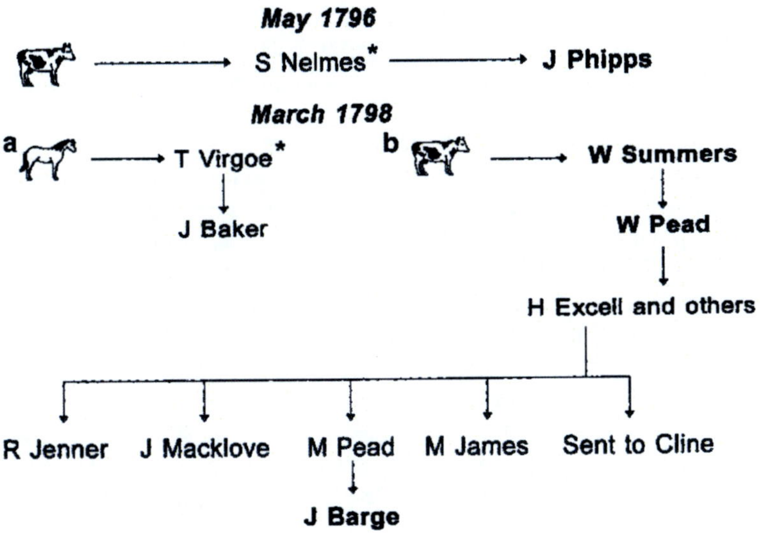

Fig. 2 Jenner's vaccination described in the *Inquiry*. *Asterisk* accidental infection. Patients name in *bold* type resisted variolation

Of all the information in the *Inquiry*, the series of vaccinations starting with William Summers is most easily understood (Fig. 2). Summers was vaccinated with material taken from a cow, later variolated and did not develop smallpox; we therefore know that Jenner's starting material was genuine. William Pead and Barge, the second and last of the series, were subsequently variolated by Henry Jenner. A transient local lesion developed, which began to subside by the 4th day. The same material was successfully used to variolate someone who had never had smallpox, an important check on the activity of the smallpox virus used. These results showed that the vaccine could be passed at least four times by arm-to-arm transfer without altering its effects. This would considerably reduce dependence on animals as a source of vaccine [4].

However, the results still left many unanswered questions. The number vaccinated was still small; ten named individuals plus those vaccinated with Hannah Excell (Fig. 2). Only four were variolated, and this was done quite soon after the vaccinations and gave no realistic idea of how long immunity lasted. Jenner sought to overcome these and other deficiencies by including in the *Inquiry*, the circumstantial information on the 28 cases briefly mentioned above. In general, Jenner proposed that smallpox could be prevented by cowpox inoculation, which could be passed arm-to-arm. Also, unlike variolation, vaccination produced a lesion only at the site of inoculation, did not cause serious illness or death, and could not be transmitted to contacts by "effluvia."

One controversial claim was that the origin of his vaccine was an equine disease called "grease," transferred to cows when horse handlers occasionally helped with milking. Jenner believed that those infected from cows were more reliably protected that those infected from horses. His evidence was circumstantial, but the idea caught his imagination and he persisted in the belief. Jenner's most extravagant claims "that the person who had been thus affected is for ever after secure from the infection of the Small-Pox" was wishful thinking. The evidence in which this was based was the few individuals whose natural cowpox infection had occurred many years before they resisted variolation or naturally-acquired smallpox. Jenner was familiar with smallpox and variolation and would have known that variolation was no absolute protection, and that rare second attacks of smallpox could occur. He had no good reason to believe cowpox would be more effective.

Jenner made his findings freely available, but for anyone hoping to confirm them, it was essential that Jenner should provide good description and/or illustrations of bovine and human cowpox, and of inoculated cowpox. Here, the *Inquiry* was deficient because, although the description of bovine cowpox was good, no illustration was provided. Jenner also created confusion by stressing the similarity between the lesions of cowpox and inoculated smallpox. Although this might have been an attempt to attract the variolators, the appearance of the lesions was in fact quite different [4].

Jenner, recognising that his work was incomplete, ended the *Inquiry*: "I shall myself continue to prosecute this inquiry, encouraged by the hope of its becoming essentially beneficial to mankind." He was later to foresee its use in the eradication of smallpox.

True and Spurious Cowpox

The following year (1799), attempting to overcome deficiencies in the *Inquiry*, Jenner published his *Further Observations on the Variolae Vaccinae*, and came very close to deserving the status of genius. Jenner used the term "true cowpox" to describe material from the cow and which gave the expected result, i.e., a relatively mild infection which remains localised, healed uneventfully, and conferred immunity to smallpox. In contrast, "spurious cowpox" did not give the expected result and by the time *Further Observations* was published, he had recognised four distinct varieties [4].

The first referred to pustules on the cow: "which pustules contain no specific virus." This was Jenner's recognition of other bovine infections transmissible to humans, but which did not produce immunity to smallpox. One such infection (Milker's nodes) is still a common occupational hazard of farm workers.

The second described "matter (although originally possessing the specific virus), which has suffered a decomposition through putrefaction." Here, Jenner recognised that true cowpox could be rendered ineffective by improper storage, and that any lesion would be cause by other "spurious" matter. Jenner knew nothing about bacteria, but in modern terms, he was recognising inactivation of cowpox virus by contaminating bacteria and the production of bacterial abscesses.

The third variety was "matter taken from an ulcer in an advanced state, which ulcer arose from a true Cow Pock." In modern terms, this was recognition that material from an old cowpox lesion was likely to contain little or no cowpox virus, but was probably contaminated with bacteria.

Jenner's fourth variety was produced by "peculiar morbid matter generated by a horse." Here again is Jenner's belief that true cowpox originated in horses, but had to be passed via cows to be fully effective. It was his attempt to explain the variability of some of his results without being aware that different viruses might be involved.

The Jenner's analysis of true and spurious cowpox was almost universally criticised, in some cases, to the end of the nineteenth century [6], shows how far advanced this concept was. In particular, his recognition of the first three varieties of spurious cowpox shows powers of observation, insight, and deduction of a very high order. CW Dixon, smallpox authority and no particular supporter of Jenner, described the analysis of true and spurious cowpox as "quite masterly", and in 1962 wrote: "His (Jenner's) capacity to visualise the properties of a specific infective element distinguished from that producing the ordinary septic lesions and capable of loss through defective storage or not being present in a lesion too advanced, is quite remarkable" [7].

Jenner in Perspective

Jenner's seminal studies were completed by 1799, but he was soon involved in an argument with William Woodville, whose vaccination trial could have been definitive [4]. However, it took place in a Smallpox Hospital and many vaccines acquired smallpox during the trial, either from contaminated vaccine or naturally. Jenner's later activities were largely confined to supplying vaccine and advice, and defending vaccination in general and priority in particular. Unfortunately however, although he had private doubts, he maintained in public to the end that vaccination would confer complete protection, and misused his concept of spurious cowpox to explain failures [4].

Others such as Pearson who challenged Jenner's priority; Sacco of Italy, de Carro of Vienna, and Ballhorn and Stromeyer of Hanover who all supported Jenner did much to extend and confirm Jenner's initial observations. Vaccines were established from cattle and horses in circumstances where smallpox virus was not yet involved [4]. These, and modern laboratory studies, indicated that smallpox vaccine was not simple attenuated smallpox virus, and that Jenner's work marked a radical new departure. In fact, cowpox virus also differs from modern smallpox vaccine (vaccinia virus) [8], and it is possible that the latter represents laboratory survival of the now naturally extinct horsepox [4].

Strong anti-vaccination movements soon developed, but Jenner's crucial achievement was acknowledged by such distinguished contemporaries as Lettsom, Dimsdale, Haygarth and Willan, and also by British and foreign governments and organisations. Later, Pasteur's tribute was to propose that "vaccination" be used to describe any immunisation, a practice still common. Jenner's work was not immediately followed by vaccines for other diseases for obvious, but largely ignored, reasons. Until Pasteur developed laboratory methods for producing vaccines, suitable material could only be obtained from easily recognised skin lesions. Smallpox and cowpox were eminently suitable, but other diseases provided a stark contrast in this respect.

Conclusion

Although Jenner's first vaccinations were few, they were without known precedent, carefully controlled and based on careful epidemiological observations. He was dogmatic and made errors, but he established the potential of arm-to-arm vaccination with a safe animal virus. Much remained to be done to establish the safety and efficacy of vaccination, but Jenner made public the crucial first observations and experiments. Genius is described as "extraordinary capacity for imaginative creation, original thought, invention or discovery," and Jenner's investigations of the milkmaid's story and analyses of true and spurious cowpox were surely this "extraordinary." Perhaps the final words should be left to the authors and *Smallpox and its Eradication* [8]: "promulgation by Jenner of the idea of vaccination with a virus other than variola virus constituted a watershed in the control of smallpox, for which he more than anyone else deserved the credit."

References

1. Edward Jenner: *The History of a Medical Myth*. Firle: Caliban, 1977
2. Horton R. Jenner did not discover vaccination. *Br Med J* 1995;310:62
3. Jenner Centenary Number. *Br Med J* 1896;i:1246–314
4. Baxby D. *Jenner's Smallpox Vaccine: the Riddle of Vaccinia Virus and its Origin*. London: Heinemann Educational Books, 1981

5. Baxby D. The genesis of Edward Jenner's *Inquiry* of 1798: a comparison of the two unpublished manuscripts and the published version. *Med Hist* 1985:29:193–9
6. Creighton C. *Jenner and Vaccination.* London: Sonnenschein, 1889
7. Dixon CW. *Smallpox.* London: Churchill, 1962
8. Fenner F, Henderson DA, Arita I, Jezek Z, Ladnyi ID. *Smallpox and its Eradication.* Geneva: World Health Organization, 1988

Edward Jenner, Benefactor to Mankind

Ian Bailey†

The story of vaccine begins with "Blossom", a mahogany-brown old Gloucester cow, one of the oldest dairy breeds in England, and now one of the rarest. Her portrait is in the Jenner Museum, Berkeley, Gloucestershire, and her hide is on the wall of the library of St. George's Hospital, London. Blossom and her milkmaid, Sarah Nelmes, were infected with cowpox. On 14 May 1796, Edward Jenner (Fig. 1), an apothecary-surgeon in Berkeley, took material from the sore on her hand and inoculated James Phipps, a healthy 8-year-old boy, through two superficial incisions, each about half an inch in length. There was a mild illness between the 7th and 9th day. A vesicle formed and died away without giving the least trouble. On 1 July, the boy was inoculated with smallpox through several slight punctures and incisions. No disease followed. Jenner submitted a paper to the Royal Society, but was given to understand that he should be cautious and prudent, and that he ought not to risk his reputation by presenting anything, which appeared so much at variance with established knowledge. Cowpox recurred in the diaries in the spring of 1798. Just over ten people were inoculated with cowpox and the first and three others were tested by inoculation with smallpox. Arm-to-arm transfer was shown to be effective. Jenner's findings were published privately later in 1798 [1]. On the title page is a quotation from the poem by Lucretius "On the Nature of the Universe" written in 55 BC, which can be translated "What can be a surer guide to the distinction of true from false than our own senses?"[2].

Jenner wrote that man's familiarity with animals might lead to disease. He describes that condition of the heel of the horse with farriers had termed "The Grease". He thought that infection was carried to the dairies by men who looked after horses and who "without due attention to cleanliness might incautiously milk cows with some particles of the infectious material adhering to their fingers". In rural Gloucestershire, smallpox occurred occasionally and was easily recognized. Jenner had observed that those who suffered cowpox, but never smallpox, failed to contract smallpox whether by inoculation or exposure, and that this protection

†Deceased

S.A. Plotkin (ed.), *History of Vaccine Development*,
DOI 10.1007/978-1-4419-1339-5_4,

Fig. 1 Edward Jenner (1742–1823)

could last up to 53 years. The crucial experiment of 14 May 1796 gave proof of immunity, and this was given support by further inoculations of 1798.

Jenner recognized another mild condition of the nipple of the cow, which only rarely spread to milkmaids and which did not lead to immunity. He called this "spurious cowpox". In a second publication of 1799 [3], he reported that failure might also be due to decomposition of the virus by putrefaction, from taking material from too old a lesion, and that there was sometimes failure on direct inoculation from a horse. In a third publication of 1800 [4], Jenner reported that more that 6,000 people had been inoculated and most of them showed to be immune to smallpox. In a final publication of 1801 [5], Jenner gave a concrete history of his observations and wrote: "A hundred thousand persons, upon the smallest computation, have been inoculated in these realms. The numbers who have partaken of its benefits throughout Europe and other parts of the Globe are incalculable: and it now becomes too manifest to admit of controversy, that the annihilation of the Small Pox, the most dreadful scourge of the human species, must be the final result of this practice".

In his publications, Jenner use the word "virus" to mean, as it had from Roman times, a noxious agent or poison. He referred to the "vaccine virus, vaccine disease, vaccine inoculation and vaccine matter". He described security against smallpox and protection or shielding of the human constitution, but did not explain how this might arise and did not use the word immunity. Nor did he use the word vaccination, which was introduced in 1800 by Richard Dunning, a Plymouth surgeon, with Jenner's approval [6]. He never referred to vaccinology for there was only one vaccine until Louis Pasteur introduced the vaccines for fowl cholera, anthrax, and rabies.

Pasteur, in an address given on the inauguration of the Faculty of Science, University of Lille, on 7 December 1854, had said: "Where observation is concerned, chance favors only the prepared mind". Jenner's mind had been prepared over many years. He was born on 17 May 1749 in the Old Vicarage, Berkeley, the son of Reverend Stephen Jenner. He was orphaned at the age of five. While at school at Wotton under Edge, he was inoculated with smallpox, a procedure introduced to England from Turkey by Lady Mary Wortley Montagu in 1721. It had been practiced for centuries before this in China where smallpox was given by nasal inhalation; in India, in Africa from where it was introduced to North America and in West Wales were it was referred to as "buying the smallpox". For 6 weeks, Jenner was bled and purged, then "haltered up" in the inoculation stables. Fortunately, he escaped with only a mild attack of smallpox. Later at Cirencester Grammar School, he met Caleb Hillier Parry and Joseph Heathfield Hickes and made other lifelong friends [7]. At the age of 13, he was apprenticed as an apothecary to Daniel Ludlow and later to George Hardwicke in Chipping Sodbury, a market town of just under a thousand people in a dairy farming area 10 miles northeast of Bristol. Louis Valentin of Nancy met Jenner in 1803, who said to him that the Duchess of Cleveland in the court of Charles II had been told she might lose her beauty from the ravages of smallpox, but replied she had no fear since she had been preserved by the cowpox. A milkmaid in Chipping Sodbury told Jenner she could not take the smallpox because she had had the cowpox. He may have heard the nursery rhyme: "Where are you going to my pretty maid? I'm going a-milking, sir, she said… What is your fortune, my pretty maid? My face in my fortune, sir, she said".

Jenner left Chipping Sodbury in 1770 and became one of John Hunter's first pupils at his home in Jermyn Street and at St. George's Hospital, London. He mentioned the cowpox story to Hunter and later, showed him a painting of the cowpox lesion, but Hunter never took up the matter nor referred to it in correspondence, probably because of his success with inoculation and its low mortality in his hands [8]. Hunter recommended Jenner to Joseph Banks and Jenner helped to classify the material brought home in July 1881 from the voyage on Endeavour with Captain James Cook. Jenner was offered the post of botanist with the second expedition, but he preferred to return to Berkeley in 1772. He had been interested in natural history since childhood. Hunter in his many letters encouraged the study of the hibernation of the hedgehog, the breeding of toads, and the migration of birds and wrote: "Why think, why not try the experiment?" In a paper read to the Royal Society in March 1788, Jenner described how the young cuckoo ejected the eggs or nestlings of the host hedge sparrow by means of a depression on its back, present for only the first 12 days of its life. For this work he was elected Fellow of the Royal Society in February 1789.

In 1772, Jenner joined a medical society, which met at the Ship Inn, Alveston near Bristol, and called it the Convivio-Medical Society. A fellow member, John Fewster, had observed that those who had had cowpox did not take smallpox by inoculation. He mentioned this to Jenner, who, according to Baron [9], "often recurred to the subject of these meetings; at length… it became so distasteful to his companions that they threatened to expel him if he continues to harass them with so unprofitable a subject".

In 1788, Jenner was a founder member of another medical society, which met at the Fleece Inn, Rodborough, near Stroud, and which he called the Medico-Convivial Society. At the first meeting, Caleb Hillier Parry presented a case of angina and published his observations 11 years later [10]. He records that Jenner had found calcification of the coronary arteries at a post-mortem examination several years before his paper, but to avoid distress to John Hunter, did not publish his findings. At the time, he mentioned this to Mr Clines, a London surgeon, and Mr Home, Hunter's brother-in-law. After Hunter's death, Mr Home wrote to Jenner about the dissection to tell him he was right. Jenner gave a paper describing disease of the heart following rheumatism, and with Hickes, described inoculation with swinepox and proposed an experimental enquiry into the nature of the disease and the associated immunity. This was not taken further, probably because Jenner thought it was a mild form of smallpox.

Jenner was awarded a doctorate in medicine by St Andrews University in July 1792 on the recommendation of Hickes and Parry. This enabled him to practice as a physician in Cheltenham. The influence of prominent residents and distinguished visitors to the spa town helped to spread knowledge of vaccination and played a part in its introduction to the armed services in the early 1800s. Jenner received and wrote a large number of letters, and described himself as the "vaccine clerk to the world". He visited London each year and for a short time praciticed in the city, but anyone could now vaccinate and few people came to see him. When his wife died in 1815, he returned to Berkeley where he died on 26 January 1823, 1 month after Louis Pasteur was born in Dole. There were centenary celebrations in 1922–1923 for the death of Pasteur and the bicentenary of the first vaccination.

Valentin [11] recalled that France received in 1800, Jenner's discovery as an inestimable benefaction: it was the subject of all conversation; poets sang the praises of Jenner and his vaccine. In England, Coleridge wrote to Jenner proposing a poem and an essay on vaccination. "Fame is a worthy object for the best men… it is, in truth, no other than benevolence extended beyond the grave, active virtue no longer cooped in between the cradle and the coffin". Neither the poem nor the essay was written, but the Poet Laureate, Robert Southey, wrote of Jenner in "The Tale of Paraguay" in 1825:

> Jenner! For ever shall thy honour'd name
> Among the children of mankind be blest,
> Who by thy skill hast taught us how to tame
> One dire disease…
> For that most fearful malady subdued
> Receive a poet's praise, a father's gratitude.

At Jenner's request and on a number of occasions, Napoleon released English prisoners who had been detained in France [12]. He is reported to have said to the Empress Josephine: "What that man asks is not to be refused". Despite the war, Jenner was elected a foreign associate of *l'Académie Royale des Sciences de l'Institut de France*.

Others had proposed inoculation with cowpox before Jenner. The best known, Benjamin Jesty, a Dorset farmer, had inoculated his family with cowpox in 1774, but did not make this generally known. Valentin considered Jenner "*incontestablement I'linventeur de la découverte*".

The *Société de Sciences Industrielles, Arts et Belles Lettres*, Paris, had voted for a statue in honour of Jenner in 1857. A site was found at Boulogne-sur-Mer where

Woodville and Nowell had brought vaccination to the people of France in June 1800. The statue by Eugéne Paul was unveiled on 11 September 1865. The plinth records that the vaccine was sent to Paris where Woodville performed further vaccination between 20 July and 18 August 1800.

For 83 years, until 1879, when Pasteur described immunization by attenuated chicken cholera, there was only one vaccine. In 1881, Pasteur immunized sheep and cattle against anthrax. Later that year, at the International Medical Congress in London, under the Presidency of Sir James Paget, Pasteur extended the term vaccination to other immunising agents "as homage to the merit of and to the immense services rendered by one of the greatest of Englishmen, you Jenner".

The last natural case of smallpox occurred in Somalia in 1977, and in 1980 the World Health Assembly declared that smallpox had been eradicated. Fifty years after the development of polio vaccine, poliomyelitis was eliminated from the Americas with the last case in Junin, Peru, in 1991, and there is hope that there may be global eradication early next century, Both Jenner and Pasteur would be fascinated by the prospect of the elimination of measles, and of new vaccines to meet the challenge of malaria, schistosomiasis, AIDS, tuberculosis, diarrhoeal diseases, and *Helicobacter* among others, by the possibility that hepatitis B vaccine may reduce the frequency of some cancers, and by the hope that science will eventually triumph over infectious disease.

Jenner's former home, The Chantry, Berkeley, now the site of the Jenner Museum and Conference Centre, was opened in 1985. There Jenner is remembered as a naturalist, scientist, and father of immunology, and about all as one of the greatest benefactors to mankind.

References

1. Plotkin SA, Mortimer EA. *Vaccines*. W.B. Saunders Company, Philadelphia, 1998.
2. Latham RE, translation. *Lucretius. On the Nature of the Universe*. London: Penguin Books, 1994:27
3. Jenner E. *Further Observations on the Variolae Vaccinae*. London: Sampson Low, 1799
4. Jenner E. *A Continuation of the Facts and Observations Relative to the Variolae Vaccinae or Cow Pox*. London: Sampson Low, 1800
5. Jenner E. *The Origin of the Vaccine Inoculation*. London: DN Shury, 1801
6. Dunning R. *Some Observations of Vaccination or the Inoculated Cow-Pox*. London: March and Teape, 1800
7. James FE. Cirencester Grammar School, the Revered Dr John Washbourn, and some medical pupils. *Trans Bristol and Glos Archael Soc* 1993;111:191–9
8. Turk JL, Allen E. The influence of John Hunter's inoculation practice on Edward Jenner's discovery of vaccination against smallpox. *J R Soc Med* 1990;83:266–7
9. Baron J. *The Life of Edward Jenner*, vol I. London: Hnery Colburn, 1827:48
10. Parry CH. *A Inquiry into the Symptoms and Causes of the Syncope Anginosa*. Bath: R Cruttwell, 1799:3–5
11. Valentin L. *Notice Historique sur le Docteur Jenner*. Nancy: Hissette, 1824
12. Nixon JA. British prisoners released by Napolean at Jenner's request. *P R Soc Med* 1939; 32:877–83

Smallpox Eradication: The Vindication of Jenner's Prophesy

Frank Fenner†

In 1802, Jenner [1] published a short pamphlet in which he made the statement: "…it now becomes too manifest to admit of controversy, that the annihilation of the Small Pox, the most dreadful scourge of the human species, must be the final result of this practice." Jenner was correct, but it took another 176 years to achieve this result.

There were many reasons for this long delay, Jenner was mainly thinking of European countries; he did not accept the necessity for revaccination, and more generally, he underestimated the difficulty of delivering potent vaccine and of achieving satisfactory vaccination levels, especially in rural areas. Further, although heat inactivation of the vaccine had not proved a problem in Europe, experience in European colonies in Africa, India and Indonesia showed that liquid vaccine was not very efficient in tropical conditions [2]. From the early 1900s, attempts were made by Dutch and French scientists concerned with vaccination in their colonies to produce more heat-stable vaccine, but it was not until the early 1950s that Collier [3] produced freeze-dried vaccine on a commercial scale. This led the first Director-General of the World Health Organization (WHO), Brock Chisholm, to propose a global smallpox eradication programme to the World Health Assembly in 1953. However, the delegates considered smallpox eradication to be "too vast and complicated" to be considered, although 2 years later they approved the vastly, more difficult and expensive proposal of malaria eradication [4].

The First WHO Global Smallpox Eradication Campaign

In 1958, the delegates of the Soviet Union led by V.M. Zhdanov, outlined to the World Health Assembly, a detailed proposal to achieve global smallpox eradication within 4–5 years by mounting mass vaccination campaigns to the endemic

†Deceased

S.A. Plotkin (ed.), *History of Vaccine Development*,
DOI 10.1007/978-1-4419-1339-5_5,

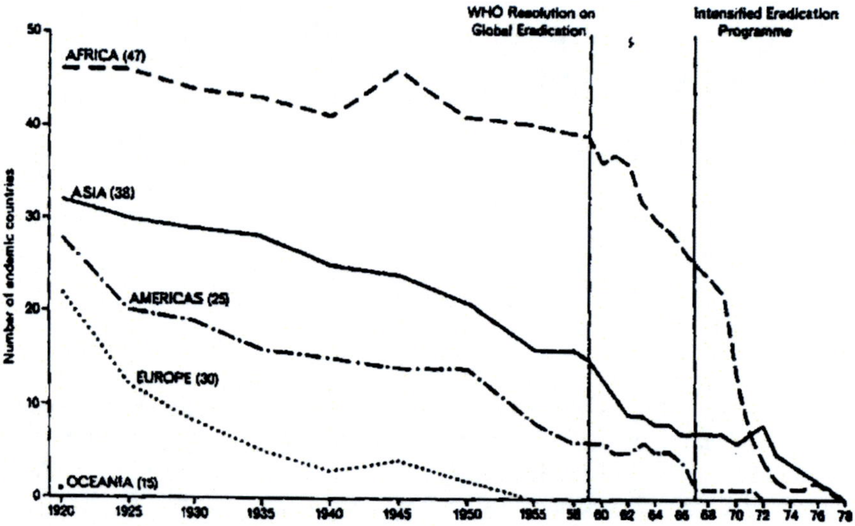

Fig. 1 Number of countries and territories in which small pox was endemic between 1920 and 1978, arranged by continent. From [6], courtesy of the World Health Organization

countries, in the assumption that if 80% of the population was vaccinated, transmission would be interrupted [5]. This concept was accepted by the Assembly in 1959. The Soviet Union undertook to supply large amounts of freeze-dried vaccine to endemic countries, as did a number of other industrialised countries. This campaign resulted in the elimination of smallpox from many of the smaller endemic countries (Fig. 1). However, by 1965, it was clear that the mass vaccination at the suggested level on 80% of the population was not going to achieve interruption of the transmission of smallpox in the larger countries in Africa and Asia, and certainly not in the Indian subcontinent [6].

The Intensified Smallpox Eradication Programme

Something more was required if smallpox was to be eradicated globally [7]. WHO asked DA Henderson, Chief of the smallpox programme at the Center for Disease Control (CDC) in the USA, to come to Geneva. There he worked with Karel Raska, the Director of WHO's Division of Communicable Diseases, and the WHO medical officer responsible to smallpox, Isao Arita, to develop a plan for Intensified Smallpox Eradication Programme. In 1966, their plan was accepted by the World Health Assembly; the finance requested being approved by two votes, the narrowest margin for the acceptance of a budget in the history of WHO. Henderson agreed to come to Geneva as Chief of the WHO Smallpox Eradication Unit for the 10 years that the programme was expected to take.

Vaccination in the Intensified Smallpox Eradication Programme

It takes over 100 pages of the book *Smallpox and its Eradication* to describe just the planning of the programme [8]. For this conference, the focus is on vaccination, details of which are set out in another chapter of the book [9]. An early priority was that none of the WHO regular budget of $2.4 million a year should be spent on procurement of vaccine, since the cost of the vaccine required would have substantially exceeded the total budget. Hence, it was decided that the vaccine would have to be provided by donation or by local production in the smallpox-endemic countries. This was not an easy task. Disillusioned with "eradication" by its experiences in the malaria eradication campaign, UNICEF (United Nations Children's Fund) provided minimal help, and Henderson himself had to spend a substantial part of his time persuading industrialised countries to make a donation of the vaccine.

Although by the mid-1960s, tissue culture production of vaccines was a well-established procedure, virtually all of the smallpox vaccine production laboratories in the world, in industrialized as well as developing countries, were producing it by scarification of the skin of calves, sheep or buffalo. Because it would take years to set up efficient tissue culture production, it was decided to let production laboratories continue with their existing methods, crude though these were.

Many endemic countries wished to embark on the production of freeze-dried smallpox vaccine, but vaccine production solely for local use was uneconomic in a country with a population less than ten million, and the WHO rarely supported it in small developing countries. However, local production was essential in the large developing countries. These countries produce very large amounts of vaccine and some of them were able to provide vaccine to other countries.

Although the bulk of smallpox vaccine came through local production or bilateral aid programmes, the vaccine donated to the WHO, which was distributed through the Smallpox Eradication Unit, was critical in ensuring that emergency requirements could be met. The unit maintained a stock of some half million doses in Geneva for emergency use. In all, 465 million doses of vaccine were donated to the WHO for the Intensified Smallpox Eradication Programme, by 27 countries, and each year from 1967 to 1969, between 15 and 45 million does of vaccine were dispatched by the WHO to endemic countries.

Just as important as ensuring adequate quantities of vaccine was the need to ensure that all vaccine used was potent. To achieve this, Henderson insisted that all the vaccine used in the programme should meet standards for potency, heat stability and freedom from pathogenic bacteria that had been set up by the WHO some years earlier. Arita took the responsibility for this programme of quality control [10]. A questionnaire relating to the mode of production, the strain of virus and the method of freeze-drying, and the results of potency testing, heat stability and bacterial content was sent to 77 laboratories in 52 countries and elicited replies from 59 laboratories in 44 countries. Only 52% reported satisfactory results for potency testing and only 27% for heat stability [11]. The Smallpox Eradication Unit therefore established a WHO Reference Centre for Smallpox Vaccine for the

Americas in the Connaught Laboratories in Toronto, Canada and an International WHO Reference Centre for Smallpox Vaccine at the National Institute of Public Health in Bilthoven, the Netherlands, for the rest of the world. A panel of experts from the reference laboratories and other production laboratories in industrialised countries was called together and produced a report [12] entitled *Methodology of Freeze-dried Smallpox Vaccine Production*, which went into great detail in practical aspects of vaccine production, and was widely distributed. Subsequently, several expert consultants visited production laboratories in some 35 countries in the developing world, and the two Smallpox Vaccine Reference Laboratories developed a system of testing samples of smallpox vaccine that was to be used in the Intensified Programme. This was an unprecedented step, which some WHO officials said was impossible because it constituted a breach of national sovereignty. But by combination of report, the visits by consultants and the regular testing of production bathes right through the programme, the percentage of satisfactory batches rapidly rose from the initial figure of 36% to about 75% in 1968 to 1971 and about 95% thereafter (Table 1). This was a most important step, for nothing is more calculated to bring a campaign into disrepute than to be distributing vaccine that is not potent.

New Methods of Vaccination

Before 1967, vaccination was carried out either by a scratch method, or by a multiple pressure technique. Two new methods of vaccination were developed during the Intensified Programme. A "high-tech" method, the jet injector [13], was used by the United States CDC staff in the campaign in west and central Africa and later in Brazil. However, it could not be used effectively in sparsely settled rural areas, and in developing countries maintenance was a problem.

Far more effective was an invention by Ben Rubin of Wyeth Laboratories that was donated by them to the smallpox programme – the bifurcated needle [14]. Dipped into a vial of reconstituted vaccine, it held a dose between its prongs. After this had been deposited on the skin, 15 vertical pricks with the bifurcated needle through the droplet resulted in successful vaccination. A reusable plastic container was designed by Ehsan Shafa, and produced in Pakistan, which could be filled with sterile needles each morning, used throughout the day, with a fresh needle for each vaccine, and collection of used needles for sterilization by boiling and re-use the next day. Apart from its efficiency and simplicity, the bifurcated needle resulted in the use of only one quarter as much vaccine as was needed for conventional multiple pressure vaccination.

However, the Intensified Smallpox Eradication Programme had been launched because it was clear from the results of the early 1960s that vaccination alone, even with potent vaccine, was not enough to achieve global smallpox eradication. Subsequently, Arita et al. [15] showed that even with 80% vaccination rates, in India, there remained a density of unvaccinated persons higher than that of the total

Table 1 WHO quality control of freeze-dried vaccine: results of tests carried out in the WHO Reference Centres for Smallpox Vaccine in Bilthoven, the Netherlands and Toronto, Canada on experimental and production batches from producers in various parts of the world.[a]

Year	No of producers	No of batches	No of satisfactory batches (%)		No of satisfactory batches (%)		Unsatisfactory		
							Initial potency	Heat stability	Bacterial count
1967	20	74	27	(36)	47	(64)	32	12	8
1968	23	136	74	(54)	62	(46)	26	36	5
1969	30	164	128	(78)	36	(22)	23	12	5
1970	27	380	312	(82)	68	(18)	27	35	13
1971	32	206	154	(75)	52	(25)	31	23	5
1972	27	311	241	(77)	70	(23)	32	39	1
1973	30	392	367	(94)	25	(6)	5	20	0
1974	28	231	199	(86)	32	(14)	11	20	1
1975	21	167	139	(83)	28	(17)	15	10	6
1976	16	213	203	(95)	10	(5)	2	7	3
1977	11	114	101	(89)	13	(11)	1	12	1
1978	9	59	57	(97)	2	(3)	0	2	0
1979	10	85	82	(96)	3	(4)	3	1	0
1980	5	46	46	(100)	0	(–)	0	0	0
Total	–	2,578	2,130	(82.6)	448	(17.4)	208	229	48

[a]From [11], courtesy of the World Health Organization.

population of any country in west or central Africa, except Nigeria and Gambia; and in Bangladesh, the population density of subjects susceptible to immunisation after 80% vaccination was about three times higher than that in India.

Surveillance and Containment

It had been clear from the outset of the Intensified Programme that some other strategy was required as well. This strategy, surveillance and containment [16], was introduced by Henderson as the key procedure right at the start of the programme. Briefly, it consisted of the active search for cases and the determination of where they had acquired the infection, followed by ring vaccination of affected houses and villages.

The two key elements of the success of the Intensified Smallpox Eradication Programme were the provision of sufficient high quality vaccine and the assiduous application of the principle of surveillance and containment. Within a few months of the stipulated 10 years, Jenner's prophesy was fulfilled, and last case of smallpox to have occurred in the field was diagnosed in Somalia in October 1977.

References

1. Jenner E. *The Origin of the Vaccine Inoculation.* London, UK: DN Shury, 1801
2. Tanganyika Territory. *Annual Reports of the Principal Medical Officer and the Senior Sanitary Officer. Period. November 1918 to November 1920*, 67–68
3. Collier LH. The development of a stable smallpox vaccine. *J Hyg [Lond]* 1955; 53:76–101
4. Fenner F, Henderson DA, Arita I, Jezek Z, Ladnyi ID. *Smallpox and its Eradication*, Geneva, Switzerland: World Health Organization, 1988; 392–393
5. Fenner F, Henderson DA, Arita I, Jezek Z, Ladnyi ID. *Smallpox and its Eradication*, 366–371
6. Fenner F, Henderson DA, Arita I, Jezek Z, Ladnyi ID. *Smallpox and its Eradication*, 171
7. Fenner F, Henderson DA, Arita I, Jezek Z, Ladnyi ID. *Smallpox and its Eradication*, 393–419
8. Fenner F, Henderson DA, Arita I, Jezek Z, Ladnyi ID. *Smallpox and its Eradication*, 422–538
9. Fenner F, Henderson DA, Arita I, Jezek Z, Ladnyi ID. *Smallpox and its Eradication*, 540–592
10. Arita I. The control of vaccine quality in the smallpox eradication programme. In: *International Sympossium on Smallpox Vaccine, Bilthover, the Netherlands, 11-13 October 1972; Symposia Series in Immunobiological Standardization.* 1973; 19:79–87
11. Fenner F, Henderson DA, Arita I, Jezek Z, Ladnyi ID. *Smallpox and its Eradication.* Geneva: World Health Organization, 1988;560
12. WHO documents, SE series SE/68.3 Rev 2. *Methodology of Freeze-Dried Smallpox Vaccine Producation.* Geneva, Switzerland: World Health Orgainzation. Listed in Fenner F. Henderson Da, Arita I, Jezek, Ladnyi ID. *Smallpox and its Eradication*, 1407
13. Fenner F, Henderson DA, Arita I, Jezek Z, Ladnyi ID. *Smallpox and its Eradication*, 573–580
14. Fenner F, Henderson DA, Arita I, Jezek Z, Ladnyi ID. *Smallpox and its Eradication*, 567–573
15. Arita I, Wickett J, Fenner F. Impact of population density on immunization programmes. *J Hyg [Lond]* 1986;96:459–466
16. Fenner F, Henderson DA, Arita I, Jezek Z, Ladnyi ID. *Smallpox and its Eradication*, 473–7, 493–515

Pasteur and the Birth of Vaccines Made in the Laboratory

Hervé Bazin

Introduction

Louis Pasteur was born in Dole, a town in the French Jura, in a family of craftsmen. His father was a tanner. His mother was kept busy with her family of five children, Louis being the only boy. At first they lived in rather poor conditions in Dole, but soon moved to the nearby town of Arbois, where his father found a small tannery along the Cuisance River. Pasteur grew up in a supportive and loving familial environment and was successful in the local school and then in the secondary school of Besançon, a much bigger town not very far from Arbois. With the help of his father and his mentors in Arbois and Besançon, Pasteur spent 2 years preparing for the examination to enter the Ecole Normale Supérieure (ENS) in Paris, to which the best candidates were accepted with an annual grant. He was successful and was

H. Bazin (✉)
Emeritus Professor University of Louvain and 4 rue des Ecoles, 92330 Sceaux, France
e-mail: herve-marie.bazin@wanadoo.fr

S.A. Plotkin (ed.), *History of Vaccine Development*,
DOI 10.1007/978-1-4419-1339-5_6,

admitted to the ENS, from which students were selected for teaching positions in French secondary schools and even in universities [1–3].

During the 3-year spent in the ENS, Pasteur fell really in love with laboratory work in Physics and Chemistry. This great interest became well recognized at the ENS and he was given a position of demonstrator for one extra year, during which time he discovered the highly interesting phenomenon of molecular dissymmetry. This point is important, as due to this exceptional discovery, Pasteur obtained large and early support for his career from powerful French scientists, giving him the opportunity as he moved from position to position to carry on laboratory work, later with assistants.

Pasteur was predisposed to study biological phenomena from his earliest results as he believed that dissymmetric molecules were made by living organisms. This idea pushed him to study fermentation, to develop the *in vitro* culture of "ferments," and then to study spontaneous generation. At each step, he proposed a new hypothesis which he clearly verified, often with new devices of his invention. At a certain point he was asked by one of his mentors to study a disastrous problem of the silk industry in the south of France: the silk worms were dying from an unknown cause. Pasteur worked for a long time on the subject, acquiring basic understanding of a contagion from worm to worm and from adult insects to larvae. He gave rather simple recommendations to reduce this epizootic through hygienic measures.

From that time (1877), Pasteur started to be interested on contagious disease of animals and humans. He was about 54 years old, in rather poor health, but still very active. His laboratory was located in the ENS, rue d'Ulm in Paris. It was about 400 m^2 in size with large basements and some possibilities to maintain outside animal labs containing rodents, chicken, rabbits, sheep, monkeys, etc. Moreover, Pasteur and his team rented spaces in houses in the neighborhood of the main laboratory and had at their disposal parts of the closed *Collège Rollin* within walking distance. Pasteur himself lived in the main building of the ENS, while his two closest assistants, Emile Roux and Charles Chamberland, lived respectively in a small room inside the main laboratory and in an apartment of three small rooms in the former *Collège Rollin*.

Studies on Anthrax

The reputation of Pasteur was increasing and the French government therefore asked him to study the anthrax disease that was a scourge for agriculture in France and elsewhere in Europe. Pasteur, always very interested in practical applications of science, learned how the disease was transmitted from animal to animal and from year to year in some places called "champs maudits" (cursed fields). At that time, very little was known about pathogens. Parasites like intestinal worms or surface flea or lice had been correctly described, but the origin of anthrax as well as other contagious diseases was unknown. Casimir Davaine had proposed that there was a parallel between the butyric ferment discovered by Pasteur and some tiny sticks seen in the blood of cattle or sheep dying from anthrax, but this highly interesting idea was not fully accepted. Pasteur knew that after a first attack of contagious diseases such as smallpox, measles, sheep pox, and bovine peripneumonia, although all caused by unknown factors, humans or animals were rarely susceptible to a second attack of the same contagious disease. Pasteur discovered that a first attack of anthrax, a contagious disease due to a visible pathogen, was similarly capable of inducing a non-susceptible state against a second attack of the same pathogen. This highly interesting observation, the first made on a disease due to a known and visible pathogen, provided him with the possibility to study, *in vitro*, microorganisms capable of inducing a state of insusceptibility.

Pasteur and the Chicken Cholera

On 30 October 1878, Pasteur received a strain of bacteria that caused chicken cholera from Henry Toussaint, a professor of the Veterinary School of Toulouse. He soon learned how to grow the chicken cholera microbe in chicken broth. At that time, they were only few bacteria known to be pathogenic for animals (or humans) that induced contagious diseases, including anthrax, septic vibrio, and chicken cholera. Pasteur was pleased to have them at his disposal but was not particularly excited by the latter. However, during the spring of 1879, he started some experiments on virulence, trying to infect chickens with food contaminated with a culture of chicken cholera microbes. Many chickens were killed by these polluted meals, but some recovered and were resistant to a second exposure to the same pathogen given by an inoculation of a lethal dose of the germ. Thus, immunity could be induced in chicken against chicken cholera.

Pasteur left Paris at the end of July 1879 to spend some holidays in Arbois. During that period he left his collaborator Emile Roux in charge of the laboratory. In mid-October, probably anxious to keep alive his strain of chicken cholera microbes, he asked Roux to put in culture some of the bacteria from July. The culture did not grow after 24 h in the incubator, so then Roux inoculated two chickens with the most recent July culture, but both chickens survived the injection. Roux, probably worried that the organisms were no longer viable, re inoculated the two chickens with a slow growing subculture of a culture from July. The birds died in 3 and 4 days, instead of 1 or 2, which was the normal time of death after a lethal challenge. From one of these chickens, a culture was done on the 28 October 1879, called X. This was the origin of the culture of microbes used later by Pasteur [4]. Clearly on 28 October 1879 it was still a virulent strain, perhaps not fully virulent but certainly rather virulent. Although it has been proposed that Roux was a major contributor to the birth of the idea of attenuation of virulence [5], it does not appear that attenuation was perceived to have taken place before Pasteur's experiments in December, 1879, described below.

At the end of his holidays, Pasteur suffered of an intestinal problem and then was busy with the wedding of his daughter at the beginning of November. Probably for those reasons he did not return to chicken cholera before the 5 December 1879, using a culture of 22 November 1879 called X_1, which was a subculture of X. This strain had been left 39 days in a medium that was more or less acid (through prolonged contact with the air, an indication which was often specified by Pasteur, who did not believe that the same result could be obtained with an oxygen draft as suggested by Roux [3]). Pasteur inoculated five "new" (never inoculated) chickens and they did not die. Six days later, Pasteur reinoculated the same animals with a subculture of X_1 and again they survived. Pasteur was clearly surprised by these results. By luck or by chance at the same time he studied the action of cold (1 h at −30 to −38°C) on the same bacteria and obtained a virulent strain capable of killing a chicken in 50 h [6]. Now Pasteur had in his hands two strains of the same microbe, one non virulent and the other virulent. On 18 December 1879, Pasteur thought of immunization (he used the verb "vacciner" for the first time in his notes) [7] by the X_1 strain, as the birds inoculated with that strain resisted his virulent one. By the end of January 1880, Pasteur was convinced that he was able to vaccinate against chicken cholera. The 9 February 1880, Pasteur already described his results to the "Académie des Sciences," in Paris, but without details of his technique of virulence attenuation. In April 1880, Pasteur described more details about the immunity after inoculation by attenuated chicken cholera. For the first time he used in public the word "vacciner" with its new extended meaning. He also presented the fact that vaccination by inoculation gave a total body immunity and not just immunity at the point of injection. In addition, he proposed an exhaustion theory of immunity: that any invasion of a microbe in a susceptible organism removed essential nutrient(s) for that microbe, leading to a state of protection if the microbe did not kill the host [8]. As a consequence of this hypothesis for a mechanism of immunity, the contact of a living microorganism was necessary to obtain a state of protection ("non-recidive" as Pasteur called it). This exhaustion theory was propounded by Auzias-Turenne, already in 1865, and perhaps before by several other scientists. Pasteur did not give reference to others, particularly to Auzias-Turenne [3].

By March 1880, Pasteur knew correctly how to immunize chickens against their cholera, but only described his technique in October 1880, citing the role of prolonged contact with oxygen in the atmosphere [9].

This vaccine is the first one called Pastorian, i.e. made with a strain of pathogen of attenuated virulence made by human hands in the laboratory. However, this chicken cholera vaccine was not so good: it gave a limited period of immunity and possible severe secondary effects. It was never really employed in extensive breeding of chickens and suffered clear criticisms, particularly from Pierre Galtier, professor at the Veterinary School of Lyon who specialized in infectious diseases! [10].

Pasteur and Toussaint and Their Competition for an Anthrax Vaccine

During his studies on chicken cholera Pasteur never forgot anthrax, his main subject and source of funding of that time. In July 1880, Toussaint, a competitor in the field of anthrax, disclosed a vaccine to immunize sheep and dogs. The concept of vaccine was from Pasteur, but the vaccine technique, from Toussaint: defibrinated blood from a sheep freshly dead of anthrax heated at 55°C for 10 min with or without a filtration through paper or the addition of carbolic acid. Toussaint did experiments on sheep in Toulouse, then in Alfort with sheep of a herd belonging to the Ministry of Agriculture, so with official support. The results of the Toussaint vaccine inoculation were rather good with just a low percentage of sheep deaths from the vaccine injections, not really different from those sometimes observed in the Pasteur's subsequent experiments.

Pasteur, very rapidly criticized Toussaint's vaccine, which was in principle dead and thus not in accordance with the exhaustion theory of immunity that he believed at that time.

Pasteur and his assistants tried to develop their own vaccines. Chamberland wrote a note dated the 18 February 1881, in which he described the culture of anthrax bacteria in a chicken broth mixed with a small percentage of potassium bichromate and successful immunization of animals including sheep [11]. Who had the original idea for this experiment is unknown, but, the Pasteur's laboratory was accustomed to grow microorganisms in presence of various substances to study their effects, at least after May 1879 [12]. It is obvious that the vaccines used for the Pouilly-le-Fort experiment were derived from the results described in the Chamberland's note of February 1881.

Pasteur described on 21 March 1881 at the "Académie des sciences" that he was in the possession of a new technique of immunization against anthrax. A few days later, on 2 April, Hippolyte Rossignol, a veterinarian opposed to the Pasteur germ theory [13] asked the "Société d'agriculture de Melun" to organize a great and public experiment. On the 28 April, a protocol was signed by Pasteur [14, 15]. The protocol called for the vaccination of 24 sheep, 1 goat, and 6 cows by two injections of attenuated strains (no details were given) of anthrax microbes, the first one on 5 May and the second one on 17 May. The lethal challenge to the vaccinated and not vaccinated groups (a total of 48 sheep, 2 goats, and 10 cows) was to be given on the 30 June. From Pasteur's notes we learn that the first vaccine was the Chamberland strain, kept for a long time in the bichromate, without virulence for sheep, but of which the virulence was reinforced by three successive passages through mice; the second vaccine was an anthrax strain kept only few days in a culture with bichromate. As Pasteur described in his notes, the last anthrax strain (the second vaccine of Pouilly-le-Fort) killed one of two sheep in one experiment but when preceded by the first Chamberland strain (the one reinforced by three passages in mice) had often protected very well against a challenge with a highly virulent strain, the "bactéridie" of 4 years [16]. This "bactéridie" of 4 years, an anthrax microbe that grew easily in culture, was selected by Pasteur on 10 January 1881 from his collection of anthrax cultures kept in his laboratory [17]. Clearly, the vaccines of Pouilly-le-Fort were prepared by Pasteur with the help of Chamberland and perhaps Roux, but not very well tested (Fig. 1).

The results of the Pouilly-le-Fort experiment were excellent, even exceptional: all animals vaccinated were in perfect condition on 2 June 1881; others animals were dead, dying, or in the case of the bovines, in bad condition but surviving. One vaccinated ewe died 1 day later but was found to be carrying a dead fetus. Pasteur

Fig. 1 Animal immunization against anthrax: a popular representation of the Pouilly-le-Fort experiment, celebrating the hundredth anniversary of the birth of Pasteur. Drawing of Damblans (Le Pèlerin, n° 2333 of 5 November 1922)

was still persuaded that the best method to obtain an attenuated strain of bacteria was to expose it to the action of the oxygen from the atmosphere. He never officially wrote or said that the vaccines used in Pouilly-le-Fort were made with the help of an antiseptic, no more than with the oxygenation method that he repeatedly described. Geison has written: "More than that, Pasteur was surely motivated in part by a well-founded concern that a full disclosure of the events at Pouilly-le-Fort would lead his more hostile critics to award Toussaint credit for the discovery of vaccination against anthrax, despite the very real technical differences between their procedures and results." [18]. Pasteur was attached to oxygen exposure, a beautiful theory to explain the mechanism of attenuation of a highly virulent and epidemic pathogen and also probably feared the competition of Toussaint.

What should one think about the Toussaint and Pasteur vaccines? The two vaccines were clearly different as admitted by Geison [18]. The Toussaint vaccine, as already mentioned, was composed of sheep blood dead of anthrax, heated at 55°C for 10 min, with or without the addition of a small percentage of carbolic acid. It was, in principle, a dead vaccine, without *in vitro* culture, injected only one time. The Pastorian live vaccine consisted of two injections at 13 days interval and as has been described consisted first of a culture in a medium mixed with a low quantity of potassium bichromate, then reinforced with three successive passages on mice; the second dose was a rather virulent strain after just a few days in culture with added bichromate. It does not appear that the two approaches inspired one another.

In any case, the hypotheses of Toussaint and Pasteur were different, Toussaint's idea being that vaccination made the lymph nodes impermeable to the bacteria and blocked the diffusion in the body of microbes from the nodes, whereas for Pasteur vaccination with a "live" vaccine exhausted some needed chemical in the body. Both ideas were wrong [3].

The Toussaint vaccine method was employed by Piot in the Middle East [3] and later improved by Chauveau, which allowed the production of enough batches of vaccine to immunize thousands of animals [19]. The value of Pasteur's vaccine was demonstrated in many places in France and elsewhere.

Pasteur and Thuillier and the Swine Erysipelas

On 15 March 1882, Louis Thuillier, an assistant of Pasteur was sent by him to Bollène in the south of France. This was in response to a request of Pasteur by a veterinarian named M. Maucuer who was having a great problem with an epizootic of swine erysipelas. Quickly, Thuillier found the microbe (also identified by Detmers in Chicago) of the disease. By successive passages in pigeons, Thuillier and Pasteur obtained a highly virulent microbe for these birds and also for swine. Then they turned their efforts to rabbits and got, after some repeated passages, a poorly virulent strain of swine erysipelas, which was used for swine immunization. This was the first vaccine developed by serial passages on an animal species different from the normal host.

Pasteur and the Treatment of Rabies

In December 1880, Pasteur was informed of the presence, in a Paris hospital, of a young boy dying of rabies. He tried to isolate the microbe of rabies from the saliva of the boy; however, the inoculation of the saliva did not give rabies to rabbit, the animal species recommended by Victor Galtier, professor at the Veterinary school of Lyon, to study rabies in laboratory. Galtier also had shown that it was possible to induce immunity in sheep with rabies contaminated saliva inoculated by the intravenous route. Pasteur and Roux rapidly improved the techniques used to transfer rabies from animal to animal: first by using inocula of nervous tissue in large quantities to avoid the necessity to discriminate between an increase of the rabies incubation period due to small quantities of virus and intrinsic attenuation of its virulence, instead of saliva. Secondly, they (at least Roux, for the actual manipulations) used an intracranial route of inoculation. These two improvements gave them 100% transmission of the disease, a much shorter incubation period and much less secondary microbial infections, providing the opportunity to do serial passages of a strain of rabies virus in rabbits and thus to keep the rabies virus at disposal in the laboratory.

Pasteur then tried to modify the virulence of the rabies virus by serial passages in different species of animals: in dogs, he obtained a shorter incubation time; in rabbits, he obtained an increase of virulence for them and for dogs, and considered that strain as the "virus fixe" (fixed virus) which was later used for many purposes in his laboratory; in monkeys, he got an interesting decrease of virulence for dogs, but he failed to reproduce these results in ulterior series [3].

On the 2 May 1885, Pasteur was informed of the presence of a patient named Girard in a hospital with a diagnosis of rabies following a bite by a stray dog in March. With the agreement of Dr Rigal, the head of service where Girard was admitted, Roux or a hospital assistant injected Girard with a preparation of an "attenuated rabies virus," in fact, it was an emulsified spinal cord extracted from a rabbit dead of experimental rabies left to dry for 5 days in a "dry bottle." The treatment of Girard was stopped after the first injection by the hospital authorities and after some days, Girard left the hospital apparently in good health [20]. Very probably, Girard had never been affected by rabies as the recovery from that disease is extremely rare and the treatment given by Pasteur was short both in quantity and in time. However, it is difficult to be sure of what really happened, except that Pasteur tried to save that patient with a treatment, which still had live virus as the two rabbits inoculated by the intracranial route with the Girard inoculum died of rabies. However, Girard was inoculated by the subcutaneous route which is much less aggressive and efficient than the intracranial route to transmit rabies. Nevertheless, Pasteur did not publish on that case. The disappearance of Girard after he came out from the hospital seems not exceptional and it is unlikely that Girard died sometimes later from rabies as such deaths were most often well characterized, although not always. Another incident involved an 11-year girl, Julie-Antoinette Poughon, admitted to the hospital Saint Denis on 22 June 1885, suffering from "declared" rabies. Although Pasteur knew, like everyone else, that "declared" rabies is lethal, she received two injections of rabbit spinal cords kept 7 and 5 days in dry flasks but unfortunately died the next day. Two rabbits were inoculated with the bulb of the girl and died from rabies after 17 days (a monkey inoculated at the same time did not suffer from rabies) showing that the death was due to a street rabies virus [21]. One can perhaps find a reason to explain why Pasteur chose to act in such case: "We have met cases of spontaneous recovery of rabies after only the first rabid symptoms were developed, never after the appearance of the acute symptoms."[22]. Pasteur might have hoped that the patients were still in a prodromic state like those he observed in his inoculated dogs, but possibly he was simply interested in testing his new technique. As Geison wrote: "Pasteur's desperate attempts to save Girard and Antoinette Poughon from 'declared' rabies did not violate any accepted ethical standards" [18].

Pasteur and his team tried to culture or to just to maintain the rabies agent in nervous tissue kept in glass tubes in various atmospheres without great success. However, on the 13 January 1885, he started to employ bottles with two necks, one on the top and the other close to the bottom in order to get an air draft inside the receptacle. This technical device was used by Roux in the incubator room of the laboratory as described by Loir, Pasteur's nephew [23, 24]. Pasteur saw in this system a way to dry spinal cord of rabbits' dead of rabies. As modified by him to use room temperature and dry air called "dry bottle," it was a clear improvement to his glass tube. During the months between January and May 1885, Pasteur learned how much time was necessary to eliminate the virulence of rabies virus in the spinal cords of rabbits or other species, trying to define intermediate steps of decreased virulence. However, he was still balancing the attenuation of rabies virus versus the quantities of remaining live virus. This important point was never clearly solved by Pasteur, the general knowledge of viruses at that time being insufficient.

On the 28 May 1885, Pasteur started to immunize four groups of ten dogs with a new protocol. Since 16 May, he had kept, in dry bottles at room temperature, pieces of spinal cords of rabbit killed by high passage rabies virus (fixed virus, very virulent for rabbits). Then every day he injected each dog with an identical quantity of more or less dried rabbit spinal cord, beginning with a cord of 12 days, then of 11 days, up to a spinal cord directly taken from a dead rabid rabbit. During and after this treatment, the dogs remained healthy.

Unexpectedly, on the 6 July 1885, three persons presented themselves in Pasteur's laboratory. They came from Alsace, hoping to receive help from Pasteur. One of them was a young boy who was bitten 4 days before by an apparently rabid dog. The boy was with his mother and the owner of the dog. The latter quickly left the laboratory to return home, as he had been only pinched, not bitten. Pasteur asked the advice of two physicians, Alfred Vulpian, a well-known physiologist and Jacques-Joseph Grancher, an MD working in his laboratory, before taking the momentous decision to treat or not to treat the young boy. Grancher did the inoculations as Pasteur was not a physician and could not inject patients. Some days later, the boy Joseph Meister was able to return home in good health. A second boy, Jean-Baptiste Jupille, came in October 1885 to be treated by Pasteur by a rather similar protocol and ending with the same happy result. Very rapidly, Pasteur took the opportunity to present these cases to the *Académie des sciences*. It was certainly too early to believe in a total success for many reasons: there was no complete certitude that the dogs were rabid, and the transmission of the rabies depends on the localization and the gravity of the bites and never reaches 100% (Fig. 2).

However, patients arrived quickly at Pasteur's laboratory from all over Europe, even from Russia, and also from the USA, and new rabies institutions were founded everywhere in the world. For example, bitten people came to the new Pasteur Institute located in Saïgon (Hô Chi Minh-City) from the Dutch colonies (Indonesia) nearby to be treated by the Pastorian method. Although accidents happened and patients died after the treatment, after some years it became clear that the treatment was valuable for the great majority of treated persons.

The Diffusion of the First Generation of Vaccines and Why!

First, one must define the distinction between variolation and Pastorian vaccination, both being founded on the specific protection given to a person by a first attack of a contagious disease or by a first contact with its agent. Variolation consisted of inoculating a normal pathogenic agent in order to give the host immunity against this agent. The only trick used to avoid a purely natural disease with all the

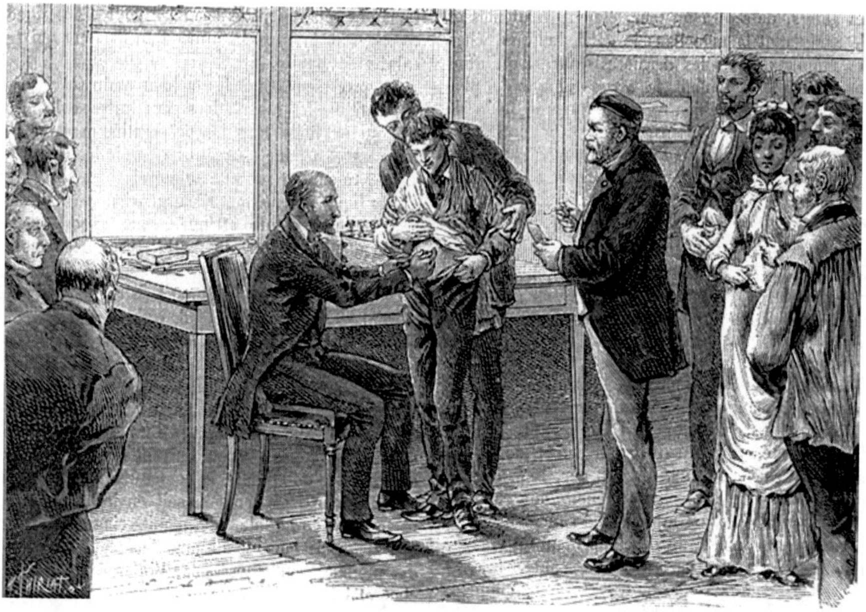

Fig. 2 In 1886, people bitten by rabid animals came from many places, but most of them were European. In Pasteur's office, rue d'Ulm, Doctor Grancher is inoculating a patient who is held by an aide (this was the first role of Alexandre Yersin in the Pasteur team!). Pasteur keeps the list of patients and the record of each treatment. At right, patients are awaiting their inoculations; they have uncovered a small part of their body, the hypochondrium, where they will be inoculated, 1 day on the left, the following day on the right. At left of the picture, the spectators who were invited by Pasteur to watch the inoculations. Pasteur never forgot publicity

sometimes disastrous consequences, was to inoculate at a very young age, to diminish the dose of pathogen injected (for example, variolation with cutaneous punctures) or to choose an abnormal route of administration (for example, the end of the tail for the bovine peripneumonia). In occidental countries, variolation was used during a period of approximately 80 years for many diseases but mostly for smallpox, sheep pox, and bovine peripneumonia. In contrast, Pastorian vaccination induced immunity by using a pathogenic agent diminished in its virulence by an artificial technique to give a very mild disease. Later, the meaning of vaccination was extended to killed pathogens or to artificially modified toxins, etc.

It is interesting to consider the influence of Pasteur's idea of attenuated vaccines, disclosed with the first publication concerning the vaccine against the chicken cholera, in the context of his times, when many specific pathogens were being discovered.

9 February 1880. First publication by Pasteur concerning his chicken cholera vaccine, although without a description of methods.

12 July 1880. Toussaint was the first to exploit Pasteur's concept of vaccination disclosed in February and to publish the description of his own anthrax vaccine.

1882. Arloing, Cornevin, and Thomas started to successfully fight a disease called bovine symptomatic anthrax or blackleg, which is actually caused by

Clostridium chauvoei, by a variolation technique using serous fluid from subcutaneous tissue of infected cows to immunize other animals. The fluid was injected by the intravenous route [25]. Then, in 1882, they quickly improved their method and prepared, by heating the dried serous substance for some hours, a two injection vaccine protocol which was well employed in France and also in many European countries [3, 19].

December 1884. Jaime Ferran, a Spanish medical doctor, started to inoculate humans with a culture of the apparently normal cholera microbe. It is unclear whether this was a case of variolation or vaccination. His results are still subjects of long discussion [3].

1884–1886. Daniel Elmer Salmon and Theobald Smith developed the first inactivated vaccine against the hog cholera bacteria injected in pigeons. The model was very artificial, but the results were interesting and promising for the future.

1892. Waldemar Haffkine, at the Institut Pasteur in Paris, developed a dead cholera vaccine for humans.

1896. On the suggestion of Haffkine, Almroth Wright used the technique employed for cholera to make an anti typhoid vaccine, which was ready in 1896. In late 1896, Pfeiffer and Kolle published on a similar vaccine, causing Wright to accuse them of copying his technique.

Toussaint, Arloing, Ferran, Haffkine, and Wright all specified in writing, that they were indebted to Pasteur and his work for the idea of vaccine development. Salmon and Smith were aware of the works of Pasteur when they developed the first chemically inactivated vaccine.

Pasteur is, without any doubt, the originator of vaccines made in the laboratory and particularly of the idea of attenuated vaccines.

Conclusion

Pasteur was an outstanding scientist in advance of his contemporary colleagues in many respects. He moved from subject to subject up to the rabies vaccine with an exceptional clairvoyance.

However, he used and abused the publicity and the media of his time in order to boost his position in the Academic world and to get funds for his laboratory studies. So, in May 1884, only having in hand some results of dog immunizations with a virus attenuated for monkeys, he declared that he was in a position to save people bitten by rabies animals with “three small injections”[3]. That was perfectly unjustified.

It is still difficult to clearly understand his comportment in all circumstances. He was very eager for honors that would enhance his dignity and that of his family, but also believing strongly in the beneficial role of science for the future and the necessity to support it (Fig. 3).

At the end of his life he was exceptionally famous in France and elsewhere. However, he was not a rich man. He was granted a pension of 25,000 francs (more or less the triple of his salary when he was university professor [26]) accorded as a

Fig. 3 Pasteur and his wife in 1888 (archive Romi). That year, Pasteur left his laboratory in the Ecole Normale Supérieure, rue d'Ulm in the Quartier Latin of Paris, to go to the new Institut Pasteur located at some distance. On the one hand, this institute is the achievement of his scientific life but, on the other hand, it was the end of his personal experimentation and a great change in his life

national recompense, to revert first to his widow and then to his children. He owned a family house in Arbois and a small vineyard, not very much by comparison to his contribution to the capacity of production of the fermentation firms (beer, wine, vinegar, etc.), to the silk industry, to the fight against infectious diseases through hygiene (in surgery, for example) and to the development of vaccines.

Pasteur's contribution to the development of vaccines and even to the welfare of humanity is clearly exceptional.

Abbreviations used in this article are BN: (french) Bibliothèque Nationale the references are given directly in the text: (NAF., xxxxx, f.P.x) means (B.N., Nouvelles Acquisitions Françaises, item X or f.P. which means folio Pasteur).

Most of the articles written by Pasteur have been published by his grandson Louis Pasteur Vallery-Radot.

Pasteur Louis Oeuvres de (O.C.) réunies par Pasteur Vallery-Radot, Masson et Cie éditeurs, Paris, 1922–1939, 7 tomes.

Pasteur Louis Correspondance (Cor.), réunie et annotée par Pasteur Vallery-Radot, Flammarion, 1940–1951, 4 tomes.

Acknowledgments My greatest thanks to Stanley Plotkin who helped me to improve this text with competence, kindness, and patience.

References

1. Vallery-Radot R, *M. Pasteur, histoire d'un savant par un ignorant.* Paris: J. Hetzel et Cie, 1884 translated in English by Lady Claude Hamilton, *Louis Pasteur his life and labours by his son-in-law.* New-York: D. Appleton and Company, 1885
2. Vallery-Radot R. *La vie de Pasteur.* Paris: Flammarion, 1900 translated in English by Mrs R. L. Devonshire *The life of Pasteur.* New-York: Garden city Publishing co., Inc.
3. Bazin H. *L'Histoire des vaccinations.* Paris: John Libbey Eurotext, 2008
4. NAF 18014, f.P. 20 verso
5. Cadeddu A. Pasteur et le choléra des poules: révision critique d'un récit historique. *Historical Philosophical Life Science* 1985; 7: 87–104
6. NAF 18014, f.P.23
7. NAF 18014, f.P. 29
8. Pasteur, O.C., t.6, p. 291 & 303
9. Pasteur, O.C., t.6, p. 323
10. Galtier V. *Traité des maladies contagieuses et de la police sanitaire des animaux domestiques.* Lyon: Imprimerie de beau jeune, 1880
11. NAF 18092, item 129
12. NAF 18013, f.P. 2, 8, 51, 55
13. Pasteur, O.C., t.6, p. 112
14. NAF 18016, f.P. 106 & 107
15. Roux E. Louis Pasteur (1822–1895); l'œuvre médicale de Pasteur. In: *L'agenda du chimiste.* Paris: librairie Hachette et Cie, 1896; supplément 1896: 527–48
16. NAF 18016, f.P. 113
17. NAF 18016, f.P. 36
18. Geison GL. *The private science of Louis Pasteur.* Princeton, New-Jersey: Princeton University Press, 1995
19. Arloing Dr S. *Les virus.* Paris: Félix Alcan, 1891
20. NAF 18019, f.P. 62
21. NAF 18019, f.P.79
22. Pasteur, O.C., t.6, p. 575
23. NAF 18019, f.P.3
24. Loir A. *A l'ombre de Pasteur.* Paris; Le mouvement sanitaire, 1938
25. Arloing, Cornevin et Thomas O. Recherches expérimentales sur l'inoculation du Charbon symptomatique et sur la possibilité de conférer l'immunité par injection intra-veineuse. *Journal de médecine vétérinaire et de zootechnie* 1880; 31: 561–9
26. Belèze G. *Dictionnaire universel de la vie pratique.* Paris: librairie Hachette et Cie, 1873

Antituberculosis BCG Vaccine: Lessons from the Past

Marina Gheorgiu

The next decade will celebrate the 100th anniversary of the discovery of the *Mycobacterium bovis* BCG. As a matter of fact, it was on December 28, 1908 that Albert Calmette and Camille Guérin reported to the *Académie des Sciences* that they had obtained a new "race of biliated tuberculosis bacilli" [1]. BCG was not discovered by chance. Based on previous knowledge of tuberculosis and earlier tuberculosis vaccines, BCG was about all the result of the dedicated work of these two personalities, who thoroughly studied the changes occurring during their experiments.

Previous Knowledge

Calmette and Guérin were working in a scientific context of hectic research on tuberculosis (TB), focusing on its transmission and on the production of a vaccine. The period ranging from 1880 to 1904 was a key period for the acquisition of knowledge concerning tuberculosis. Koch's bacillus, the casual agent of the disease, was discovered in 1882. Koch's phenomenon illuminated the sensitization and acquired resistance on guinea pigs when exposed to a secondary tuberculosis infection, and knowledge was acquired from the failure of earlier tuberculosis vaccines. Inspired by Pasteur's successes in obtaining protective vaccines with attenuated microorganisms, such as poultry cholera virus and *Bacillus anthracis* (1882), and by the protection conferred by antibody-inducing toxins, such as those obtained by Behring–Kitasato against diphtheria, numerous scientists tried to obtain tuberculosis vaccines using similar methods. The failure of Koch's tuberculin to prevent the diseases and its disastrous effects on tuberculosis (TB) treatment led to the search for vaccines using different nonpathogenic, attenuated, or killed tubercle bacilli of bovine, human, or of equine origin, instead of components of the organism. All these earlier tuberculosis

M. Gheorgiu (✉)
Laboratoire du BCG, Institut Pasteur, 25 rue du Dr Roux,
75724 Paris cedex 15, France

S.A. Plotkin (ed.), *History of Vaccine Development*,
DOI 10.1007/978-1-4419-1339-5_7,

vaccines were tried in animals and were described by Calmette in his book on tuberculosis [2], but two of them deserve to be mentioned here as we believe they could have inspired him. One of them was the "bovo-vaccin" obtained in 1902 by Behring. It was prepared with human tubercle bacilli attenuated by maintenance, i.e., aging in the laboratory for six and a half years and then desiccating under vacuum. The "bovo-vaccin," which had the merit of being the first to protect against TB, was largely used for "Jennerization" of bovines. It induced protective, short-lasting immune responses, but was variable in its attenuation and presented the risk of human contamination from vaccinated animals. The second anti-TB vaccine was the so-called "tauruman" prepared by Koch. It is worth noting that the method he used to attenuate the human and bovine tubercle bacilli consisted of successive passages on glycerinated broth medium, followed by desiccation. The results were similar to those obtained with the "bovo-vaccin," and the vaccine was therefore abandoned. These early vaccines were obtained using Pasteur's methods, but attenuation of tubercle bacilli failed to be safe and protective.

After returning home from Indochina in 1895, Calmette dedicated himself to research and to the fight against tuberculosis in the town of Lille where TB mortality reached 43%. He created another Pasteur Institute and the first dispensary in which tuberculosis patients were taken care of. Guérin joined him in 1897, as Calmette needed a veterinary surgeon for animal experiments. They first discovered that animals infected with low virulence or very few bacilli resisted a virulent reinfection. Thereafter, their studies were aimed at verifying Behring's hypothesis that pulmonary tuberculosis was acquired not only after respiratory, but also after oral contamination. This was the starting point of the development of BCG [3].

Discovery of BCG

The *M. bovis* strain isolated by Nocard from the milk of a heifer with tuberculosis mastitis, known as "Nocard's milk," was transferred in 1904 to Calmette who used it for his studies. Very fine homogenous bacillary suspensions were needed for oral administration of this strain because it was the only way to facilitate the translocation of bacilli from the intestinal lumen across the mucosa and their dissemination via lymphatics and blood before reaching the lungs. Guérin reported that it was "very difficult to homogenize the dry, tightly clumped bacillary mass grown on glycerinated potato in and agate mortar." They observed that adding a drop of sterile beef bile to the bacilli in the mortar made the homogenization "remarkably easy." Then, they tried "to grow Koch's bacillus on strongly alkaline medium simulating bile." After many unfruitful assays, a bile medium was produced as follows: "potato slides were immersed into 5% of glycerinated beef bile and heated in a water bath at 75°C for 3 h. They were then placed into a tube with a narrow waist at the base of which a new glycerinated bile was added to make contact with the potato slides without submerging them, and autoclaved 30 min at 120°C" (Fig. 1). Successive passages every 21 days on this culture medium modified the initial

Fig. 1 Calmette is seated, Guérin standing. November 1932

M. bovis strain. Guérin described the strain changes as follows: "the first passages were poor and then became abundant. The morphology changed from a hard, rich and scaly mass to a smooth, glassy, pasty bacillary mass" [4]. At the origin, the virulence of the strain was known: 10^{-4} mg killed guinea pigs in 40–60 days. Guérin reported that the initial virulence increased after 1 year of successive passages on this culture medium as 10^{-5} mg killed the guinea pigs after 45–60 days. During the following years, the virulence regularly decreased until the 39th passage which was unable to kill animals. Calmette and Guérin then presented a note at the *Académie des Sciences* in which they described it as a new "race of biliated tuberculosis bacilli." They called it "bacille tuberculeux bilié," which later became "bacilli bilié Calmette–Guérin" and was finally simplified to Bacille Calmette–Guérin (BCG).

Antituberculosis Immunity

Between 1908 and 1921, the strain showed no reversion to virulence over 230 passages on bile-potato medium. On the contrary, after 30 days it conferred resistance to challenge with virulent bovine and human tubercle bacilli. This was found in almost all animal species: bovines, guinea pigs, mice, horses, rhesus monkeys, and chimpanzees, using different doses and routes of administration. As Calmette reported [2]: "immunity existed only if the vaccine-bacille was viable and able to disseminate through the lymphatic system into the host." The relationship between tuberculin sensitization and immunity of infected hosts was also confirmed in BCG-vaccinated animals.

Calmette and Guérin noticed that hypersensitivity to tuberculin injection preceded the acquired resistance and that "only sensitized animals were immunized." Nevertheless, they realized that immunity and hypersensitivity were two distinct and independent states of infected or vaccinated organisms. Calmette also reported that this immunity, first called "pre-munition" from the Latin "premunire," protected against "severe tuberculosis and contributed to the decrease in mortality." At the first BCG International Congress held in 1948, it was agreed that a vaccinated subject submitted to massive and repeated, and/or highly virulent bacilli challenge could develop a tuberculosis reinfection. Thus, immunity to tuberculosis was "relative."

Various researches contributed to the slow progress of the knowledge of the mechanisms involved in tuberculosis [5, 6]. The protective immune response to the disease implies the activation of infected microphages by antigen-specific sensitized T cells and the subsequent killing of intracellular tubercle bacilli. Both CD4 and CD8 T cells are involved. The killing capacity of activated macrophages seems to depend on the production of reactive oxygen and nitrogen intermediates. Their activation is related to multiple factors such as the production of lymphocytes by sensitized lymphocytes, e.g., gamma-interferon (γ-INF) IFN and interleukins IL-2 and IL-12 [7, 8]. Tumor necrosis factor (TNF) also plays a role in granuloma formation which limits bacilli diffusion or, on the contrary, facilitates tissue destruction followed by spread. These contradictory events depend on either the efficacy or the lack of efficacy of immune responses [9]. The contradictory results obtained with the new subunit vaccines [10] show that the conclusions from the past experiences are still valid, i.e., BCG has to be viable, to be able to multiply, and to persist in the target organs (draining lymph nodes, spleen, lungs) to induce immune responses [2, 10–12].

Delayed-type hypersensitivity (DTH) to injected tuberculin still remains a skin test indicator of the cellular immune responses of infected or vaccinated persons. Again, the elders' observation that DTH was distinct from immunity has been confirmed by clinical trials showing a dissociation between protection and the percentage of DTH. Dissociation between T cells inducing one or the other state [10] was also reported in mice.

BCG Vaccination

Once the safety and protective efficacy of BCG were demonstrated in animals, human vaccination became imperative because tuberculosis epidemiology showed 3% morbidity and 20–43% mortality of symptomatic cases. The first baby was thus vaccinated in 1921 by Weill-Hallé at the *Hôpital de la Charité* (presently the *Hôtel Dieu*) in Paris. BCG was given orally in a little spoon and followed by ingestion of milk. Three doses of 2 mg each were administered (at the 3rd, 5th, and 7th days of life), i.e., a total of 6 mg BCG representing 240,000,000 "bacillary units" (colony-forming units, CFU). After 1 year, the dose was increased to reach 10 mg per inoculation, i.e., a total of 30 mg (1,200,000,000 "bacillary units") were administered.

After a very careful follow-up of the first 30 vaccinated babies who were safe and protected against family contact with tuberculosis, BCG vaccination rapidly spread over France and Europe. On the occasion of the Conference of the League of Nations held in Paris in 1928, Calmette reported results covering the period 1921–1926 on more than 50,000 BCG-vaccinated children. It was shown that the mortality rate of previously immunized tuberculosis contacts decreased to 1.8% as compared to 25–32.6% seen in Paris among the nonvaccinated. The conference ended with the unanimous recognition of the safety of BCG, and its use was encouraged [13].

Soon after, in 1929, a catastrophe occurred that cast a cloud over the reputation of the vaccine. In Lübeck (Germany), 252 infants received locally prepared BCG. Seventy-two of them died from tuberculosis, 43 remained healthy, and others had chronic forms of the disease. Subsequent investigations carried out by Bruno and Ludwig Lange, two German experts in tuberculosis, revealed that the BCG vaccine given to children had been locally and accidentally contaminated by a human tubercle bacilli strain (the Kiel strain) under study in the same laboratory [14]. The morphological aspects of the cultures and their virulence in guinea pigs helped to differentiate the attenuated bovine BCG strain from virulent human Kiel strain. The safety of BCG was finally recognized, but the public confidence in BCG was damaged for a while and Calmette died in 1933, disheartened by a long lawsuit [15].

Oral vaccination was discontinued during the 1960s, although a few countries continued to recommend it until recently. The reasons for discontinuation were as follows: postimmunization DTH was low, variable (30–80%), and short-lasting (1 year); BCG oral administration was associated with cervical lymphadenitis; and parenteral routes such as Rosenthal's multiple puncture, Bretey's scarification, and especially Wallgren's intradermal route permitted the injection of a small, precise does of BCG [16] and also had the advantage of inducing a high (90%) and long-lasting (5 years) DTH. However, there is today a renewed interest in mucosal-induced immune responses. Tubercle bacilli inhalation had been first tested by Calmette, using a safe apparatus he had designed, and aerosol BCG vaccination was subsequently taken up by many others. BCG by aerosol induced better immunization than that by the intradermal route [17, 18].

Reasons for the increased interest in oral immunization are multiple: oral BCG is simple to administer; it avoids diseases which could be transmitted by injections such as hepatitis B, AIDS, etc.; and the mucosally associated lymphoid tissue (MALT) is now known to respond with local and systematic immune responses in animals after mucosal stimulation with specific antigens. Wild or recombinant BCG vaccines induced specific local and systematic cellular, humoral, and protective immune responses whether they were administered by the respiratory or oral route [18, 19].

BCG Strain Diversity

The first difference in "vitality" between the Pasteur and the Copenhagen strains was reported by Orskov in 1948 at the first BCG Congress [16]. It was supposed that the differences were due to the successive passages into the Sauton liquid medium without intermediate passages onto bile potato and/or glycerinated potato as was the case at the Pasteur Institute. Later on, in 1956, Dubos reported difference between BCG strains with respect to morphology and multiplication capacity in mice and showed that residual virulence correlated with the degree of protection [20]. Furthermore, numerous individual or collaborative studies under the aegis of the World Health Organization (WHO) and the International Association for Biological Standardization (IABS) confirmed that the current BCG strains were not identical. The WHO therefore recommended the maintenance of BCG strains as freeze-dried primary and secondary seed lots for vaccine production to avoid further mutations through successive passages on culture media [21, 22]. At present, the BCG strains more commonly used in vaccine production are designated by the laboratory or country name where they are kept, e.g., Danish 1331, Glaxo 1077, Connaught (formerly Montreal), Japanese 172, Pasteur 1173P2, Tice (USA), etc. Biochemically, BCG strains can be divided into two groups: the Japanese, Brazilian, Russian strains, which produce a 23-kDa protein, have two copies of IS 986, and contain methoxymycolate. In contrast, the Pasteur and Danish strains carry a single copy of IS 986, do not produce the 23-kDa protein, and do not contain methoxymycolate [12]. The few clinical trials performed so far emphasized differences in reactogenicity, but the impact of a given BCG strain on protection is more difficult to estimate. Differences in immunogenicity are also of major importance for the development of efficient recombinant BCG strains expressing foreign genes [23, 24]. Considerable differences in immunogenicity and protective responses were recently observed, with the Pasteur, Glaxo, Russian strains being better than the Japanese or Prague BCG strains [12].

BCG Vaccine Manufacture

The tiglet and sticky growth of mycobacteria was of great concern to Calmette. Attempts to improve homogenization contributed to the discovery of BCG. The growth characteristics remain a concern for vaccine preparation and standardization. Since Guérin, some laboratories have continued to grow BCG as veils on Sauton's medium. The bacillary mass is obtained by filtration, dispersed by ball milling, and resuspended in protective solutions in order to stabilize the vaccine. The bacillary content in colony-forming units largely differs because of the variability in water content, the degree of dispersion, the killing of bacteria by ball milling homogenization, and the differences in manufacturing processes.

Dispersed grown cultures were first used by Dubos and Fenner, and subsequently by others [25]. At present, many BCG vaccine producers use dispersed grown BCG. These cultures permit obtention of a high ratio of live to dead bacilli per dose and offer better resistance to stabilization by freeze-drying. Thus, the dispersed grown freeze-dried BCG vaccines contain about 50% more viable units per moist weight unit than conventionally obtained vaccines [26, 27].

Because of the diversity of BCG vaccine manufacturing processes, the WHO drew up a booklet on "the requirements for freeze-dried BCG vaccine" to ensure good quality and well-stabilized vaccines. For the same reasons, the quality of the vaccines delivered through the UNICEF is controlled by independent laboratories designated by the WHO.

BCG Efficacy

The efficacy of BCG in the prevention of tuberculosis in children was usually estimated in the past by observing TB mortality, which reached about 25% among unvaccinated subjects and only 2% in vaccinated ones. Mortality was as high as 53% in unvaccinated TB-contact babies from birth to 1 year of age [13]. The relative protection conferred by BCG is subject to continuous controversies. Although problems with efficacy have been recognized from the beginning, the urgency of fighting a public scourge imposed BCG vaccination. However, a number of prospective clinical trials have been carried out, from the first by Aronson among the North Indian population to the most recent one in India [28]. Table 1 shows that protective efficacy has varied widely in different parts of the world and its impact on the control of tuberculosis worldwide remains unclear. The analysis of these divergent results demonstrated that methodological bias had contributed to conflicting data. Moreover, factors interfering with immune responses, such as contamination with environmental mycobacteria, problems with follow-up of TB cases in young children, genetic diversity of vaccinated populations, living conditions, and the quality of freeze-dried vaccine strains, could have interfered with BCG efficacy [29]. A recent meta-analysis of published literature on the efficacy of the BCG vaccine in the prevention of tuberculosis showed that it prevented 80%

Table 1 Prospective BCG human trials

Trials	TB deaths/ 100,000	BCG (mg)	No vaccinated	TB	No non-vaccinated	TB	Protection (%)
Aronson 1946–49 US	200	0.1	123	4	139	11	80
UK BMC 1950–72	35	0.1	20,500	18	19,600	97	78
US Comstock Palmer	30	0.2	16,200	26	17,854	32	14
India 1968–70	200	0.1	88,200	162	44,135	44	0

of severe forms of tuberculosis, including meningitis and miliary, and some 50% of mild forms [30].

Tuberculosis still remains a public scourge in all the developing countries, while its incidence is increasing in the industrialized countries. Multidrug resistance and HIV-related tuberculosis are of great concern. Even if imperfect, BCG remains one of the least expensive vaccines with few self-limited side effects and is a generally efficient tool of prevention. This is why it is still in the WHO Expanded Program of Immunization [31]. New and better vaccines are not yet available, and even if soon produced experimentally, it will take a long time before they are accepted for use in humans.

We can therefore expect that the BCG vaccine will celebrate its 100th anniversary while still in use.

References

1. Calmette A, Bocquet A, Nègre L. Contribution à l'étude du bacille tuberculuex bilié. *Ann Inst Pasteur* 1921;9:561–70
2. Calmette A. *L'infection bacillaire et la tuberculose.* Paris: Masson, 1928:771–864
3. Calmette A, Guérin C. Origine intestinale e la tuberculose pulmonaire. *Ann Inst Pasteur* 1905;19:601–18
4. Guérin C. Le BCG et la prévention de la tuberculose. *Rev Atomes* 1948;27:183–8
5. Lurie MB. *Resistance to Tuberculosis.* Cambridge, MA: Harvard University Press, 1964
6. Mackaness GB, Blanden RV. Cellular immunity. *Prog Allergy* 1967;11:89–140
7. Dannenberg AM Jr., Rook GAW. Pathogenesis of pulmonary tuberculosis: a interplay of tissue-damaging and macrophage-activating immune responses. In: Bloom RR, ed. *Tuberculosis Pathogenesis, Protection and Control.* Washington, DC: ASM Press, 1994;459–83
8. Kaufman SHE. Immunity to intracellular bacteria. In: Paul WE, ed. *Fundamental Immunology,* third edition. New York: Raven Press, 1993
9. Orme IM. Immunity to mycobacteria. *Curr Opin Immunol* 1993;5:497–502
10. Roberts AD, Sonnenberg MG, Ordway DJ et al. Characteristics of protective immunity engendered by vaccination of mice with purified culture filtrate protein antigens of Mycobacterium tuberculosis. *Immunology* 1995;85:502–8
11. Collins FM. The immune response to mycobacterial infections: development of new vaccines. *Vet Microbiol* 1994;797:1–16
12. Lagranderie M, Balazuc AM, Deriaud E, Leclerc C, Gheorghiu M. Comparison of immune responses of mice immunized with five different *Mycobacterium bovis* BCG vaccine strains. *Infect Immun* 1996;64:1–9
13. Calmette A, Guérin C, Nègre L, Bocquet A. Prémunition des nouveau-nés contre la tuberculose par le vaccin BCG (1921–1926). *Ann Inst Pasteur* 1926;2:89–120
14. Lange B. Nouvelles recherches sur les causes des accidents de Lübeck. *Rev Tuberc* Extrait 1931;XII:1142–70
15. Bernard ML. Le drame de Lübeck. *Bull Acad Natl Méd* 1931;106:673–82. Premier Congrés International du BCG, Inistut Pasteur, Paris 1948
16. *Premier Congrés International du BCG.* June 18–23, 1948. Paris: Institut Pasteur, 1948
17. Rosenthal SR. Routes and methods of administration. In: Rosenthal SR, ed. *BCG Vaccine:Tuberculosis-Cancer.* Littleton MA: PSG Publishing Company Inc, 1980;146–75
18. Gheorghiu M, BCG-induced mucosal immune responses. *Int J Immunopharmacol* 1994;16:435–44

19. Lagranderie M, Murray A, Gicquel B, Leclerc C, Gheorghiu M. Oral immunization with recombinant BCG induces cellular and humoral immune responses against the foreign antigen. *Vaccine* 1993;11:1283–90
20. Dubos RJ, Pierce CH. Differential characteristics in vitro and in vivo of several substrains of BCG. *Amer Rev Tuberc* 1956;74:655–717
21. WHO. Requirements for dried BCG vaccine. *Tech Report* 1979;Ser 638:116–47
22. Gheorghiu M, Augier J, Lagrange PH. Maintenance and control of the French BCG strain $1173P_2$ (Primary and secondary seed-lots). *Bull Inst Pasteur* 1983;81:281–8
23. Stover CK, de la Cruz VF, Fuerst TR et al. New use of BCG for recombinant vaccines. *Nature* 1991;351:456–60
24. Winter N, Lagranderie M, Rausier J et al. Expression of heterologous genes in *Mycobacterium bovis* BCG: induction of a cellular response against HIV1 Nef Protein. *Gene* 1991;109:47–54
25. Dubos RJ, Fenner F. Production of BCG vaccine in a liquid medium containing Tween 80 and a soluble fraction of heated human serum. *J Exp Med* 1950;91:261–84
26. Gheorghiu M, Lagrange PH, Fillastre C. The stability and immunogenicity of a dispersed-grown freeze-dried Pasteur BCG vaccine. *J Biol Standard* 1988;16:15–26
27. Gheorghiu M, Lagranderie M, Balazuc AM. Stabilization of BCG vaccines. *Der. Biol. Stand* 1996;87:251–61
28. WHO. Tuberculosis prevention trials: Madras. Trial of BCG vaccines in South India for tuberculosis prevention. *Bull WHO* 1979;57:819–27
29. Bloom BR, Fine PEM. The BCG experience: implications for future vaccines against tuberculosis. In: Bloom RR, ed. *Tuberculosis Pathogenesis, Protection and Control.* Washington DC: ASM Press, 1994;531–57
30. Colditz GA, Brewer TF, Berkey CS et al. Efficacy of BCG vaccine in the prevention of tuberculosis. Meta-analysis of the published literature. *JAMA* 1994; 271:698–702
31. WHO. BCG vaccination politics. *Tech Rep* 1980; Ser 652

A History of Toxoids

Edgar H. Relyveld

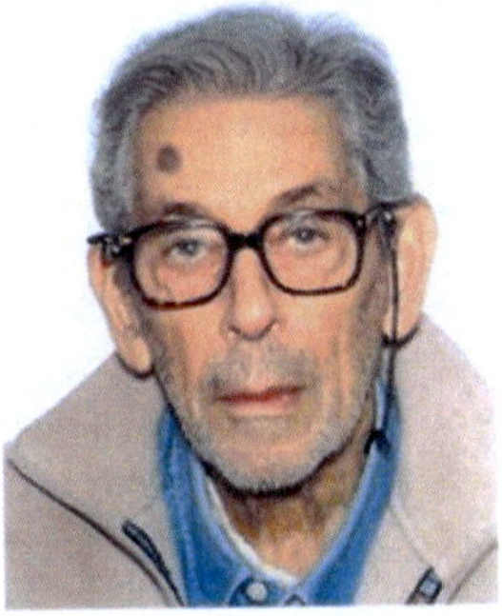

Introduction

Vaccination is the ultimate way to prevent infectious diseases. Many microbial toxins have been isolated, purified, and toxoided to prepare immunizing agents that are well-tolerated and provide long-lasting protection.

The description of toxins and their production, purification, and detoxification to prepare single, combined fluid or adsorbed vaccines has already been presented in several scientific publications and handbooks [1–8], but research is still going on to develop new preparations without side-effects and exhibiting solid immunity.

Some preparations were introduced many years ago without significant improvements, but advances have also been achieved to obtain vaccines devoid of untoward reactions that use economical procedures necessary for large scale production and, therefore, protection of man in developing countries all over the world, as well as for veterinary use.

E.H. Relyveld (✉)
6 rue du, Sergent Maginot, 75016 Paris, France

S.A. Plotkin (ed.), *History of Vaccine Development*,
DOI 10.1007/978-1-4419-1339-5_8, © Springer Science+Business Media, LLC 2011

Despite the discovery of safe and potent tetanus and diphtheria toxoids more than 70 years ago, both diseases are still frequent, even in some industrialized countries. We even had new diphtheria epidemics reported during the last 5 years in the New Independent Russian Republics that are now spreading over Europe and other foreign countries, e.g., Algeria and Turkey. Reasons for the resurgence of diphtheria epidemics may include low immunization coverage and the use of adult-type vaccine (Td) in the vaccination of children to prevent side-effects [9, 10]. Even though tetanus and diphtheria vaccination is compulsory in France, where the number of cases went down drastically since the introduction of toxoids, recent studies, especially using a new simple hemagglutination test [11], showed that 19% of the population was not protected against tetanus. Most of the subjects were over 60 years old, mostly women. A recent serological survey [12] also showed significant difference in protection against diphtheria according to the subject's age: 32.3% of people 65 years old, mostly women, were not protected.

Tetanus still remains a major health problem in developing countries, and the number of estimated neonatal cases was over 500,000 in 1992. A revised plan of action for eradication of this disease has been initiated by the World Health Organization (WHO) [13].

The major objective in writing this chapter has been, therefore, to not only compile information on the history of the two most frequently used diphtheria and tetanus vaccines in the world, but also to provide new guidelines for the preparation and application of these vaccines in a near future.

Discovery of Diphtheria and Tetanus Toxins, Toxoids and Antitoxins

Diphtheria is a widespread disease, known since ancient times, and epidemics with high mortality rates were common. However, the disease was not differentiated from other throat distempers until its description by Bretonneau of Tours, France in 1826, who gave the disease its name according to the leathery appearance of the diphtheritic membranes.

The bacillus was seen for the first time in 1883 by Klebs [14] in smears of the membranes and was isolated in a pure culture by Loeffler in 1884 [15], who also showed that the bacteria produced that fatal disease in animals. The bacilli were, however, only found at the level of the local lesions and were absent from remote organs which exhibited the typical histopathological appearance of the disease. Loeffler has already postulated the presence of a toxin secreted by the bacilli. In 1888, Roux and Yersin [16] established a correlation between the localized infection and the remote lesions, which could be produced by heat-labile exotoxin present in culture filtrates of the bacteria. Two years later, in 1890, von Behring and Kitasato [17, 18] showed that specific antitoxin was present in the serum of animals that had received sublethal doses of the toxin. This antitoxin serum had the capacity to neutralize the toxin present in culture filtrates. Serum therapy in children was initiated 1 year later, using antitoxins raised in sheep and horses.

Boston Ramon
1886–1963

Due to serum therapy, diphtheria mortality in Paris dropped from 147 per 100,000 persons to 35 per 100,000 persons. A neutral mixture of diphtheria toxin and horse antitoxin was used by Theobald Smith [19, 20] to immunize guinea pigs and horses, and later by von Behring for children [21].

After the introduction of the Schick test in 1913 [22–24], many unprotected, Schick-positive children were successfully immunized with toxin–antitoxin mixtures. A large scale immunization program of school children in New York City was undertaken by Park in 1922 [25]. This hazardous method was used until 1924 and then replaced by toxoid. In 1914, Loewenstein added 0.2% formalin to tetanus toxin and, after incubation at 34°C, nearly succeeded in producing toxoid [26]. He also studied the action of formalin on diphtheria toxin, and later studied toxin-antitoxin mixtures.

It was shown in 1923 by Glenny and Hopkins [27] that diphtheria toxin could be converted into toxoid by the action of formalin. Its toxicity was thus reduced, but the resulting product was useful as an immunizing agent in animals only in combination with antitoxin [28].

A stable, atoxic diphtheria toxin was finally produced by Ramon in 1923 [29] by the action of formalin on crude toxin, followed by incubation at 37°C for several weeks. This same procedure was applied to prepare tetanus and several other toxoids used for the immunization of people worldwide.

Through experimentation in animals, Nicolaier found in 1884 [30] that tetanus was associated with bacilli in the soil. He thought that tetanus was not caused by bacilli, but rather by diffusion of a poison produced in the soil-contaminated wound. In 1889, Kitasato isolated the causative agent in a pure form, based on the resistance of the organism to heating [31]. The existence of tetanus toxin was demonstrated in 1890 by Knud Faber [32], and tetanus antitoxin was produced the same year by von Behring and Kitasato after immunizing rabbits with the attenuated toxin [17].

Later on, tetanus antitoxin was prepared in large quantities by immunization of horses with small quantities of toxin followed by increasing doses, or with attenuated toxin. Many attempts were made, without success, to prepare a safe, stable and potent toxoid for human use.

A procedure to obtain highly antigenic toxoids by treatment of crude toxins with formalin and incubation at 37°C for several weeks, was reported by Ramon in 1923 [29] and is still used today for toxoiding tetanus toxin. The first human vaccinations with tetanus toxoid were successfully carried out by Ramon and Zoeller in 1926 [33].

Preparation of Diphtheria Toxin

Diphtheria toxins were originally prepared by growing the strains in complex liquid media of meat extracts and peptone in Fernbach or Roux bottles for at least 10 days. Semisynthetic media containing casein hydrolysate, or even synthetic media, were later used. Production of toxin under static conditions required a large number of bottles, and the titers were low, about 50–100 Lf (flocculation units) per mL. The titers could vary, depending mostly on the iron concentration of the medium, and these crude filtrates were rather difficult to purify [34, 35].

Diphtheria toxin is presently prepared by first growing a lyophilized strain of *C. diphtheriae* (PW no 8) on Loeffler's serum plate for 40 h are 37°C. From this plate, a loop is transferred to several 500-mL bottles containing 100–200 mL of sterile Linggood medium [36] and shaken for 33 h at 35°C. Cultivation is carried out in glass, aluminum or stainless steel fermenters by submerged culture for 40 h at pH 7.8 under a continuous stream of air, according to methods described by the WHO [37] and van Hemert [38]. Titers of up to 350 Lf units per mL are obtained, and the toxin prepared in fermenters is easily purified to prepare highly immunogenic vaccines.

Preparation of Tetanus Toxin

Tetanus toxin has been produced for many years by cultivation of *Clostridium tetani* in 5-L flasks on complex medium consisting of enzymatic digests of beef meat and liver [39]. Cultivation is carried out for 10 days under anaerobic conditions, and the crude toxin is harvested by filtration. Semisynthetic media are presently used: they are made of tryptic casein digests supplemented with beef heart infusion [40, 41].

Highly toxigenic *C. tetani* strains have been selected, and special digest have been developed (NZ Case, Tryptose T or TS) for large scale production of the toxin in high yields on protein-free media [42]. Toxin production was initially carried out on casein digest media in beakers or stainless steel containers, but is now produced in high capacity fermenters under strictly controlled conditions of temperature, pH, agitation, gas flow, etc. Bacterial growth is prolonged for about 1 week to allow intracellular toxin (tetanospasmin) to be discharged into the culture medium as a result of cell lysis [38, 43–45].

Tetanus toxin is synthesized intracellularly as a single protein of 150,500 Da. The toxin released into the medium is cleaved by proteases into a nicked protein derivative with an NH_2-terminal light chain A of 52,300 Da and a COOH-terminal heavy chain B-C of 98,300 Da, which are linked by a disulphide bridge. One milligram of tetanus toxin contains about 25×10^6 mouse lethal doses [46].

Preparation of Purified Toxoids

Toxoids were first prepared by addition of formalin to the crude toxin. The formaldehyde reaction results in a cross-linkage between the ε-amino groups of lysine and a second amino, imidazole, or phenol group, leading to a stable methylene bridge. These reactions can take place between amino acids of the same toxin molecule, between two toxin molecules, or between a toxin molecule and other bacterial proteins (accessory antigens) or peptides of the culture medium.

Although these toxoids have proven to be both effective and safe immunogens, their administration to man can give rise to adverse reactions. Untoward reactions have been related to the artificial incorporation into the toxoid proper of foreign compounds present in the complex media as a result of the cross-linking action of formaldehyde. Composite derivatives that are not eliminated by purification are thus formed. Therefore, purified toxins should be used for toxoid preparation [35, 47–51].

Purification and crystallization of diphtheria toxin produced in fermenters is easily achieved by the successive steps of ultrafiltration and salt fractionation [52]. Procedures for toxoiding which have been described in detail, must be done under very specific conditions, pH 8.55 in the presence of 0.01 M lysine or another amino acid, to avoid any risk of toxin reversal [53].

The properties of highly purified diphtheria and tetanus toxins have been reported elsewhere [35]. Millions of doses of stable vaccines devoid of untoward reactions in babies, children, and adults have been produced. Controlled production methods also guarantee safety to the vaccinated employees.

The Use of Adjuvants

Purified fluid toxoids are poor immunogens and adjuvants must be added to enhance their immunogenicity. The use of adjuvants to stimulate the immune response and increase the level of circulating antibodies was discovered by Ramon in 1925 [54–57]. Ramon observed that the highest titers were obtained in animals having an abscess of inflammation at the side of inoculation. After some preliminary trials, he showed by injection of horses with a mixture of toxoid and sterilized tapioca that good results were achieved through the provocation of an inflammatory reaction at the inoculation site. He also insisted on the slow release of the toxoid, and published his findings on the "Adjuvanting and stimulating action of agents for immunity" in 1925. Different types of starch were tested, but the best results were obtained with tapioca, which elicited a local reaction, was slowly resorbed by the organism and devoid of unfavorable side-effects. Other products had also been tested such as calcium, magnesium, aluminum salts as well as lanolin, tannin, kaolin, carbon, and even bread crusts.

About 1 year later, Glenny et al. showed that neutralized, alum-precipitated, crude diphtheria toxoid gave a much higher immune response after only one injection than the same fluid toxoid [58]. To diminish local reactions, the precipitate was washed with saline; the adsorbed vaccine completely lost its irritating effects, while still having a high immunostimulating capacity [59]. The precipitate obtained by adding alum to crude diphtheria toxoid was, in fact, a mixture of aluminum hydroxide [$Al(OH)_3$] and aluminum phosphate ($AlPO_4$) because of the presence of phosphates in the culture media used for toxin production. Both adjuvants have, since then, been used for the preparation of billions of doses of single and combined vaccines, particularly T, diphtheria tetanus pertussis (DTP) and DTP Polio, and have been inoculated into adults, children and babies. The procedures for preparation of aluminum-adjuvanted toxoids have been published [37, 45, 60] and even recently reviewed [61, 62], and are therefore not reported again here. However, several studies have provided evidence that animals, as well as in humans, aluminum adjuvants increase the level of antigen-specific and total IgE antibodies, and may promote IgE-mediated allergic reactions [63–70, 71].

Acknowledgements The author is indebted to Prof. B. Bizzini, Drs. M. Huet and L. Lery for their helpful discussions, to Dr. M.A. Fletcher for his editorial assistance and to C. Raoul and S. Hermann for typing and editing the text.

References

1. Ajl SJ, Ciegler A, Kadis S, Montie TC, Weinbaum G. *Microbial Toxins* Vol I-III. New York: Academic Press, 1970–1972
2. Alouf JE, Fenrenbach FJ, Freer JH, Jeljaszewicz J. *Bacterial Protein Toxins*. London: Academic Press, 1984
3. Falmage P, Alouf JE, Fenrenbach FJ, Freer JH, Jeljaszewicz J, Thelestam M. *Bacterial Protein Toxins*. Stuttgart: Gustav Fisher, 1986
4. Germanier RG. *Bacterial Vaccine*. Orlando: Academic Press, 1964
5. Harshman S. *Microbial Toxins: Tools in Enzymology, Methods in Enzymology,* vol 165. San Diego: Academic Press, 1988
6. Woodrow GC, Levine MM. *New Generation Vaccines*. New York: Marcel Dekker, 1990
7. Singh H, Bhatia R. *Vaccines, Prospects and Perspectives*. Delhi: Forward Publishing Company, 1993
8. Plotkin SA, Mortimer EA. *Vaccines.*Philadelphia: WB Saunders Comp, 1994
9. CDC. Diphtheria epidemic – new independent states of the former Soviet Union, 1990–1994. MMWR 1995;44:17–81
10. Maurice J. Diphtérie: L'épidémie se léve à l'Est. *J Int Med* 1955;343:12–13
11. Relyveld EH, Huet M, Lery L. Passive haemagglutination tests using purified antigens covalently coupled to turkey erythrocytes. *Dev Biol Stand* 1995;86:225–41
12. Vincent-Ballereau F, Schrive I, Fisch A et al. La population adulte française est-elle protégée de la diphtérie en 1995. Résultats d'une equête sérologique multricentrique. *Méd Mal Infect* 1995;25:622–6
13. OMS. Global immunization of children once again on the rise, saving 3 million lives annually. Press Release WHO/73; 3 October 1995

14. Klebs E. Ueber Diptheria. *Verh Deutsch Kongr Inn Med* 1883;2:139–54
15. Loeffler F. Untersuchungen über die Bedeutung der Mikroorganismem für die Entsthung der Diptherie beim Menschen, bei der Taube and beim Kalbe. *Mitt Klin Gesund Berlin* 1884,2:629–61
16. Roux E, Yersin A. Contribution à l'étude de la diphtérie. *Ann Inst Pasteur* 1888;2:421–99
17. Behring von E, Kitasato S. Ueber das Zustandekekommen de Diphtherie – Immunität und der Tetanus – Immunität bei Tieren. *Dtsch Med Wschr* 1890;16:1113–4
18. Behring von E. Untersuchungen über das Zustandekekommen de Diphtherie - Immunität bei Tieren. *Dtsch Med Wschr* 1890;16:1145–8
19. Smith TH. The degree and duration of passive immunity to diphtheria toxin transmitted by immunized females guinea-pigs to their offspring. *J Med Res* 1907;16:359–79
20. Smith TH. Active immunity produced by so-called balanced or neutral mixtures of diphtheria toxin and anti-toxin. *J Exp Med* 1909;11:241–56
21. Behring von E. Ueber ein neues Diptherieschützmittel. *Dtsch Med Wschr* 1913;39:873–6
22. Schick B. Kutanreaktion bei Impfung mit Diptherietoxin. *Münch Med Wschr* 1908;55:504–6
23. Schick B. Die Diptherietoxin – Hautreaktion des Menschen als Vorprobe de prophlaktischen Diptherieheil-seruminjektio. *Münch Med Wschr*1913;60:2608–10
24. Relyveld EH, Hénocq E, Raynaud M. Etude de la reaction de Schick à l'aide d'une toxine pure. *Bull Acad Méd* 1962;146:101–9
25. Park WH. Toxin-antitoxin immunization against diphtheria. *Amer Med Ass* 1922;79:1584–90
26. Loewenstein E. Euber Immunisierung mit antischen Toxinen und mit übercompensierten Toxin-antitoxin Muschungen bei Diptherie. *Z Exp Path Ther* 1914;15:279–321
27. Glenny AT, Hopkins BE. Diphtheria toxoid as an immunizing agent. *Brit J Exp Pathol* 1923;4:283–8
28. Ebisawa I. Three to four instead o one millilitre of formalin. *Vaccine* 1996;14:247
29. Ramon G. Sur le pouvoir floculant et sur le propriétés immunisantes d'une toxine diphtérique rendue anatoxique (anatoxine). *CR Acad Sci* (Paris) 1923;177:1338–40
30. Nicolaier A. Ueber infectiösen Tetanus. *Dtsch Med Wschr* 1884;10:842–4
31. Kitasato S. Ueber den Tetanus bacillus. *Z Hyg Infektionskr* 1889;7:225–34
32. Faber K, Die Pathogenie des Tetanus. *Berl Klin Woch* 1890;27:717–20
33. Ramon G, Zoeller C. De la valeur antigénique de ;'anatoxine tétanique che l'homme. *CR Acad Sci* (Paris) 1926;182:245–7
34. Relyveld EH. *Toxine et antitoxine diphthériques – Étude immunologique.* Paris: Editions Scientifiques Hermann, 1959
35. Relyveld EH. Current developments in production and testing of tetanus and diphtheria vaccines. In: Mizrahi A, Hertman T, Klungberg MA, Hohn A, eds. *New Developments with Human and Veterinary Vaccines.* New York Alan R Liss, 1980;51–76
36. Linggood FV. Purification of diphtheria toxin and toxoid made from tryptic digest broths. *Br J Exp Pathol* 1941;22:255–61
37. WHO (BLG) UNDP(77.1 Rev 1) Manual for the production and control of vaccines; diphtheria toxoid
38. van Hemert P. *Vaccine Production as a Unit Process.* Delft: Thesis, 1971
39. Prévot AR, Boorsma HJ. Répartition d l'azote et métabolisme azoté dan la toxinogénèse tétanique. *Ann Inst Pasteur* 1939;63:600–10
40. Mueller JH, Miller PA. Production of tetanal toxin. *J Immunol* 1945;50:377–84
41. Mueller JH, Miller PA. Variable factors influencing the production of tetanus toxin. *J Bact* 1954;67:274–7
42. Latham WC, Bent DF, Levine L. Tetanus toxin production in the absence of protein. *Appl Microbiol* 1962;10:146–52
43. Bizzini B, Turpin A, Raynaud M. Production et purification de la toxine tétanique. *Ann Inst Pasteur* 1969;116:686–712
44. Neilsen PA. Large scale production of tetanus toxoid. *Appl Microbiol* 1967;15:453–4
45. WHO (BLG/UND P/77.2 Rev 1) Manual for the production and control of vaccines:tetanus toxoid
46. Bizzini B. The chemistry of tetanus toxin. In: Veronesi R, ed. *Tetanus: Important New Concepts*. Amsterdam: Excerpta Medica, 1981:8–17

47. Bizzini B. Tetanus. In: Germanier R, ed. *Bacterial Vaccines*. Orlando: Academic Press, 1984;1–36
48. Pappenheimer AM. Diphtheria. In: Germanier R, ed. *Bacterial Vaccines*. Orlando: Academic Press, 1984;38–68
49. Rappouli R. New improved vaccines against diphtheria and tetanus. In: Woodrow CG, Levin MM, eds. *New Generation Vaccines*. New York: Marcel Dekker, 1990;251–68
50. Relyveld EH, Hénocq E. Bizzini B. Studies on untoward reactions to diphtheria and tetanus toxoids. *Dev Biol Stand* 1979;43:33–7
51. Relyveld EH. Recent developments with vaccines against bacterial protein toxins. In: Alouf JE, Fehrenbach FJ, Freer JH, Jeljaszewicz J, eds. *Bacterial Protein Toxins*. London: Academic Press, 1948;217–26
52. Relyveld EH, Raynoud M. Préparation de la toxine depthérique pure cristallisée à partir de cultures en fermentuer. Caractéres de la toxine pure. *Ann Inst Pasteur* 1964;107:618–37
53. Relyveld EH. Studies on the detoxification of purified diphtheria toxin. *Progr Immunobiol Standard* 1969;3:258–63
54. Ramon G. Sur la production de l'antitoxine diphtérique. *CR Soc Biol* 1925;93:508–9
55. Ramon G, Descombey P. Sur l'immunisation atiténtique et sur la production de l'antitoxine tétanique *CR Soc Biol* 1925;93:508–9
56. Ramon G. Sur la production des antitoxins. *CR Acad Sci* 1925;181:157–9
57. Ramon G. *Quarante Années de Recherches et de Travaux*. Toulouse: L'Imprimerie Régionale, 1957
58. Glenny AT, Pope CG, Waddington H, Wallace U. The antigenic value of toxoid precipitated by potassium alum. *J Path Bact* 1926;29:38–9
59. Glenny AT, Barr M. Alum-toxoid precipitates as antigens. *J Path Bact* 1931;34:118
60. Holt LB. *Developments in Diphtheria Prophylaxis*. London: Heinemann, 1950
61. Gupta RK, Rost BE, Relyveld E, Siber GR. Adjuvant properties of aluminum and calcium compouds. In: Powell MF, Newman MJ eds. *Vaccine design: The Subunit and Adjuvant Approach*. New York: Plenum Press, 1995;229–48
62. Lindbald EB. Aluminum adjuvants. In: Stewart-Tull DES, ed. *The Theory and Practical Application of Adjuvants*. Chicester: John Wiley, 1995;21–35
63. Edelman R, Hardegree MC, Chedid L. Summary of an international symposium on potentiation of the immune response to vaccines. *J Infect Dis* 1980;141:103–12
64. Food and Drug Administration, Center for Drugs and Biologics. Summary Minutes Allergenic Products Advisory Committee and Report of Safety Considerations for the Aluminum Component of Alum-Precipitated Allergenic Extracts. CDB, Office of Biologics Research and Review, Biologics Information Staff (NFN-20) FDA, Bethesda, MD 20892, USA 1987
65. Goto N, Akama K. Histopathological studies of reactions in mice injected with aluminum adsorbed tetanus toxoid. *Microb Immunol* 1982;26:1121–32
66. Gupta RK, Relyveld EH. Adverse reactions after injection of adsorbed diphtheria-pretussis-tetanus (DTP) vaccine are not only due to pertussis organisms or pertussis components in the vaccine. *Vaccine* 1991;9:699–702
67. Matuhasi T, Ikegami H. Elevation of levels of IgE antibody to tetanus toxin in idividuals vaccinated with diphtheria-pertussis-tetanus vaccine. *J Infect Dis* 1982;146:290
68. Nagel JE, White C, Lin MS, Fireman P. IgE sythesis in man. II. Comparison of tetanus and diphtheria IgE antibody in allergic and nonallergic children. *J Allergy Clin Immunol* 1979;63:308–14
69. Paulwels R, Bazin H, Platteau B, van der Straeten M. The influence of different adjuvants on the production of IgD and IgE antibodies. *Ann Immunol (Inst Pastuer)* 1979;130C:49–58
70. Vassileb TL. Aluminum phosphate but not calcium phosphate stimulates the specific IgE response in guinea-pigs to tetanus toxoid. *Allergy* 1978;33:155–9
71. Nagel JE, White C, Lin MS, Fireman P. IgE sythesis in man. II. Comparison of tetanus and diphtheria IgE antibody in allergic and nonallergic children. *J Allergy Clin Immunol* 1979;63:608–14

Vaccination Against Typhoid Fever: A Century of Research. End of the Beginning or Beginning of the End?

Philippe Sansonetti

Introduction

As its name implies, typhoid fever is a septicemic illness associated with alteration in consciousness (the Greek word "typhos" means "stupor"). It is cause by the Gram-negative bacillus *Salmonella typhi* which belongs to the enterobacteriaceae family. In its untreated form, the disease is marked by increasing fever, headache, insomnia, and general malaise. More serious complications include intestinal perforations and bleeding, severe alteration in consciousness, and possibly shock, which can be fatal. Similar but usually less severe cases, called paratyphoid fever, are caused by *Salmonella paratyphi* A, B, and C. Together, *S. typhi* and *S. paratyphi* infections are referred to as enteric fevers.

P. Sansonetti (✉)
Unité de Pathogénie Microbienne Moléculaire, INSERM U389, Institut Pasteur, 28 rue du Docteur-Roux, 75724 Paris, France
e-mail: Psanson@pasteur.fr

S.A. Plotkin (ed.), *History of Vaccine Development*,
DOI 10.1007/978-1-4419-1339-5_9, © Springer Science+Business Media, LLC 2011

Theobald Smith
1859–1934

Human beings appear to be the only reservoir of *S. typhi*, which is usually acquired by ingestion of food or water that have been contaminated by fecal material. The risk of infection is correlated in the developing world, compared to other endemic enteric infections such as cholera and related infections, bacillary dysentery and diarrhea of viral origin. About 16–33 million estimated cases still occur each year in the developing world, with more than 600,000 deaths: a toll, in terms or mortality, equivalent to bacillary dysentery. The peak incidence is between 5 and 20 years of age.

Historical Overview of Typhoid Fever

In an epic 50-year period in the late nineteenth century, typhoid fever was separated from other febrile fevers, its mode of contamination was described and its causative agent indentified. This led the way to a vaccine that became available by the end of the same century. Considering the limited means of clinical and paraclinical investigations available and the embryonic stage of microbiology, the speed at which the entire problem was solved is amazing. Fifty more years were necessary before chloramphenicol became available.

In 1829, PCA Louis in Paris described typhoid and related the clinical symptoms to lesions observed in the intestines, mesenteric lymph nodes, and spleen [1]. Schoenlein, in 1839, clearly differentiated two distinct forms of typhus, thus confirming the typhoid entity ("typhus exanthematicus" versus "typhus abdominalis"). Bretonneau [2] and Smith then recognized the contagious nature of the disease as well as the immunity conferred by the disease. In 1873, Budd showed that bowel discharge was the major mode of water contamination [3], and in 1880, Eberth finally indentified the etiologic agent from a patient's tissue sample [4]. In 1884, this discovery was rapidly confirmed by Gaffky, who cultivated and isolated *S. typhi* in pure culture from the spleen of typhoid patients [5]. Finally, while AE Wright and F Smith developed an agglutination test for Malta fever, Widal reported that convalescent serum agglutinated and immobilized *S. typhi* in vitro [6], and, with Gruber, Durham and Pfeiffer, this led to the discovery of the agglutination test which became the classic serological test for diagnosis of infection by *S. typhi*.

It is likely that a discussion between R Pfeiffer and A Wright, the former reporting in the latter that similar agglutinins were present in the serum of a man injected subcutaneously with a heat-killed preparation of typhoid bacilli, set the stage for the development of a typhoid vaccine and for a great scientific controversy.

Introduction of Typhoid Vaccination: A Wright or R Pfeiffer?

It now seems clear that this controversy should no longer continue and that both authors be given equal credit for the discovery of typhoid vaccine [7].

In September 1896, in the *Lancet* article, Almoth Wright, a British researcher, reported on the use of oral calcium chloride for the treatment of serous hemorrhages in patients suffering from defective blood coagulopathies, and also demonstrated its usefulness to control local side-effects after subcutaneous injection of typhoid bacilli [8]. In this article, he quoted experiments with heat-kill, phenol-preserved typhoid vaccine in a horse, and subsequently in two officers of the Indian Medical Corps, one of whom then received an inoculum of the wild type *S. typhi* and appeared protected. Wright was then sufficiently obstinate and convincing to obtain evaluation of his vaccine in 2,835 volunteers in the Indian Army [9]. Despite local and generalized adverse reactions that were carefully examined by several panels of experts, results were considered sufficiently convincing to decide to vaccinate British troops engaged in the Boer War. As a consequence of this constant pressure, by World War I, typhoid vaccination had almost become routine in the British Army.

In November 1896, German scientists Richard Pfeiffer and Wilhelm Kolle published an article about prophylactic vaccination of humans against typhoid fever [10]. They demonstrated, following similar experiments conducted with the cholera bacillus, that the sera of patients convalescing from typhoid fever and of animals infected experimentally with *S. typhi* contained protective antibodies in the guinea-pig protection and agglutination test. They did not quote Wright's paper, but instead referred to earlier work of other scientists who had used typhoid vaccines for therapeutic purpose.

Years of debate followed, personal ambition, and scientific emulation being exacerbated by growing political tension between England and Germany on the eve of World War I. It now seems that many groups were independently trying to develop a typhoid vaccine and that Wright and Pfeiffer were the first to succeed.

Typhoid Fever Morbidity and Mortality in Relation to Vaccination: A Debate

It is generally considered that the large introduction of various forms of killed cellular typhoid vaccine in Europe, North America, Australia, and Japan as the turn of the twentieth century, and the sustained immunization policies that followed, contributed to the tremendous drop in the incidence of typhoid fever in these regions. This view may be considered rather simplistic since typhoid fever morbidity had already started to decrease at the end of the nineteenth century, due to the constant improvements in hygiene, particularly in urban areas.

The best evidence that vaccination was effective was the spectacular drop in the number of cases of typhoid fever among the military personnel who were involved in the various conflicts that marked our century. From September 1914 to May 1915, 65,748, cases of typhoid fever were reported in the French army, with 11,000 deaths. After immunization became systematic, the incidence of typhoid fever dramatically dropped. During the last year of the war, only 615 cases were reported, despite poor hygiene conditions in trenches. Whether typhoid vaccination "saved" World War I is of course another issue. Likewise, from September 1939 to May 1940 the number of reported cases was 144 (with 5 deaths) in a largely immunized French military population. Similar experiences have been reported in other armies with some limitations however. In both wars, many outbreaks of typhoid fever were reported in properly immunized troops, raising doubts among some sanitary authorities about the actually efficacy of typhoid vaccination. It is likely that in military operations the inocula encountered can be so high that immunity is overwhelmed. During World War I, Friedberger in Germany even doubted the value of typhoid vaccination, claiming that in many wars a spontaneous reduction in the incidence of enteric infections was observed with time, long before typhoid vaccination was introduced [11]. This important debate at some points very passionate, could not be solved in retrospective studies. Controlled studies were required, but were not conducted until the early 1960s under auspices of the World Health Organization (WHO).

At Last a Response

Between 1960 and 1970, the WHO sponsored a series of controlled field trials in an attempt to definitely assess the efficacy of parenteral typhoid killed cell vaccines that had been available for 70 years. These trials are summarized in Table 1. In Yugoslavia, a heat-inactivated, phenol-preserved parenteral vaccine seemed to confer better protection when compared to an alcohol-inactivated and preserved vaccine. Subsequently, two lyophilized vaccines were prepared. One was heat-phenol inactivated (L); the other was acetone-inactivated (K). The K vaccine was reproducibly more protective that the L vaccine in the various field trials, including two well-designed randomized, controlled, double-blind trials conducted in Yugoslavia [12] and in Guyana, respectively. In summary, the efficacy varied between 79 and 88% for the K preparation, and between 51 and 66% for the L preparation. This can be considered as the definitive demonstration of efficacy of whole-cell parenteral vaccine against typhoid fever. However, despite indisputable efficacy, these vaccine formulations remained poorly tolerated. Attempts therefore were made at preparing extracts of *S. typhi* that would be better tolerated.

Table 1 Field trials of two doses of lyophilized acetone (K) and heat-phenol (L) inactivated vaccine

Site and date	Age in years	Vaccines	Number of vaccinees	Duration of survey (years)	Incidence/ 10^5/year	Efficacy (%)
Yugoslavia	2 to 50	K	5,028	2.5	318	79
1960/63		L	5,068	2.5	727	51
		Control	5,039	2.5	1,488	
Guyana	5 to 15	K	24,046	7	67	88
1960/67		L	23,431	7	209	65
		Control	24,241	7	602	
Poland	5 to 14	K	81,534	3	7	88
1961/64		Control	83,734	3	47	
Russia	7 to 15	L	36,112	2.5	55	66
1962/65		Control	36,999	2.5	162	

From Ivanoff et al [26].

Toward a Significant Improvement: Vi Polysaccharide Vaccines

To improve the tolerance of whole-cell killed typhoid vaccines, numerous attempts were made over the years to prepare more or less purified extracts. Among these, the VI antigen seems to be the most promising and best chemically defined candidate.

The Vi polysaccharide of *S. typhi* is composed of a homopolymer of *N*-acetylgalacturonic acid expressed as a capsule around the bacterial body. This capsular polysaccharide was recognized as a virulence factor and as a possible immunogen as early as 1932 by Felix and Pitt [13]. In 1954, Landy proposed to study the Vi antigen as a vaccine formulation by injecting it in a highly purified form [14]. Unfortunately, the technique they employed was denaturing for the polysaccharide and the vaccine did not provide significant protection for volunteers [15].

Subsequently, in 1972, Wong and Feeley were able to purify the Vi antigen in nondenaturing conditions with hexadecyltrimethylammonium bromide [16]. This work opened the way to further development, resulting in the first chemically defined parenteral vaccine candidate against typhoid fever as initially assessed in volunteers [17]. This vaccine was then evaluated in randomized, placebo-controlled, double-blind field trials in Nepal [18], and in South Africa [19]. Its good local and general tolerance was confirmed after intramuscular injection of a single dose of 25 μg which, on average, conferred about 65% protection against typhoid fever for at least 2 years. The actual duration of protection needs to be evaluated. This vaccine (Typhim Vi®) is now available in several countries.

Despite these promising results, it appeared that booster doses of this vaccine did not provide antibody titers superior to those obtained with a single dose. This is essentially due to lack of T cell-dependent response and immunological memory of inherent to pure polysaccharide antigens. Recently, the Vi antigen has been conjugated to carrier proteins such as the tetanus toxoid [20]. A typical secondary response is observed in animals with this conjugated vaccine. Human trials are currently underway.

Live Attenuated Mutants of *S. typhi* Given Orally: An Alternative Option?

Over the last 20 years, an alternative approach to typhoid vaccination has been evaluated. The rationale for using live attenuated mutants of *S. typhi* administered orally is that for a disease whose portal of entry is the intestinal mucosa, optimal immunization should be obtained at this level. The only currently available approach to eliciting strong mucosal immunity is to administer live strains that are able to replicate, even at a limited level, inside the inductive sites of the gut-associated lymphoid tissues. This approach is not without potential problems. Among these, finding the proper balance between insufficient attenuation; making the vaccine candidate reactogenic; and excessive attenuation; making it nonimmunogenic; are of primary importance.

Germanier and Furer in Switzerland were the pioneers in this area when, in 1975, they proposed the *gal*E mutant of *S. typhi* Ty21a as an orally administered vaccine candidate [21]. This strain, obtained by chemical mutagenesis and selected as a *gal*E mutation, results in the absence of activity of uridine diphosphate (UDP) galactose-4-epimerase which accounts for the conversion of UDP-galactose into UDP-glucose. In the absence of galactose, Ty21a does not express the smooth O-antigen and is not immunogenic. In the presence of galactose, proper immunogenic lipopolysaccharide (LPS) is synthesized, but galactose-1-phosphate and UDP-galactose accumulate, leading to bacterial lysis. This is the theoretical rationale for virulence attenuation. However, other mutations are also present in this strain, probably accounting for its remarkable tolerance in volunteers and in field trials [22]. Several controlled field trials have been conducted with variable results. It appears that the formulation of the vaccine and the number of doses administered are the two major parameters to consider. In the first field trial conducted in Egypt, 6- to 7-year-old children received three oral doses of the vaccine over a week after a table of $NaCHO_3$ to neutralize gastric acidity. During 3-year surveillance, an outstanding 96% protective efficacy was observed [23].

Subsequent studies using a lyophilized formulation in enteric-coated capsules were carried out in Chile. Three doses administered over a week provided only 67% protection during the first 3 years [24]. Further studies in Chile and Indonesia have emphasized the importance of the formulation, liquid suspension being more efficacious that enteric-coated capsules [25]. The needs for several doses and the lack of consistent results with Ty21a have recently stimulated the search for new attenuated vaccine candidates.

Strains of *S. typhi* with precise genetically engineered attenuating mutations are currently being constructed and evaluated as recently reviewed [26]. A variety of possibilities are being tested, with two categories of mutations currently being evaluated. Mutations affecting regulatory pathways such a *cya/crp* are promising but, even in association with a *cdt* mutation which affects deep tissue colonization in mice, febrile adverse reactions accompanied by vaccinemia were observed in some volunteers receiving this strain as a live vector carrying a plasmid encoding a HBV antigen. Mutants in the *pho*P/*pho*Q senso-regulatory system are also being currently assessed.

Finally, the most advanced candidates at the moment are auxotrophic mutants such as *aro*C and *aro*D of *S. typhi* strain Ty2. These mutants cannot grow in the absence of para-amino benzoic acid (PABA) and dihydroxy acid (DHB), components which are absent or present in a very low concentration in human tissues. Based on these two mutations, the most recent strain CVD908 has shown promising results in volunteers. It is clinically well-tolerated, although vaccinemia is consistently observed [27]. More work needs to be done to improve this category of candidate vaccines.

Is It the End, the Beginning of End or the End of the Beginning?

After a century of intensive research, with the now available support of molecular biology, biochemistry, and immunology, a fully satisfactory typhoid vaccine is still missing. Therefore, it is certainly not the end of the search that started with Wright and Pfeiffer. However, we have certainly passed the "end of the beginning" milestone. A very promising conjugated Vi antigen candidate is available for parenteral immunization. A choice of live attenuated vectors for oral immunization is also available, and it seems that the groups involved in the development of these mutants have reached the point of "fine tuning" of their strains. In addition, a successful live vaccine could also represent a tremendous vector expressing heterologous antigens for immunization against other mucosal pathogens. Maybe the "beginning of the end" milestone is in view. However, several hurdles and various traps may still be anticipated. Will we be able to do better than nature? In other words, will we be able to reach close to 100% protective efficacy, given the fact that the natural disease is not fully protective? Will we need to combine parenteral and oral vaccines for full protection? How long will protection be maintained? Will available vaccination schedules be affordable by citizens of the developing world? Will vaccination allow eradication of typhoid fever? At what cost? Under which protocol?

History unfortunately does not help us because eradication of smallpox does not represent a relevant model. At best, an invitation to a "return to the future" sounds reasonable. Polio should be eradicated soon. Typhoid fever and polio are similar in many aspects. Let us hope that the program of global polio eradication will establish bases for typhoid eradication.

References

1. Louis PCA, *Recherches antomiques, pathlogogiques, thérapeutiques, sur la maladie connue sous le nom de gastroentérite, fièvre putride, a dynamique, typhoïde, etc, compareé avec les maladies aiguës les plus ordinaries.* Paris: JB Baillière, 1829
2. Bretonneau P. Notice sur la contagion de la dothinentérie. *Arch Gen Méd* 1829; 21:57

3. Budd W. Typhoid fever: its nature, mode of spreading, and prevention. London: Longmans, 1873
4. Eberth CJ. Organisms present in the organs during abdominal lymphoid infection. *Virchows Arch Path Anat* 1880; 58:74–81
5. Gaffky G. On the etiology of abdominal typhoid infection. Berlin: *Mitteilungen aus dem kaiserlichen Gesundheitsamte* 1884;2: 372–420
6. Widal GFI, Sicard A. Rescherche de la reaction aggluntinate dans le sang et le sérum desséchés typhiques et dans la sérosité des vésications. *Bull Soc Méd Paris* (3rd ser) 1896; 13:681–682
7. Gröschel DHM, Hornick RB. Who introduced typhoid vaccinations: Almoth Wright or Richard Pfieffer? *Rev Infect Dis* 1981;6:1251–1254
8. Wright AE. On the association of serous hemorrhages with conditions of defective blood-coagulopathy. *Lancet* 1896; 2:807–809
9. Wright AE, Semple D. Remarks on vaccination against typhoid fever. *Br Med J* 1897;1: 256–259
10. Pfieffer R, Kolle W. Experimentelle Untersuchungen zur Frage der Schutzimpfung des Menschen gegen Typhus abdominalis. *Disch Med Wochenschr* 1896;22:735–737
11. Friedberger E. Zur Frage der Typhus-intestinalis Choleraschuzimpfung. *Z Immunol* 1919; 28:119–185
12. Yugoslavia Typhoid Commission. A controlled field trial of the effectiveness of acetone-dried and inactivated and heat-phenol activated typhoid vaccines in Yugoslavia. *Bull WHO* 1964; 30:623–630
13. Felix A, Pitt RB. A new antigen of *Bacillus typhosus*. Its relation to virulence and to active and passive immunization. *Lancet* 1932; ii:186–191
14. Webster ME, Landy M, Freeman ME. Studies on Vi antigen. II. Purification of Vi antigen from *Escherichia coli* 5396–38. *J Immunol* 1952; 69:135–142
15. Landy M. Studies on Vi antigens: immunization of human beings with purified Vi antigen. *Am J Hyg* 1954; 60:52–62
16. Wong WH, Feeley JC. Isolation of Vi antigen and a simple method for its measurement. *Appl Microbiol* 1972; 24:28–33
17. Tacket CO, Ferriccio C, Robbins J. Safety and immunogenicity of the *Salmonella typhi* Vi capsular polysaccharide vaccines. *J Infect Dis* 1986; 154:342–345
18. Acharya IL, Lowe CU, Thapa RL. Prevention of typhoid fever in Nepal with the Vi capsular polysaccharide of *Salmonella typhi. N Engl J Med* 1987;18:1101–1104
19. Klugman KP, Gilbertson IT, Koornhof HJ. Protective activity of Vi polysaccharide vaccine against typhoid fever. *Lancet* 1987; ii: 1165–1169
20. Szu SC, Li X, Schneerson R. Comparative immunigenicities of Vi polysaccharide-protein conjugates composed of cholera toxin or its B subunit as a carrier bound to high- or lower-molecular weight Vi. *Infect Immun* 1989; 57:3823–3827
21. Germanier R, Furer E. Isolation and characterization of *gal*E mutant Ty21a, of *Salmonella typhi*: a candidate strain for live oral typhoid vaccine. *J Infect Dis* 1975; 141:553–558
22. Levine MM, Taylor DN, Ferriccio C. Typhoid vaccines come to age. *Pediat Infect Dis* 1989; 8:374–381
23. Wahdan MH, Serie C, Cerisier Y. A controlled field trial of live *Salmonella typhi* strain Ty21a oral vaccine against typhoid: three year results. *J Infect Dis* 1982; 145:292–295
24. Levin MM, Ferriccio C, Black RE. Large-scale field trial of Ty21a live oral typhoid vaccine in enteric-coated capsule formulation. *Lancet* 1987; i: 1049–1052
25. Levin MM, Ferreccio C, Cryz S, Ortiz E. Comparison of enteric-coated capsules and liquid formulation of Ty21a a typhoid vaccine in a randomized controlled field trial. *Lancet* 1990; 336:891–894
26. Ivanoff B, Levine MM, Lambert PH. Vaccination against typhoid fever: present status. *Bull WHO* 1994; 72:954–971
27. Tacket CO, Hone DM, Losonsky G. Clinical acceptability and immunogenicity of CVD908 *Salmonella typhi* vaccine strain. *Vaccine* 1992; 10:443–446

The History of Pertussis Vaccination: From Whole-Cell to Subunit Vaccines

Marta Granström

Bordetella pertussis was first cultured by Jules Bordet, pictured in Figure 1. Cultivation of the organism led to the development of two types of vaccines: whole cell and acellular

The Efficacy of Whole-Cell Vaccines

A review of all studies and trials cannot be made in this context, but two sources are recommended, i.e., Whooping cough by Joseph E. Lapin [1] and a review by Fine and Clarkson [2]. In the early years of the twentieth century, attempts were made to develop whole-cell vaccines, but for these early studies, the current definition of vaccination may not be valid since many workers have considered prophylaxis to include preventive treatment in the incubation period of the disease, as noted by Lapin.

M. Granström (✉)
Department of Clinical Microbiology, Karolinska Hospital, 171 76 Stockholm, Sweden
e-mail: marta.granstrom.ki.se

S.A. Plotkin (ed.), *History of Vaccine Development*,
DOI 10.1007/978-1-4419-1339-5_10, © Springer Science+Business Media, LLC 2011

An example of the problem – and the value of checking original sources – is the 1933 report of Thorvald Madsen, Director of the Danish State Serum Institute, describing the use of whole-cell vaccine during the 1923–1924 and 1929 epidemics in the Faroe Islands [3]. The paper is often quoted as an important contribution for the acceptance of the notion of prevention of whooping cough by vaccination, as it represented the first pertussis vaccine prepared in a standardized fashion and submitted to clinical trials. According to Madsen, vaccination during the 1923–1924 epidemics had no effect in preventing the disease, as "the majority of both vaccinated and unvaccinated individuals contracted whooping cough." The major effect was said to have been a reduction in mortality and that "the course of the disease was as a whole much more severe in the nonvaccinated individuals." Consequently, the protective efficacy of the vaccine in the 1923–1924 trials was quoted as nil [2]. Madsen concluded that "in the latter epidemic the prophylactic value of the vaccine was much better than in the first one."

A critical review can only be made of the 1923–1924 epidemic, for which Madsen gives as references the original Danish reports from general practitioners working in different districts in the Faroe Islands [4, 5]. Rasmussen [4] divided the cases of pertussis among the vaccines into four subgroups, i.e., those with clinical symptoms prior to the first dose, those with symptoms arising during vaccination, within 8 days of the last dose, or in more than 8 days after the last dose. His data do not allow calculation of the vaccine efficacy (VE), but he states in his conclusion that "vaccination was clearly inefficient after onset of whooping cough" but that "16 to 23% of those children, who were not isolated and who had not started to cough on the day of the last vaccination, escaped the disease."

In 1926, Kofoed [5] also concluded that vaccination during the incubation phase or during the clinical disease did not provide protection, but was efficacious if it was completed prior to exposure. In a group of 195 children, only 8 had a severe disease, 45 had a rather mild disease, 65 had an uncharacteristic cough, and 77 entirely escaped the disease. In the control group composed of 30 unvaccinated children, none escaped pertussis, which was severe in 18 cases, moderately severe in 10 cases and uncharacteristic in two cases. Kofoed stated that whooping cough was difficult to differentiate from other coughs if not accompanied by whoops. He therefore pooled the data from patients without any cough with those of patients presenting with uncharacteristic cough. Altogether, these data provided a 142/195 ratio of patients that did not present with pertussis. A VE estimate taking into account this ratio along with the two uncharacteristic cases observed in the control group would thus indicate that the vaccine efficacy would roughly reach 75%. It was thus likely that the favorable results recorded during the 1929 epidemics were based on these experiences and that, during this outbreak, most of the vaccinations were done prior to exposure.

Other key reports are those from the British Medical Research Council trials in England, conducted between 1942 and 1954 [6–8]. These trials led to the introduction of general vaccination against pertussis in industrialized countries and provided the laboratory correlate to protection, the potency assay with intracerebral challenge of immunized mice. The assay was shown to correlate with protection in children [7] and minimal requirements (four protective units) were formulated. It should be noted that the test has been extremely useful for almost 50 years.

The problems caused by this test emanate from the misconception that a correlation implies a causal relationship, i.e., protection of mice is a model for protection in children. This conclusion has for many years hampered the development of subunit vaccines, as the demand that such vaccines would have to pass the test was raised. The limitations of the assay were also shown in recent trials including a whole-cell vaccine with low VE [9].

Indications that the mouse potency assay does not always correlate with protection in children have been present for years. One such example was the whole-cell vaccine produced in Sweden in 1978, which had less than two mouse protective units, but conferred a high degree of protection in children [10]. An example of the opposite situation, a vaccine of ≥4 mouse protective unit but not conferring protection, has been presented from Nova Scotia, Canada. Fully vaccinated children were shown to develop pertussis with a clinical picture indistinguishable from that seen in unvaccinated children, indicating that the vaccine did not even induce an immunological memory. This vaccine has been shown to produce practically no antibody responses to pertussis toxin (PT) and filamentous hemagglutinin (FHA), the major protective antigens of *Bordetella pertussis*. Another whole-cell vaccine, with only weak immune responses to PT and FHA, had low vaccine efficacy estimates in recent trials [9]. Other whole-cell vaccines, with good immune responses to these antigens, were shown to have high VE efficacies in other trials [9].

The Adverse Effects of Whole-Cell Vaccines

The history of the risks of whole-cell pertussis vaccine is as old as the history of its protective efficacy, since two deaths in newborns were reported by Madsen [3]. The whole issue of the risk versus benefit debate of whole-cell pertussis vaccines cannot be presented here, but an extensive review of the risk for permanent damage was done by Griffith [11], and by Wardlaw and Parton [12]. The current status of the issue is that whole-cell vaccine has not been shown to cause permanent brain damage or death. The vaccine does give rise to reversible, neurological reactions, causes fever with febrile convulsions, and induces local reactions at a rate approaching 100% [13]. The high rate of adverse effects and the fear of serious damage have resulted in many countries in a low vaccine uptake – and a concomitant increase in the disease incidence – and stimulated a renewed interest in basic research, greatly neglected after the successful introduction of whole-cell vaccine in general vaccination programs.

Development of Subunit Vaccines

Again, a full review of the subject cannot be made, but the textbook of Wardlaw and Parton [12] provides background information about today's acellular and subunit vaccines. To summarize the current status of "protective antigens" in pertussis, several antigens, i.e., pertussis toxin, filamentous hemagglutinin (with some active PT), fimbriae and pertactin (a 69 kD antigen) have been shown to be protective in experimental respiratory models. Only one antigen, a PT toxoid, has been shown to be protective by itself in humans. Since many important contributions have been made in the past 20 years, all of them cannot be mentioned here, but the contributions of three scientists must be given a special acknowledgement.

A key paper for our current understanding of pertussis was a review [14] by Margaret Pittman, published in 1979, in which she hypothesized that pertussis was a toxin-mediated disease – by analogy with diphtheria and tetanus – and that pertussis toxin (PT) was the major cause of the harmful effects of the disease. She therefore thought that a PT toxoid should be sufficient for protection. Margaret Pittman, who died in 1995, made important contributions to the pertussis field for 50 years. She maintained to the end of her days the firm belief that additional antigens in subunit pertussis vaccines – if not shown to be absolutely necessary – or impurities in the case of the other toxoid vaccines – should be avoided as they had the potential to cause harmful side-effects [15].

Two other scientists, Yuji Sato and his wife, Hiroko Sato, contributed to our basic understanding of the disease, and developed animal models relevant for test of subunit vaccines. Yuji Sato developed the first acellular pertussis vaccine, as understood today and in current use, which was introduced in Japan for general vaccination of children of 2 years of age in 1981 [12, 16]. Pertussis vaccine uptake in Japan had sharply decreased, due to two deaths in the mid-1970s and was followed by a subsequent rise in the disease incidence (with 46 deaths). The new vaccines were introduced by the Japanese authorities who faced a major public health problem, prior to their being subjected to efficacy trials. Yuji Sato, eager to have the efficacy of acellular vaccines estimated and wanting to test a monocomponent vaccine of pertussis toxoid, actively supported the first Swedish efficacy trial (trial 1 in Table 1).

The Efficacy of Subunit and Acellular Pertussis Vaccines

Table 1 summarizes the vaccines tested in clinical trials, their composition, the efficacy estimates and a personal best opinion about the likelihood of the vaccine's protective efficacy.

The first efficacy trial of the Japanese vaccines, JNIH-6 and JNIH-7, employed a novel approach to case definition – based on positive culture or positive serology – for VE calculation. It was claimed that this method would be "the most specific for pertussis" [17]. However, if *B. pertussis* were replaced by *S. pneumoniae* and if it was claimed that bacterial culture from a nasopharyngeal aspirate is specific for pneumococcal pneumonia in child, then the source of the problem encountered in this trial, i.e. colonization, would be evident.

In this study, children were sampled not only on the basis of clinical disease, but also on the basis of exposure within the family. If a child developed a cough, e.g. a common cold within 30 days after positive culture, then he/she became a (mild)

Table 1 Reported vaccine efficacy (VE) estimates for acellular subunit and whole-cell pertussis vaccines in recent (1986–1995) trials and a likely estimate of the protective efficacy after correction for major biases introduced by laboratory methods, use of laboratory methods and study design

Trial no	Vaccine (manufacturer)	Antigens per dose	Reported VE estimate (95 cl)	Corrected protective efficacy (%)
1	JNIH-6 (Biken)	25 μg PT 25 μg FHA	69% (47–82)	= 80
	JNIH-7 (Biken)	35 μg PT	54% (26–73)	= 80
2	DTaP (SIS/Amvax)	40 μg PT	71% (63–78)	= 80
3	$DTaP_2$ (SB experimental)	25 μg PT 25 μg FHA	59% (51–66)	not valid
	$DTaP_5$ (PMC "classic")	10 μg PT 5 μg FHA 3 μg 69 kD 5 μg fimbriae 2 5 μg fimbriae 3/6	85% (81–89)	≈ 80
	DTwP (PMC-US)	Whole cell vaccine	48% (37–58)	≈ 40
4	$DTaP_3$ (SB production)	25 μg PT 25 μg FHA 8 μg 69 kD	84% (76–90)	≈ 80
	$DTaP_3$ (Biocine)	5 μg PT* 2.5 μg FHA 2.5 μg 69 kD	84% (76–90)	= 80
	DTwP (PMC-US)	Whole cell vaccine	36% (14–52)	= 40
5	DTaP (Lederle/Takeda)	3 μg PT 35 μg FHA 2 μg 69 kD 1 mg fimbriae 2	82% (75–)	= (70–)80
	DTwP (Lederle)	Whole cell vaccine	91% (86–)	= 80
6	$DTaP_3$ (SB production)	As above	89% (77–95)	= 80
7	$DTaP_2$ (PMC) 25 μg FHA	25 μg PT	86% (71–93)	80%
	DTwP (PMC)	Whole cell vaccine	96% (87–99)	= 80
8	$DTaP_2$ (PMC/Biken)	25 mg PT 25 μg FHA	96% (78–99)	= 80
	DTwP (Behringwerke)	Whole cell vaccine	97% (79–100)	= 80

The efficacy estimates for vaccines above the dotted line were obtained in prospective, randomised double-blind, placebo-controlled trials.
*Genetically detoxified.

case of pertussis (Fig. 2 [17]). As a result, the vaccine efficacy (VE) estimates were low. The statistician later calculated VE estimates based on clinical symptoms and obtained high-point estimates for both vaccines [18]. The VE estimates for "severe" pertussis, i.e. >30 days of cough with laboratory confirmation, were 78% (95% CL 57–89%) for both vaccines, provided that the word "severe" pertussis is replaced by "true" pertussis.

The above conclusions eventually became obvious to everybody in the field and resulted in the "WHO criteria" used in subsequent efficacy trials. The WHO criteria represent a return to VE estimates based on clinical disease, i.e. paroxysmal cough of ≈21 days in duration confirmed by either culture of *B. pertussis* or positive serology, defined as a significant titer rise of IgG to FHA or a significant titer rise of IgG to PT. Blood sampling was to be triggered by a cough of ≥7 days duration. Additional criteria, e.g. titer rises in IgA (used in some trials) or a household contact within 28 days prior to onset of disease with a culture confirmed case of pertussis (used in some trials).

Thus, VE estimates based on a clinical case definition are correct. The point estimates, i.e. the exact VEs, cannot, however, be considered as "true," due to bias toward falsely high VE. For instance, it has been known for decades that culture of *B. pertussis* is more difficult from a vaccinated child than from a nonvaccinated (=bias introduced by culture). The serological criteria of significant titer rise to antigens also present in the vaccine will cause a significant rise, with seroconversions during the third week of the disease, in the nonvaccinated controls, i.e. a primary antibody response, versus the rapid rise of antibodies, within 3–7 after onset of symptoms, in the vaccinated group, i.e. a secondary antibody response. With the first sample drawn ≥7 days of cough, a majority of children in the vaccine group will already have high titers and thus will have no further rises in the antibody titer. A vast majority of the children in the nonvaccinated group, on the other hand, will show significant titer rises between samples taken during the acute phase of the disease and samples taken during convalescence.

In addition to the biases introduced by the laboratory confirmation, the case definition by itself introduces a bias towards falsely high efficacy estimates, since mild, atypical cases of <21 days duration – and without paroxysms – can be expected to be more numerous in the vaccine group than in the placebo group.

When studying the summary of Table 1, one has also to bear in mind that the results emanate from different types of studies and that only studies of the same kind can be (more or less) comparable with regard to point estimates. The studies above the dotted line were prospective, double-blind, placebo-controlled. Such trials tend to yield lower VE estimates. The others were of different types, e.g. open, not placebo-controlled, household contact. Trial 1 has been discussed above, but trial 2 also needs a comment. The serological biases introduced by the WHO criteria are neutral between the vaccines as long as they all contain both PT and FHA. This is the case for all the vaccines listed in the table, except for the monocomponent PT vaccine. The strong positive bias towards falsely high VE estimates introduced by the significant titer rises criterium is not true for this vaccine with regard to FHA, against which antigen both the vaccine and the placebo groups are "unimmunized." To make the VE estimates of this trial more comparable to those of the other trials, the cases in this trial diagnosed by titer rise to FHA alone should be eliminated. The vaccine efficacy estimate for this vaccine then rises to 78% (95% CL 71–84%).

A direct comparison with the other trials, after exclusion of the cases identified by FHA alone, cannot be made since these data have not been presented. A decrease in VE estimates without this bias is to be expected, although the size of the decrease may vary with the study design. For instance, another Swedish study (no. 3 in Table 1) had repeated sampling, and is therefore likely to have less bias introduced by serology. It should also be pointed out that in the first Swedish trial of a two-component and monocomponent vaccine (trial 1), the bias introduced by serology in favour of the two-component vaccine was not present since a significant titer rise to both FHA and PT was required for positive serology.

The VE estimates obtained in the studies below the dotted line show that whole-cell vaccines with good immunogenicity also have high protective efficacy. The estimates are, however, more than likely to represent overestimates due to the design of the trials. The only remaining feature in Table 1 that might seem unexpected is the low efficacy of a two-component vaccine in trial no. 3, in contrast to the VEs of similar vaccines in three other trials. My conclusion that the point estimate for this experimental diphtheria-tetanus-acellular pertussis (DTaP) vaccine is "not valid" for production lots from the same company or for other two-component vaccines must be taken as a (well founded) opinion.

Based on the above considerations, the remaining vaccines (except one whole-cell vaccine) have been assigned a corrected efficacy of ≥80%. This estimate does not include correction for the positive bias introduced by the clinical case definition. The size of this bias is difficult to estimate, considering that inclusion of clinical cases with cough of any duration would necessitate some unbiased laboratory method for confirmation of the disease, e.g. serology with an antigen not present in any of the vaccines. A true protective efficacy of ≥70% would thus be a qualified guess.

A relatively low protective effect requires both high vaccine coverage rates (for herd immunity) and repeated booster injections in order not to accumulate susceptibility in older age groups. In this respect, subunit vaccines have not been shown to differ from whole-cell vaccines. Some early as well as some recent trials raised claims for the long duration of immunity, based on the priming doses alone. Such claims are not valid when the disease is still endemic since children will receive (repeated) natural boosters.

A similar misconception about the long duration of immunity induced by natural pertussis seems to prevail. The conclusion is based on observations from the prevaccination era, but does not take into account the role of natural boosters. It is of interest to note that in a society where the disease was not endemic, as in the Faroe Islands, Rasmussen [4] described many cases of reinfections at the time of the 1923–1924 outbreak, and simply concluded that "reinfections are quite common."

The Adverse Effects of Subunit and Acellular Pertussis Vaccines

A description of the results of Phase 1, 2, and 3 trials for all the vaccines in Table 1 cannot be made here, but to summarize, the rates of local reactions are significantly lower for all acellular vaccines than for (any) whole-cell vaccine. The rate of local reactions to a monocomponent vaccine is also lower than that of a two-component vaccine [17]. It increases with successive doses of acellular vaccine, while the opposite is true for whole-cell vaccines. An acellular vaccine booster in a child primed with whole-cell vaccine results in almost no reactions. When acellular vaccine boosters are given to children primed with the same vaccine, larger (large) local reactions have been described (for some vaccines).

The most common systemic reaction to whole-cell vaccines, i.e. fever, is significantly decreased with acellular vaccines. As a consequence, the rate of convulsions is also decreased since the majority of cases represent febrile convulsions. As for the reversible neurological reactions, it was hoped that these reactions would disappear with acellular vaccines, but this does not seem to be the case (with one possible exception). The reactions, i.e. unusual crying, persistent crying of ≈3 h duration, and hypnotic, hyporesponsive episodes (HHE) – contraindicating further doses of whole-cell vaccine – have been described for all acellular vaccines (with one exception), but the rates seem to be lower than for (many) whole-cell vaccines.

The only possible exception may be the monocomponent vaccine in trial 2 (Table 1) which did cause any HHE in >5,000 infants or via some 15,000 injections given in the course of clinical studies. This reaction and other neurological reactions were first described in a large whole-cell vaccine study reported by Cody et al. [13]. A rate of HHE of ca 1:1,200 injections or ca 1:1,400 infants can be calculated for the first three doses. In the context of a historical review, it should be noted that Kofoed [5] described in his report "an 8-month old child, who ca 1 hour after the first injection, suddenly showed difficulties to breath and was laying like dead. The doctor was called on the phone but before the end of the conversation, the information that the child was healthy could be given." This symptom is generally considered to be specific for pertussis vaccines, but claims have been raised that this symptom is also seen after diphtheria tetanus (DT) vaccine, e.g. in the Italian trial (no. 4 in Table 1), while no cases were seen in the larger Swedish trials (no. 1–3 in Table 1). It should be noted that no case of HHE has been notified in Sweden since 1979, i.e. since DT vaccine replaced DTP vaccine. The Swedish parents, unaccustomed to this symptom, were so alarmed by it in a large, ongoing DTP (acellular) trial that many of them even called an ambulance to take their child to a hospital.

Irreversible damage or death have not been caused by any of the current acellular vaccines. Four children have, however, died of bacterial infections in the first Swedish trials [17], but none of the current vaccines is produced by the same method today. The mortality rate was much higher than expected, but unfortunately clusters can occur. The rate of hospitalizations for bacterial infections was twice as high in the two-component vaccine group than in the monocomponent or in the placebo groups during the period when the first three deaths related to bacterial infections occurred in the same vaccine group. The hospitalization rate study was made blindly, i.e. prior to breaking the code, as it was recognized that it may be difficult to identify bacterial infections. One child had ingested a lethal dose of heroin, but also had, according to the autopsy report, a pneumonia which was fatal by itself.

It is likely that it will never be clarified whether these events were a series of unfortunate coincidences or whether they were truly vaccine-associated. Yet, it is important to review the data since, if they were vaccine-associated, then the most likely cause would be reversion of the pertussis toxin (toxoid) components. Laboratory tests could not confirm a reversion for the two-component vaccine, but could do so for the monocomponent vaccine. It has, however, to be remembered that the laboratory models are not very sensitive, that pertussis toxin is not a "toxic" toxin (injection of large doses to adults caused no harmful effects), that the two-component vaccine had an extended expiration date (i.e. longer than that used in Japan) in order to be included in Phase 3 trials and that reversion after injection for an aluminum-adsorbed vaccine can continue for a long period of time.

A conclusion to be drawn from a hypothetical vaccine-association would be that an increase in bacterial infections would be the sign to look for if reversion of pertussis toxin is suspected in multicomponent vaccines. A basis for this effect could be found in the studies of Tuomamen [19] which have shown that the adhesins of *B. pertussis*, i.e. PT and FHA, bind to the surface of other bacteria, thus conferring on them the ability to bind to ciliated cells, a phenomenon that the author called the "piracy of adhesins." Another conclusion would be that both adhesins are needed to cause the disease, and that monocomponent vaccines of pertussis toxoid are safe, even when in reversion.

Laboratory Correlates with Protection for Subunit Vaccines

The issue has not been settled since the intracerebral challenge test is not applicable to subunit vaccines. Immunogenicity assays could be used, but no minimal requirements have, as yet, been formulated. The value of immunogenicity data has also been questioned, due to the lack of a serological correlate which could not be shown in the first Swedish efficacy trial [17]. Other data have been presented which confirm that a correlation must exist, e.g. specific immune globulin with antibodies against PT alone modified the clinical disease in children [18, 20]. A study in adults also found a correlation with antibody levels against both PT and FHA, although the correlation was stronger for the antibodies against PT [21]. The failure to show a correlation in the first trial can have several causes, but the most likely one seems to be the dilution of the "cases" by a large proportion "non-cases" (of colonization with *B. pertussis*). In trial 2, a preliminary trend analysis of the postvaccination antibody levels indicated lower attack rates of pertussis, with higher levels of PT-IgG (J. Taranger and B. Trollfors, personal communication).

In conclusion, whole-cell pertussis vaccines were useful for ≈50 years, decreasing pertussis morbidity rates in children. Major problems of this vaccine, i.e. its high rates of local reactions and some systemic reactions, will be eliminated with the subunit pertussis vaccines. Another major problem, the fact that the vaccine could not be safely administered to adults, should also be solved by the introduction of subunit vaccines. This latter aspect is particularly important since there is no reason to believe that subunit vaccines will induce a more long-lasting immunity than that conferred by whole-cell vaccines. Long-lasting immunity to toxin-mediated diseases, e.g. pertussis and diphtheria, was based on natural booster in the prevaccination era. Natural boosters, eliminated by infant immunization, will have to be replaced by booster injections to be given to adults in order to maintain life-long protection.

References

1. Lapin JH. *Whooping Cough.* Springfield, III: Charles C Thomas, 1943
2. Finw PEM, Clarkson JA. Reflections on the efficacy of pertussis vaccines. *Rev Infect Dis* 1987;9:866–83
3. Madsen T. Vaccination against whooping cough. *JAMA* 1933; 101:187–188
4. Rasmussen RK. Om kighoste og kighostevaccination i Ejde lægedistrikt paa Færøerne. *Bibliot f læger* 1925; 131–148
5. Kofoed SE. Nogle oplysninger om optræ af kighoste i Sandø præld (Færøerne) 1923–24 speceilt med henblik paa anvendelse af kighostevaccin. *Ugesk f læger* 1926; 88:585–588
6. Medical Research Council. The prevention of whooping-cough by vaccination. *Br Med J* 1951; 1:1463–1471
7. Medical Research Council. Vaccination against whooping-cough. Relation between protection of children and results of laboratory tests. *Br Med J* 1956; 2:454–462
8. Medical Research Council. Vaccination against whooping cough. *Br Med J* 1959; 1:994–1000
9. International Symposium on Pertussis Vaccine Trials. Trial synopsis, 1995
10. Romanus V, Jonsell R, Berquist SO. Pertussis in Sweden after the cessation of general immunization in 1979. *Pediatr Infect Dis J* 1987;6:364–371
11. Griffith AH. Permanent brain damage and pertussis vaccination: is the end of the saga in sight? *Vaccine* 1989; 7:199–210
12. Wardlaw AC Parton R. *Pathogenesis and Immunity in Pertussis.* Chichester: John Wiley & Sons Ltd, 1988
13. Cody CL, Baraff LJ, Cherry JD, Marcy SM, Manclark CR. Nature and rates of adverse reactions associated with DTP and DT immunizations in infants and children. *Pediatrics* 1981; 68:650–660
14. Pitman M. Pertussis toxin: the cause of the harmful effects and prolonged immunity of whooping cough. A hypothesis. *Rev Infec Dis* 1979; 1:401–412
15. Robbins JB, Pittman M, Trollfors B, Lagergård TA, Taranger J, Schneerson R. *Primum non noncore*: a pharmacologically inert pertussis toxoid alone should be the next pertussis vaccine. *Pediatr Infect Dis J* 1993; 12:795–807
16. Sato Y, Kimura M, Fukumi H. Development of a pertussis component vaccine in Japan. *Lancet* 1984;1:122–126
17. Ad hoc group for the study of pertussis vaccines. Placebo-controlled trial of the two acellular pertussis vaccines in Sweden – protective efficacy and adverse events. *Lancet* 1988; 1:955–960
18. Blackwelder WC, Storsæter J, Olin P, Hallander HO. Acellular pertussis vaccines. Efficacy and evaluation of case definitions. *AJDC* 1991; 145:1285–1289
19. Tuomanen E. Piracy of adhesions: attachment of superinfecting pathogens to respiratory cilia by secreted adhesins of *Bordetella pertussis. Infect Immun* 1986; 54:905–8
20. Granström M, Olinder-Nielsen AM, Holmblad P, Mark A, Hanngren K. Specific immunoglobulin for treatment of whooping cough. *Lancet* 1991; 338:1230–1233
21. Granström M, Granström G. Serological correlates in whooping cough. *Vaccine* 1993; 11:445–448

Bacterial Polysaccharide Vaccines

Robert Austrian[†]

Capsulated bacteria, Gram-positive or Gram-negative, cause a variety of infections in man. Prominent among them are streptococci of Lancefield's groups A, B, and C, staphylococci, meningococci, *Haemophilus influenzae* type b, klebsiellas, *Escherichia coli,* and *Salmonella typhi*, to name but some. Since the description of the capsule as an attribute of bacteria more than a century ago, increasing knowledge of its structure and role in interactions of these organisms with their environment has enabled development of vaccines to enhance defenses of their hosts against infection and their likelihood of recovery when it occurs. Since much what has been learned has been derived from studies of the pneumococcus, emphasis in what follows will focus upon *Streptococcus pneumoniae*, additional references pertinent to other specific vaccines are cited where relevant.

†Deceased

S.A. Plotkin (ed.), *History of Vaccine Development*,
DOI 10.1007/978-1-4419-1339-5_11, © Springer Science+Business Media, LLC 2011

Although bacterial aggregates in a gelatinous substance referred to as "zoogloea" had been described before 1880, perhaps the first reference to a bacterial capsule was that of Pasteur in reporting his initial isolation of the pneumococcus:

> Each of these little particles is surrounded at a certain focus by a sort of aureole which corresponds, perhaps, to a material substance;it is certain that in several cases where the little organism is difficult to distinguish, scrutiny of the aureole has permitted its recognition [1].

Sternberg, who isolated the pneumococcus in 1880, also recognized its capsule. In a report in 1882, Sternberg, a pioneer in photomicrography, published what was probably the first microphotograph of the pneumococcus. Of it he wrote: "The most striking morphological difference between the micrococcus is the aureole which surrounds the well-defined dark central portion in the latter figure" [2].

It may have been Carl Friedlander who was the first to use the term "capsule." In his paper citing for the first time the stain of Christian Gram, he wrote:

> We found in these investigations that, in most cases of pneumonia, either a great part of the cocci or aggregated cocci were surrounded by a more or less broad layer of substance, stained weakly blue or red respectively with gentian violet and fuchsin, which encircled the cocci with a kind of hull or capsule. A structure of this kind, a capsule, has not heretofore been described among the schizomycetes. One speaks, it is true, of a gelatinous ground substance between the micrococci which agglomerates them into "zoological masses" but a single micrococcus surrounded by a well characterized capsule is heretofore unknown [3].

The concept of the capsule as a discrete structure is probably an artefactual one, the result of one or another of the methods used to visualize bacteria which cause aggregation of high molecular weight compounds about the outer margins of the cells. The presence or absence of these polymers, which require expenditure of energy for their production [4], appears dictated in part by the environmental conditions such as nutritional requirements and defense against phagocytic cells. Little is known about their evolutionary origins. It has been observed, however, that noncapsulated pneumococci derived from wild type strains of several capsular types, when grown in the presence of antibodies to surface constituents other than C or cell wall polysaccharide, give rise to mutants endowed with a capsule of C, or soluble C-like polysaccharide [5]. These strains suggest the possibility that bacterial capsules may have evolved by such a process followed by gene duplication, translocation and further, giving rise to the various components of capsular polysaccharides. It is noteworthy that all the components of pneumococcal C polysaccharide can be found in one or another of its capsular polymers [6] and similar relationships exist among the carbohydrate constituents of the meningococcal cell wall and those of meningococcal capsular antigens [7].

In the three decades following its initial isolation, considerable knowledge of the bacteriology of the pneumococcus and the infections it causes was acquired, reviewed by White in his monograph: "*The Biology of Pneumococcus*" [8]. In 1910, the delineation of distinct pneumococcal serotypes by Neufeld and Händel led to the development of serotherapy for pneumococcal pneumonia [9]; and in 1917, Dochez and Avery reported the presence of a type specific soluble substance in filtrates of pneumococcal cultures and in the sera of infected patients and rabbits [10]. This latter observation led to the discovery by Avery and his associates [11, 12] that the capsular antigens of pneumococcal types 1, 2, and 3 were polysaccharides. Prior to that time, it was the widely held view that all antigens were proteins.

The number of recognized capsular polysaccharides produced by different bacterial species varies greatly, from the single Vi polymer of *S. typhi* to the more than 80 distinct capsular polysaccharides of pneumococcus, *E. coli* or *Klebsiella pneumoniae*. Chemical composition is a more important determinant of virulence than capsular size [13], although among strains of a given type, virulence correlates directly with capsular size [14].

Among the capsulated bacteria that infect man, some are obligate and others facultative parasites. The pneumococcus, an obligate parasite, enjoys usually a commensal relationship with man [15]. Colonization may occur on the day of birth and, later, with as many as four types simultaneously. Development of type specific antibodies may follow colonization in the absence of overt illness and levels of antibodies tend to rise from infancy to adulthood [16]. Disease follows usually injury to the respiratory epithelium to the detriment of both host and parasite.

The first experiment indicative that pneumococcal infection could be prevented by injection of killed organisms was reported in 1882 by Sternberg, who found the death of rabbits inoculated with his saliva could be prevented if they had been injected previously with saliva treated with an antiseptic [17].

The first planned attempts to prevent pneumococcal infection were those of Sir Almroth Wright and his collaborators in 1911 in an effort to control epidemic pneumonia in South African gold miners. A vaccine of killed pneumococci of undetermined type was administered in trials involving approximately 50,000 men; and although Wright concluded the vaccine was efficacious [18], subsequent analysis gave little support to this view [19].

Wright's protégé in South Africa, F. Spencer Lister, independently classified pneumococcal types and determined the number of pneumococci needed to stimulate an antibody response in man [20], a number yielding an amount of polysaccharide included in contemporary vaccines [21]. He might have demonstrated the efficacy of pneumococcal vaccines had he structured his trials properly. Lister reasoned correctly, as shown later [22], that if one vaccinated half the members of a mine compound, the immunized subjects would retard the spread of organisms to the unvaccinated controls. He chose, therefore, to vaccinate all the men in one compound, using those in another as controls. Since the attack rates of pneumonia differed in different compounds, the interpretation of Lister's trials remained controversial. The problem they posed was resolved later in efficacy trials of a meningococcal polysaccharide vaccine in which only 20% of the exposed closed population was vaccinated [23]. Despite additional trials of whole pneumococcal vaccines during World War I, their efficiacy remained moot.

Although pneumococcal polysaccharides had proved to be nonimmunogenic in rabbits, their immunogenicity in mice was reported by Schiemann and Casper in 1927 [24]; and in 1930, similar responsiveness by man was described by Francis and Tillett [25]. Confirmation of the latter findings led to trials in the

Civilian Conservation Corps of vaccines containing 1 mg each of the capsular polysaccharides of pneumococcal types 1 and 2 [26]. Although the trials were suggestive of the vaccine's efficacy, bacteriologic studies were incomplete, leaving the status of the vaccine in doubt.

Clear evidence of the efficacy of a vaccine of pneumococcal capsular polysaccharides emerged from a trial at a military installation where pneumonia was epidemic during World War II [22]. Fifty microgram each of four capsular polysaccharides were administered to 8,586 recruits and 8,449 were injected with saline. Four illnesses attributed to types in the vaccine occurred in its recipients, all within 2 weeks of injection, in contrast with 26 such illnesses in controls over a 24-week period. It was observed also that, if an individual was a carrier of a type represented in the vaccine, vaccination did not abolish the carrier state; but if an individual was not a carrier of such a type before vaccination, the likelihood of his becoming a carrier was reduced by approximately half. Lister's reasoning regarding the impact of vaccinating half a closed population on the incidence of disease among unvaccinated controls was shown also to be correct.

The advent of antibiotics and their impact on the morbidity and mortality of pneumococcal infections in the 1940s and 1950s was accompanied by profound changes in physicians' attitudes, abetted by declining recognition of pneumococci in diagnostic laboratories as they abandoned pneumococcal serotyping. The view that pneumococcal infection had largely disappeared and that what remained was of little gravity widely held, and two licensed hexavalent pneumococcal vaccines were removed from the market after several years for lack of use.

Reexamination of pneumococcal infection in an urban hospital in the 1950s with traditional methods [27] revealed no evidence of decline in the incidence of pneumococcal infections, and that, among individuals sustaining irreversible injury early in the course of infection, most often the elderly, chronically ill and/or immunocompromised, antimicrobial therapy provided little benefit. In the absence of measures that might improve prognosis, prophylaxis appeared to be the only alternative.

Redevelopment of pneumococcal vaccine required determination of those serotypes most often responsible for infection; and because of the uncertain etiologic role of isolates from respiratory secretions, was based on isolates from normally sterile body sites, predominantly blood. Examination of more than 3,000 isolates revealed that half the infections were caused by six types, three-quarters by 12 types and seven-eighths by 18 types. Although rank order might shift over time, these types tend to persist as those most frequently responsible for infection [28, 29].

Preparations of capsular polysaccharides, based upon their chemical properties and molecular size were tested in volunteers in doses from 12.5 to 1,000 μg, with foreknowledge that several grams of polysaccharide might accumulate in consolidated lung [30], and combined in polyvalent formulations. When it was learned that epidemic pneumococcal pneumonia with attack rates exceeding 100 per 1,000 per annum still occurred among gold miners in South Africa, trials involving 12,000 African males were conducted between 1972 and 1976 [31]. One-third received polyvalent pneumococcal vaccine prepared by Eli Lilly & Co., one-third Group A meningococcal vaccine and the remainder a saline placebo. Putative and proved pneumococcal infections associated with types represented in the pneumococcal vaccine were reduced 78.5% among its recipients when contrasted with controls; bacteremias caused by the same types were reduced 82.3%, the *P* values for both findings being less than 0.0001. Attack rates of infection with five types represented in the vaccine were sufficiently high to demonstrate their individual efficacy in reducing homotypic infection. Other trials conducted concurrently with polyvalent vaccine prepared by Merck & Co. in South Africa yielded concordant findings [32].

Although the results of the trial of MacLeod et al. [22] were based on a study of pneumonia, it continues to be stated erroneously that evidence is lacking demonstrating that pneumococcal vaccine prevents nonbacteremic pneumococcal pneumonia. The South African trials show clearly that, in a population in which pneumococcus predominated as the cause of pneumonia, vaccination was followed by 50% reduction in radiologically diagnosed pneumonia irrespective of cause, with a *P* value of 0.0001 for the difference between pneumococcal vaccines and controls. On the basis of the foregoing findings, a tetradecavalent vaccine prepared by Merck & Co. was licensed in 1977, Eli Lilly having withdrawn from the market, and the formulation expanded to 23 antigens in 1983.

Acceptance of the vaccine has been slow. Two randomized, double-bind controlled trials in the USA, although consistent in their findings with those from abroad, were inconclusive because of the low attack rates of illness; and a large trial in the Veterans Administration [33], despite being widely misperceived as showing the vaccine's inefficacy, was equally unrevealing for the same reason [34]. Because definitive randomized double-blind trials in the USA would entail enrollment of very large populations and problems in assuring the validity of microbiologic specimens as well as great expense, trials of two other designs, case control studies [35] and indirect cohort studies [36] have been employed. The results of such studies have shown the aggregate efficacy of the vaccine to be between 56 and 67% [37–40].

In evaluating pneumococcal vaccine, the impact of its polyvalency has been largely overlooked. The few data available on the efficacy of monovalent polysaccharide vaccines in adults [23, 41] suggest they approximate 90%. If it assumed that individual antigens in pneumococcal vaccine are comparably effective and a vaccine is exposed to two pneumococcal type represented in the vaccine, the likelihood of his being infected with neither is 0.9×0.9 or 81%. Limited data on the turnover of pneumococcal types in adults suggest that one or two new types may be acquired annually. If, over a period of 4 years, one were exposed to four pneumococcal types represented in the vaccine, the antigens of which were each 90% effective, the likelihood of being infected with none would be 0.9^4 or 64%, a value corresponding closely to the observations cited.

Two other aspects of bacterial polysaccharide vaccines are currently incompletely understood. First, it is unclear how long after their administration protection persists. Most bacterial polysaccharides are not degradable by mammalian enzymes. Pneumococcal polysaccharides have been detected in the tissues of mice by immunofluorescent microscopy months after their injection [42], and Felton observed positive precipitin reactions with extracts of human tissues obtained at autopsy and antisera to pneumococcal types 1 and 2 [43]. Antibodies in young adults plateaued approximately 6 months after vaccination, and revaccination 5 years later caused only a transient rise in antibodies which returned soon to levels antedating readministration of vaccine [44]. A large case control study showed some decline in protection after 5 or more years in vaccinated adults under age 55 and that protection declined more rapidly with increasing age [39]. Further studies are needed to determine if and when revaccination is indicated and to discover also if initial vaccination early in adult life will afford comparable or better protection than if it delayed to age 65. The rising incidence of drug resistant pneumococci makes answers to these questions increasingly relevant [45].

Pneumococcal infection is a major problem in early life, both as a cause of otitis media and of bacteremia. As indicated by earlier studies [46, 47], those of contemporary polysaccharide vaccines have shown them to be poorly immunogenic with negligible impact on the carrier state [48] or otitis media [49]. There are reasons to anticipate that this deficiency of isolated polysaccharide antigens can be overcome by their chemical linkage to proteins, a topic considered elsewhere.

To date, vaccines of capsular polysaccharides of four bacterial species: *S. pneumoniae*, *Neisseria meningitidis*, *H. influenzae,* and *S. typhi* composed respectively of 23, 41, and 1 capsular antigens, have been licensed. That of *H. influenzae* has been superseded by conjugate preparations, and as such may be the ultimate form of Vi antigen.

Finally, it should be appreciated that not all bacterial polysaccharides can be the basis of vaccines suitable for humans. The hyaluronic acid capsule of Lancefield's group A and C streptococci cannot be distinguished from self by man. In similar fashion, the capsular polysaccharide of Group B meningococci and of *E. coli* K1, composed of disaccharide units present in neural tissue [50] is a poor antigen, and the possibility that antibodies, following its injection, might cause neural injury has been a cause of continuing concern.

In summary, vaccines of bacterial capsular polysaccharides have proved useful in preventing infections in adults caused by pneumococci, meningococci, *H. influenzae,* and typhoid bacilli. When infections are caused by a limited number of serotypes and when childhood infection is common, they are likely to be replaced by conjugate vaccines. In the prevention of pneumococcal disease, polysaccharide vaccines may still have a role in a paradigm in which a limited number of conjugated antigens is given in infancy and a more highly polyvalent preparation of unconjugated polysaccharides is administered at or after puberty.

References

1. Pasteur L, Chamberland, Roux MM. Sur une maladie nouvelle, provoquée par la salive d'un enfant mort de la rage. *C R Acad Sci* 1881;92:159–65
2. Sternberg GM. A fatal form of septicaemia in the rabbit, produced by the subcutaneous injection of human saliva. *Studies Biol Lab Johns Hopkins Univ* 1882;2:183–200
3. Friedlander C. Die Mikrokokken der Pneumonie. *Fortsch Medicin* 1883;1:715–33

4. Bernheimer AW. Synthesis of Type III pneumococcal polysaccharide by suspensions of resting cells. *J Exp Med* 1953;97:591–600
5. Bornstein DL, Schiffman G, Bernheimer HP, Austrian R. Capsulation of pneumococcus with soluble C-like (C_s) polysaccharide. I. Biological and genetic properties of C_s pneumococcal strains. *J Exp Med* 1968;128:1385–1400
6. Van Dam JEG, Fleer A, Snippe H. Immunogenicity and immunochemistry of *Streptococcus pneumoniae* capsular polysaccharides. *Antonie van Leeuwenhoek* 1990;58:1–47
7. Jennings HJ. Capsular polysaccharides as vaccine candidates. *Curr Top Microbiol Immunol* 1990;150:97–127
8. White B. *The Biology of Pneumococcus.* New York, NY. The Commonwealth Fund, 1938: 2nd Printing, Harvard University Press, 1979
9. Neufeld F, Haendel L. Weitere Untersuchungen über Pneumokokken-Heilsera. III. Mitteilung. Über Vorkommen und Bedeutung atypischer Varietäten des Pneumokokkus. *Arb Kais Gesund* 1910;34:293–304
10. Dochez AR, Avery OT. The elaboration of specific soluble substance by pneumococcus during growth. *J Exp Med* 1917;26:477–93
11. Heidelberger M, Avery OT. The soluble specific substance of pneumococcus. *J Exp Med* 1923;38:73–9
12. Avery OT, Heidelberger M. Immunological relationships of cell constituents of pneumococcus. Second paper. *J Exp Med* 1925;42:367–76
13. Knecht JC, Schiffman G, Austrian R. Some biological properties of pneumococcus type 37 and the chemistry of its capsular polysaccharide. *J Exp Med* 1970;132:475–87
14. MacLeod CM, Krauss MR. Relation of virulence of pneumococcal strains for mice to the quantity of capsular polysaccharide formed in vitro. *J Exp Med* 1950;92:1–9
15. Austrian R. Some aspects of the pneumococcal carrier state. *J Antimicrob Chemother* 1986; 18 (suppl A):35–45
16. Gwaltney JM, Sande MA, Austrian R, Hendley JO. Spread of *Streptococcus pneumoniae* in families. II. Relation of transfer of *S pneumoniae* to incidence of colds and serum antibody. *J Infect Dis* 1975;132:62–8
17. Sternberg GM. Induced septicemia in the rabbit. *Am J Med Sci* 1882;84:69–76
18. Wright AE, Parry Morgan W, Colebrook L, Dodgson RW. Observations on prophylactic inoculation against pneumococcus infections and on the results which have been achieved by it. *Lancet* 1914;1:1–10, 87–95
19. Maynard GD. Memorandum on Rand Mines pneumococcic vaccine experiment. *Med J S Afr* 1913;9:91–5
20. Lister FS. An experimental study of prophylactic inoculation against pneumococcal infection in the rabbit and in man. *Pub S Afr Inst Med Res* 1916;8:231–87
21. Heidelberger M, MacLeod CM, Kaiser SJ, Robinson B. Antibody formation in volunteers following injection of pneumococci or their type-specific polysaccharides. *J Exp Med* 1946; 83:303–20
22. MacLeod CM, Hodges RG, Heidelberger M, Bernhard WG. Prevention of pneumococcal pneumonia by immunization with specific capsular polysaccharides. *J Exp Med* 1945;82: 445–65
23. Artenstein MS, Gold R, Zimmerly JG, Wyle FA, Schneider H, Harkins C. Prevention of meningococcal disease by Group C polysaccharide. *N Engl J Med* 1970;282:417–20
24. Schiemann 0, Casper W. Sind die spezifisch präcipitablen Substanzen der 3 Pneumokokkentypen Haptene? Zeitschr Hyg Infektionskr 1927;108:220–57
25. Francis T Jr, Tillett WS. Cutaneous reactions in pneumonia. The development of antibodies following the intradermal injection of type-specific polysaccharide. *J Exp Med* 1930;52: 573–85
26. Ekwurzel GM, Simmons JS, Dublin LI, Felton LD. Studies on immunizing substances in pneumococci. Public Health Rep 1938;53:1877–93
27. Austrian R, Gold J. Pneumococcal bacteremia with especial reference to bacteremic pneumococcal pneumonia. *Ann Intern Med* 1964;60:759–76

28. Finland M, Barnes MW. Changes in occurrence of capsular serotypes of *Streptococcus pneumoniae* at Boston City Hospital during selected years between 1935 and 1974. *J Clin Microbiol* 1977;5:154–66
29. Austrian R. Some observations on the pneumococcus and on the current status of pneumococcal disease and its prevention. *Rev Infect Dis* 1981;3 (suppl): S1–S17
30. Frisch AW, Tripp JT, Barrett CD, Pidgeon BE. The specific polysaccharide content of pneumonic lungs. *J Exp Med* 1942;76:505–10
31. Austrian R, Douglas RM, Schiffman G et al. Prevention of pneumococcal pneumonia by vaccination. *Trans Assoc Am Phys* 1976;89:184–92
32. Smit P, Oberholtzer D, Hayden-Smith S, Koornhof HJ, Hilleman HR. Protective efficacy of pneumococcal polysaccharide vaccines. *JAMA* 1977;238:2613–6
33. Simberkoff MS, Cross AP, Al-Ibrahim M et al. Efficacy of pneumococcal vaccine in high-risk patients. Results of a Veterans Administration cooperative study. *N Engl J Med* 1986;315: 1318–27
34. Shapiro ED. Correspondence. Pneumococcal vaccine failure. *N Engl J Med* 1987;316:272–3
35. Clemens JD, Shapiro ED. Resolving the pneumococcal vaccine controversy: are there alternatives to randomized clinical trials? *Rev Infect Dis* 1984;6:589–600
36. Broome CV, Facklam RR, Fraser DW. Pneumococcal disease after pneumococcal vaccination. An alternative method to estimate the efficacy of pneumococcal vaccine. *N Engl J Med* 1980;303:549–52
37. Shapiro ED, Clemens JD. A controlled evaluation of the protective efficacy of pneumococcal vaccine for patients at high risk of serious pneumococcal infections. *Ann Intern Med* 1984;101:325–30
38. Sims RV, Steinman WC, McConville JH, King LR, Zwick WC, Schwartz JS. The clinical effectiveness of pneumococcal vaccine in the elderly. *Ann Intern Med* 1988;108:653–7
39. Shapiro ED, Berg AT, Austrian R et al. The protective efficacy of polyvalent pneumococcal polysaccharide vaccine. *N Engl J Med* 1991;325:1453–60
40. Bolan G, Broome CV, Facklam RR, Plikaytis BD, Fraser DW, Schlech WF III. Pneumococcal vaccine efficacy in selected populations in the United States. *Ann Intern Med* 1986;104:1–6
41. Erwa HH, Haseeb MA, Idris AA, Lapeyssonie L, Sanborn WR, Sippel JE. A serogroup A meningococcal vaccine. Studies in the Sudan to combat meningococcal meningitis caused by *Neisseria meningitidis* group A. *Bull WHO* 1973;49:301–5
42. Kaplan MH, Coons AH, Deane HW. Localization of antigen in tissue cells. III. Cellular distribution of pneumococcal polysaccharides Type II and III in the mouse. *J Exp Med* 1950;91:15–30
43. Felton LD. The significance of antigen in animal tissues *J Immunol* 1949;61:107–17
44. Heidelberger M, DiLapi MM, Siegel M, Walter AW. Persistence of antibodies in human subjects injected with pneumococcal polysaccharides. *J Immunol* 1950;65:535–41
45. AustrianR. Confronting drug-resistant pneumococci. *Ann Intern Med* 1994;121:807–9
46. Davies JAV. The response of infants to inoculation with Type 1 pneumococcus carbohydrate. *J Immunol* 1937;33:1–7
47. Hodes HL, Ziegler JF, Zepp HD. Development of antibody following vaccination of infants and children against pneumococci. *J Pediatr* 1944;24:641–9
48. Douglas RM, Hansman D, Miles HB, Paton JC. Pneumococcal carriage and type-specific antibody. Failure of a 14-valent vaccine to reduce carriage in healthy children. *Am J Dis Child* 1986;140:1183–5
49. Mäkelä PH, Leinonen M, Pukander J, Karma P. A study of pneumococcal vaccine in prevention of clinically acute attacks of recurrent otitis media. *Rev Infect Dis* 1981;(Suppl):3:S124–30
50. Devi SJN, Robbins JB, Schneerson R. Antibodies to poly [(2→8)-α-N-acetylneuraminic acid] and poly [(2 9)α-N-acetylmuraminic acid] are elicited by immunization of mice with *Escherichia coli* K92 conjugates: potential vaccines for groups B and C meningococci and *E coli* K1. *Proc Natl Acad Sci USA* 1991;88:7175–9

Polysaccharide–Protein Conjugate Vaccines*

John B. Robbins, Rachel Schneerson, Shouson C. Szu, and Vince Pozsgay

Historic Background

*It is to Karl Landsteiner that we owe the notion that the immunologic properties of nonimmunogenic ligands (haptens), including saccharides, can be improved by covalent attachment to proteins [1]. His pioneering studies in the 1920s influenced Walter Goebel and Oswald Avery (his colleague at the Rockefeller Institute, NY) who sought evidence that serum antibodies to the type 3 capsular polysaccharide

*This presentation is dedicated to the memory of Margaret Pittman, who died in 1995 at the age of 94 years. We acknowledge her many original and important contributions to science with special reference to *Hemophilus influenzae* type b (Hib). It was our good fortune to have been befriended by Pittman and we are grateful to the many enriching times spent with this gifted scientist and inspiring teacher.

J.B. Robbins (✉)
National Institute of Child Health and Human Development, National Institutes of Health, Bethesda, MD 20892-2720, USA
e-mail: robbinsjo@mail.nib.gov

S.A. Plotkin (ed.), *History of Vaccine Development*,
DOI 10.1007/978-1-4419-1339-5_12, © Springer Science+Business Media, LLC 2011

(CP) of pneumococci conferred protection to that pathogen [2]. These workers showed that a synthetic disaccharide (hapten), cellubiuronic acid, bound to a protein could elicit antibodies that were both reactive with the type 3 CP and conferred protection to mice challenged with that pathogen. At that time, purification of individual components of bacteria was difficult and high-titered antisera were prepared by intravenous injections of whole bacteria: such serologic reagents were multivalent. Their studies provided convincing evidence that CPs were both essential virulence factors and protective antigens of pneumococci.

Primary Pathogens

Microorganisms that cause systemic infections in otherwise healthy individuals may be designated as "primary" pathogens. Microorganisms that cause infection only in individuals, whose resistance is compromised by either genetic or acquired mechanisms, are referred to as "opportunistic." Many primary and opportunistic bacteria have surface polysaccharides (PSs) that are both essential virulence factors and protective antigens. These types of surface antigens may be CPs of gram-positive and gram-negative bacteria or lipopolysaccharides (LPS) of gram-negatives.

CPs confer virulence by virtue of their ability to "shield" the bacterium from the protective actions of complement – they have no pharmacologic activity aside from their ability to stimulate the formation of mostly serum antibodies [3, 4]. LPSs have more complicated structures than CPs and are composed of three domains [5]. The innermost, lipid A, anchors the LPS to the surface of the outer membrane and exerts properties of "endotoxin" (fever, inflammation, vasomotor effects, leukocytosis, etc.). The outer domain, known as the O-specific polysaccharide (O-SP), has virulence properties (shielding activity) similar to CPs [6]. Purified O-SPs have no pharmacologic properties, are not able to elicit antibodies, probably because of their comparatively low molecular weight, and should be considered as haptens.

Prevention of *Hemophilus influenzae* Type b (Hib) Meningitis

There are at least four reasons for preventing meningitis and other systemic infections caused by Hib (1) in the USA, Canada, Sweden, etc., Hib meningitis was common; about one in 280 newborns were affected by the age of 5 years[7]. In some populations, such as Alaskan Eskimo and Australian aboriginal children, the attack rate of Hib meningitis ranged from 1/30 to 1/50 of newborns [8–10]. Patients with defective splenic function, hypogammaglobulinemia and children who live under crowded conditions and poverty are also highly susceptible; (2) mortality is 5–10% and about 30% of "cured" patients had central nervous system (CNS) deficits ranging from deafness, seizures to mental retardation [11]; (3) about 30% of isolates were ampicillin-resistant and resistance to other antibiotics

Margaret Pittmen

is increasing; (4) Hib meningitis is about ten times more contagious in children than meningococcal meningitis [12, 13]. Infants and young children in day-care centers and nurseries were particularly vulnerable [14]. Epiglottitis, the second most common systemic Hib infection, caused even higher mortality and morbidity than meningitis [15].

Historic Considerations

When we started to work on this problem in 1969, there were three important observations on the pathogenesis of and protective immunity to Hib. First, Margaret Pittman identified capsulated and noncapsulated strains of *Hemophilus influenzae* (Hi) [16]. She distinguished six capsular types and showed that one, type b, accounted for almost all cases of meningitis caused by this species. Furthermore, she showed that serum antibodies conferred type-specific protection to rabbits challenged with Hib [16]. Second, Fothergill and Wright showed an inverse relation between the age distribution of Hib meningitis cases and the presence of serum "bactericidal power," whether inherited by newborns from the maternal circulation or acquired as "natural" antibodies, conferred protection against meningitis and other systemic infections caused by Hib (this hypothesis was shown by Goldschneider et al. [18] to be valid also for meningococcal meningitis). Hib meningitis begins to occur at approximately 3 months of age, peaks at about 9 months and is rarely encountered over the age of 5 years. The third observation was provided by Hattie Alexander, Michael Heidelberger, and Grace Leidy, who showed that the protective moiety of therapeutic antiserum prepared by multiple intravenous injections of whole, formalin-fixed Hib, could be removed by adsorption with Hib PS [19]. To summarize, these studies established that type b caused most systemic Hi infections, that serum bactericidal antibodies conferred immunity, and that passive immunization with Hib CP antibodies had therapeutic value.

Capsular Polysaccharides of *Hemophilus influenzae*

Hi are confined mostly to the respiratory tract of humans: almost all Hi from CSF, blood, pleural, and joint fluid are capsulated. Of the six CPs (a–f), only Hib survives in serum complement and causes bacteremia in pathogen-free infant rats [4]. Prevention of Hib systemic infections does not result in the emergence of the other five types as pathogens because they are susceptible to complement alone. Immunity to Hib, in contrast to other Hi types, requires a critical level of serum CP antibodies, whether passively acquired or actively induced by vaccine or cross-reactive antigens ("natural antibodies") [17, 20]. (With this and previous information, it could be predicted that a vaccine capable of inducing bactericidal antibodies in infants would be efficacious.) We also showed that type a, the polysaccharide the most structurally related to type b, also has similar, though less virulent, properties in these in vitro and in vivo assays [4]. In some developing countries, type a strains comprise about 15% of Hi from meningitis.

Development of "Natural" CP Antibodies Inversely Related to Age Incidence of Hib Systemic Infection

Acquisition of serum anti-Hib CP is age-related and usually not induced by disease or respiratory carriage of Hib. Rather, most anti-Hib CP is elicited by cross-reacting bacteria of the respiratory and intestinal tracts in the absence of the homologous organism [20, 21]. *Escherichia coli* K100 is the best studied of these structurally related and cross-reacting polysaccharides. The age-related acquisition of anti-Hib CP is inversely related to the age incidence of Hib meningitis: newborns and older children rarely contract systemic Hib infection. Vaccine-induced prevention of Hib meningitis, therefore, is required only from the age of 3 months to 5 years; thereafter, anti-Hib CP antibodies are continually stimulated by cross-reacting antigens of the normal enteric and respiratory flora. This principle, that serum anti-PS may be elicited by the normal flora, has been shown or may be inferred for most surface PS of bacterial pathogens [21].

Protective Level of *Hemophilus influenzae* Type b Antibodies

The "protective" level of serum anti-Hib CP was estimated from the successive experience of passive immunization of boys with X-linked hypogammaglobulinemia. CP antibody levels before their next injection were derived from analyses of the antibody contents of commercial immunoglobulin, the $t_{1/2}$ of IgG, its distribution in the body, the dosage and the intervals between injections [22]. The residual ("protective") level ranged from 0.12 to 0.24 μg or approximately 0.15 μg Ab/mL, which was later confirmed in clinical trials with hyperimmune human immunoglobulin (BPIG) [23].

Age-Related Antibody Responses to Hib Polysaccharide

Neither convalescence from systemic infection [24] nor stimulation by cross-reacting antigens, vaccination or revaccination with the Hib CP alone induces regular protective levels of antibodies up to 2 years of age, the age group with the highest incidence, morbidity, and mortality caused by this pathogen [22, 23, 25, 26]. Children, 2 to about 5 years old, respond to Hib CP with protective antibody levels which are comparatively short-lived [6, 22]. This immunologic property of PS in humans has been shown for most bacterial pathogens [27].

These limitations can be overcome by covalent attachment of Hib CP to a protein to form conjugates [28–31]. As a conjugate, the properties of both increased immunogenicity and T cell dependence are conferred upon the CP as manifest by a booster response in infants and a statistically significant higher level of anti-Hib CP at all ages [32–34]. Three schemes to covalently bind Hib CP to proteins have proved successful [28–31]. The resultant conjugates have slightly different immunologic properties, but all induce protective levels of anti-Hib CP in infants. Hib are now recommended for routine vaccination of infants in western developed countries and in some developing countries [35, 36].

Conjugates Are more Immunogenic than Systemic Hib Infection

Systemic Hib infection in <2 year olds results in no or low levels of Hib CP antibodies. Hib conjugates, in contrast, elicit protective levels of antibodies in this age group [24].

Duration of Anti-Hib CP in Infants Vaccinated with Conjugates

Claesson et al. [37] have measured the Hib CP antibodies in 6-year-old children vaccinated with a Hib conjugate during infancy and compared these levels with those of nonvaccinated children. The geometric mean level of Hib CP antibodies in the 6-year-olds vaccinated in infancy was statistically higher than in the controls. More importantly, there were no vaccinated children with <0.15 μg Ab/mL compared to about 15% with less than this protective level among the controls. It is likely that Hib conjugate-vaccinated children are protected from systemic Hib infections throughout their lives.

We cannot yet predict the immunologic properties of Hib conjugates in infants by unambiguous methods. Standardization of Hib conjugates has been implemented using a variety of physicochemical and biologic assays to ensure the potency of each new lot [35, 36]. Variables that affect the immunogenicity include the size and number of saccharides bound to the protein and the nature of the carrier protein [38, 39]. Hib conjugates have only trace levels of LPS ("endotoxin") and elicit no or only trivial side reactions. Reactions elicited by diphtheria and tetanus toxoids (DT) are minor and infrequent, and we predict that the new acellular vaccines eliminate the problem of adverse reactions that follow administration of DT and conjugates.

Two of the three Hib conjugates use tetanus or diphtheria toxoids as carrier proteins. Prior or concomitant administration of the carrier protein enhanced serum anti-Hib PS in animals injected with Hib PS conjugates [40]. This "carrier" effect has been shown in infants injected with Hib conjugate vaccines [41, 42]. Finally, although the three Hib conjugates, licensed for routine vaccination of infants, have differences in their overall composition and structure, injection with any combination gives protective Hib CP antibody levels [43].

"Herd" Immunity Elicited by Anti-Hib PS and Other PS Antibodies

Serum anti-PSs, including those of Hib, were thought to confer protection by inactivating organisms as they entered the blood stream. Previous experience with pneumococcal and meningococcal polysaccharide vaccines and recent data provide another, perhaps the main, protective mechanism elicited by conjugates [44]. The evidence is that (1) serum IgG is present in respiratory tract secretions. Serum IgG, whether placentally acquired or administered as immunoglobulin, confers protection against systemic infection with capsular pathogens; (2) vaccine-induced serum anti-PS inhibits colonization by pneumococci and meningococci in adults [45, 46]; (3) Hib is found almost exclusively in the nasopharynx of young children. Vaccination with conjugate vaccines inhibits colonization of children with Hib [47]; (4) in Iceland, Sweden, and Finland, routine immunization of infants with Hib PS conjugates has resulted in virtual elimination of Hib systemic infection [48–52]. In Wales, Hib meningitis had almost disappeared in infants prior to their vaccination with Hib conjugates. In Finland, Hib epiglottitis in adults has almost disappeared. Furthermore, in the USA, vaccination of only about 60% of children with Hib conjugates has also reduced the incidence of systemic Hib infections to about 1% of their former incidence. The best explanation for this extraordinary favorable effect is that transmission of the pathogen in the vaccinated population is reduced to an extent where it is unlikely that an antibody-negative (susceptible) child encounters a Hib carrier. Since Hib is a pathogen for and a habitant of humans only, it is theoretically possible that worldwide vaccination could eliminate this pathogen as was done with smallpox. Elimination of other capsulated respiratory pathogens that are inhabitants of humans only, such as meningococci and pneumococci, may also follow widespread vaccination with conjugate vaccines.

Conjugate Technology for Other Pathogens Whose Surface Polysaccharides Are Protective Antigens

Clinical trials established the value of multivalent PS vaccines for the prevention of *Neisseria meningitidis* and *Streptococcus pneumoniae* and protective levels of anti-CP for these two pathogenic species have been proposed. Vaccine-induced PS antibodies initiate complement-dependent killing (lysis for meningococci and opsonophagocytosis for pneumococci) as well as inhibition of colonization. Moreover, the extensive experience with passive immunization of patients with hypogammaglobulinemia and of native American infants, shown to be at high risk for Hib and pneumococcal infections, provides evidence that a critical level of CP-specific IgG antibodies is sufficient to prevent systemic (including pneumonia) and local infections (otitis media) with these pathogens [23, 24]. Conjugates of these capsular PS have similar properties to those of Hib PS conjugate vaccines. Accordingly, it seems logical and reasonable to use these data for licensure of new conjugates without extensive clinical efficacy trials.

Infection of newborns with group B streptococci (GBS) is closely analogous to infection with Hib, meningococci, and pneumococci. Immunoprophylaxis of neonatal GBS infections requires that pregnant women have a critical ("protective") level of type-specific serum IgG [53]. Many factors impede clinical trials of GBS conjugates, related mostly to the problems presented by vaccination of pregnant women [54]. These problems could be avoided by assuming efficacy of GBS conjugates based solely upon their ability to induce protective levels of type-specific IgG. We draw attention to experimental observations showing that serum IgG is the major Ig in cervical secretions [44]. Routine vaccination of all children with GBS conjugates could inhibit colonization and thereby reduce transmission of this pathogen in both sexes, and thus eliminate both the pathogen as well as GBS disease [54].

We theorized that serum IgG anti-LPS of enteric bacteria exert similar protective effects to those of the CPS of capsulated respiratory pathogens. We proposed that serum IgG anti-PS confers immunity to these eliminating pathogens by lysing the inoculum as the organisms enter the small intestine [55, 56]. First, vaccination with Vi CP prevents typhoid fever (Vi is now a licensed vaccine in at least 40 countries, including the USA, France, and the UK). Clinical trials of Vi conjugates are underway [57, 58]. Second, conjugate-induced antibodies to the O-SP of *Shigella sonnei* confer protection to shigellosis caused by this pathogen [59, 60]. Third, conjugates composed of the O-SP of *Salmonella typhimurium* bound to TT protected mice against lethal challenge with this pathogen [61, 62]. We are currently evaluating O-SP conjugates of *S. paratyphi* A for the prevention of this important cause of enteric fever in Southeast Asia.

The Future of Polysaccharide–Protein Conjugate Vaccines

There are many reasons to predict that conjugates serve as vaccines for inducing protective levels of antibodies to both the polysaccharide and protein components of these new vaccines. Furthermore, it seems likely that there is a limit to the number of conjugates that use the same carrier protein. We tentatively suggest the following components of conjugate vaccines for routine infant vaccination modified by the epidemiology of the country (1) capsular polysaccharides of *H. influenzae* types b and a; Groups A, B, C, Y, and W135 of meningococci, types 1, 3, 6, 9N, 14, 18C, 19F, 19A, and 23F of pneumococci; types 1, 2, and 3 of GBS; types 1, 2, 5, 12, and 13 of *E. coli*; Vi of *S. typhi* and *Vibrio cholerae* O139;(2) O-SPs of *S. typhimurium*, *S. choleraesuis*, *S. enteritidis*, *S. paratyphi* A, *Shigella dysenteriae* type 1, *S. flexneri* type 2a, *S. sonnei,* and *V. cholerae* O1; (3) genetically toxoided proteins from *Corynebacterium diphtheriae*, *Clostridium tetani*, *C. difficile*, *C. welchii*, *Bordetella pertussis* (pertussis toxin and adenylate cyclase toxin), *E. coli* (heat-labile and heat-stable toxins), *S. dysenteriae*, *Pseudomonas aeruginosa*, and *Aeromonas hydrophilia.*

Lastly, we feel that the synthetic approach may make it possible to increase the immunogenicity of the saccharide component over that of the biosynthetic product [63]. To this end, we have been developing synthetic saccharides corresponding to the O-SPs of shigellae [64].

Conclusion

It is likely that an infant vaccine composed of conjugate vaccines for almost all "primary" pathogenic bacteria will be found (Table 1). This new formulation will induce protective levels of antibodies to both the PS and protein components of the conjugates. Widespread vaccination with these new vaccines could eliminate many of these bacterial pathogens that are inhabitants of humans only.

Table 1 Components of future conjugate vaccines.

Capsular polysaccharides of:
Haemophilus influenzae types b and a*
Meninogocoocal groups A,B,C,Y and W135
Pneumococcal types 1,3,6,9N,14,18C,19F,19A and 23F
Group B streptococcal types 1,2,3
Escherichia coli K types 1,2,5,12 and 13
Vi of *Salmonella typhi*
***Vibrio cholerae* O139**
O-specific polysaccharides of:
S typhimurium, S cholerasuis, S enteriditis, S paratyphi **A**
Shigella dysenteriae **type 1,** ***S flexneri*** **type 2a,** ***S sonnei***
E coli O157, 101, 111, ***V cholerae*** **O1**

continued

Table 1 (continued)

Genetically-toxoided proteins:
Corynebacterium diphtheriae
Clostridium tetani, C difficile, ***C welchii***
Bordetella pertussis (pertussis toxin. adenylate cyclase toxin)
E coli (heat-labile and heat-stable toxins)
S dysenteriae type 1, SLT 1 and SLT II
Pseudomonas acruginosa
Aeromonas hydrophila

*Bold type indicates components proposed for developing countries.

References

1. Landsteiner K. *The Specificity of Serological Reactions*. Revised ed. Cambridge, MA: Harvard University Press, 1970
2. Avery OT, Goebel WF. Chemoimmunological studies on conjugated carbohydrateproteins. II. Immunological specificity of synthetic sugar-proteins. *J Exp Med* 1929;50:521–33
3. Wood WB Jr, Smith MB. Surface phagocytes – its relation to the mechanism of recovery in acute pneumonia caused by encapsulated bacteria. *Trans Assoc Am Physicians*. 1947;60:77–81
4. Sutton A, Schneerson R, Kendall-Morris S, Robbins JB. Differential complement resistance mediates virulence of *Haemophilus influenzae* type b. *Infect Immun* 1982;35:95–104
5. Lunderitz O, Freudenberg MA, Galanos C, Lemann V, Rietchel ET, Shaw DH. Lipopolysaccharides of Gram-negative bacteria. *Curr Top Microbiol Transp* 1982;17:79–151
6. Liang-Takasaki CJ, Grossman N, Leive L. *Salmonella* activate complement differentially via the alternate pathway depending on the structure of their lipopolysaccharide O-antigen. *J Immunol* 1983;130:1867–79
7. Parke JC, Schneerson R, Robbins JB. The attack rate, age incidence, racial distribution and case fatality rate of *Haemophilus influenzae* type b meningitis in Mecklenburg County, North Carolina. *J Pediatr* 1972;81:765–9
8. Gilsdorf J. Bacterial meningitis in southwestern Alaska. *Am J Epidemiol* 1977;106:388–91
9. Ward JJ, Lum MKW, Margolis HS. *Haemophilus influenzae* in Alaskan Eskimos: characteristics of a population with an unusual incidence of invasive disease. *Lancet* 1981;1:121–5
10. Hanna JN. The epidemiology of invasive *Haemophilus influenzae* infections in children under five years of age in the Northern Territory: a three year study. *Med J Aust* 1990;152:243–40
11. Sell HW, Merrill RE, Doyne EO, Zimsky EP. Long-term sequellae of *Hemophilus influenzae* meningitis. *Pediatrics* 1972;49:206–11
12. Glode MP, Schiffer MS, Robbin JB et al. An outbreak of *Haemophilus influenzae* type b meningitis in an enclosed hospital population. *J Pediatr* 1976;88:36–40
13. Glode MP, Daum RS, Goldmann DA, LeClair J, Smith A. *Haemophilus influenzae* type b meningitis: a contagious disease. *Br Med J* 1980;i:1–7
14. Redmond SR, Pichichero ME. *Haemophilus influenzae* type b disease. An epidemiologic study with special reference to day-care centers. *JAMA* 1984;252:2581–4
15. Claesson B, Trollfors B, Ekstrom-Jodal B et al. Incidence and prognosis of acute epiglottitis in children in a Swedish region. *Pediatr Infect Dis* 1984;3:534–8
16. Pittman M. Variation and type specificity in the bacterial species *Haemophilus influenzae*. *J Exp Med* 1931;53:471–92; *Ibid.* The action of type-specific *Haemophilus influenzae* antiserum. *J Exp Med* 1931;58:683–706
17. Fothergill LD, Wright J. Influenzal meningitis: relation of age incidence to the bactericidal power of blood agsint the causal organism. *J Immunol* 1933;24:273–84
18. Goldschneider I, Gotschlich EC, Artenstein MS. Human immunity to the meningococcus. I. The role of humoral antibodies, *J Exp Med* 1969;129:1307–26

19. Alexander HE, Heidelberger M, Leidy G. The protective or curative element in *H influenzae* rabbit serum. *Yale J Biol Med* 1944;16:425–30
20. Schneerson R, Bradshaw M, Whisnant JK et al. An *Escherichia coli* antigen crossreactive with the capsular polysaccharide of *Haemophilus influenzae* type b: occurrence among known serotypes, and immunochemical and biologic properties of antisera towards *H influenzae* type b. *J Immunol* 1972;108:1551–62. Shneerson R, Robbins JB. Induction of serum *Haemophilus influenzae* type b capsular antibodies in adult volunteers fed cross-reacting *Escherichia coli* 075:K100:H5. *N Engl J Med* 1975;292:1093–6
21. Robbins JB, Schneerson R, Glode MP et al. Cross-reactive antigens and immunity to diseases caused by encapsulated bacteria. *J Allergy Clin Immunol* 1975;56:141–51
22. Robbins JB, Parke JC, Schneerson R et al. Quantitative measurement of "natural" and immunization-induced *Haemophilus influenzae* type b capsular polysaccharide antibodies. *Pediatr Res* 1973;7:103–10
23. Santosham M, Reid R, Ambrosino DM, et al. Prevention of *Haemophilus influenzae* type b infections in high-risk infants treated with bacterial polysaccharide immune globulin. *N Egl J Med* 1987;317:923–9
24. Trollfors B, Lagergärd T, Cleasson BA, Thornberg E, Martinell J, Schneerson R. Characterization of the serum antibody response to the capsular polysaccharide of *Haemophilus influenzae* type b in children with invasive infections. *J Infect Dis* 1992;166:1335–9
25. Smith DH, Peter G, Ingram DL, Anderson P. Responses of children immunized with the capsular polysaccharide of *Haemophilus influenzae* type b. *Pediatrics* 1973;52:637–41
26. Peltola H, Kayhty H., Sivonen A, Makela PH. *Haemophilus influenzae* type b capsular polysaccharide vaccine in children: a double blind study of 100,000 vaccinees 3 months to 5 years of age in Finland. *Pediatrics* 1977;60:730–7
27. Austrian R. Some observations on the pneumococcus and on the current status of pneumococcal disease and it prevention. *Rev Infect Dis* 1981;(Suppl):S1–S17
28. Schneerson R, Barrera O, Sutton A, Robbins JB. Preparation, characterization and immunogenicity of *Haemophilus influenzae* type b polysaccharide protein conjugates. *J Exp Med* 1980;152:361–76
29. Chu CY, Schneerson R, Robbins, Rastogi, SC. Further studies on the immunogenicity of *Haemophilus influenzae* type b and pneumococcal type 6A polysaccharide protein conjugates. *Infect Immun* 1983;40:245–56
30. Anderson P. Antibody responses to *Haemophilus influenzae* type b and diphtheria toxin induced by conjugates of oligosaccharides of the type b capsule with the nontoxic protein CRM 197. *Infect Immun* 1983;39:233–8
31. Marburg S, John D, Tolman RL, et al. Biomolecular chemistry of macromolecules – synthesis of bacterial polysaccharide conjugates with *Neisseria meningitidis* membrane protein. *J Am Chem Soc* 1986;108:5282–7
32. Anderson PW, Pichichereo ME, Insel RA. Immunization of two-month-old infants with protein coupled oligosaccharide derived from the capsule of *Haemophilus influenzae* type b. *J Pediatr* 1985;107:346–51
33. Claesson BA, Trollfors B, Lagergard T et al. Clinical and immunological responses to the capsular polysaccharide of *Haemophilus influenzae* type alone or conjugated to tetanus toxoid in 18- to 23-month-old children. *J Pediatr* 1988;112:695–702
34. Parke JC, Schneerson R, Reimer C et al. Clinical and immunologic responses to *Haemophilus influenzae* type b-tetanus toxoid conjugate vaccine in infants injected at 3, 5, 7 and 18 months of age. *J Pediatr* 1991;118:184–90
35. Robbins JB, Schneerson R. Polysaccharide protein conjugates: a new generation of vaccines. *J Infect Dis* 1990;161:821–32
36. World Health Organization. Requirements for *Haemophilus* type b conjugate vaccines. *WHO Tech Rep Ser* 1991;814
37. Claesson BA, Trollfors B, Anderson PW et al. Serum antibodies in 6-year-old children vaccinated in infancy with *Haemophilus influenzae* type b-tetanus toxoid conjugate vaccine. *Pediatr Infect Dis J* 1996;170–2

38. Anderson PW, Pichichero ME, Insel RA. Vaccines consisting of periodate cleaved oligosaccharides from the capsule of *Haemophilus influenzae* type b coupled to a protein carrier: structural and temporal requirements for priming in the human infant. *J Immunol* 1986;137:1181–6
39. Szu SC, Li X, Schneerson R, Vickers JH, Bryla DA, Robbins JB. Comparative immunogenicity of Vi polysaccharide-protein conjugates composed of cholera toxin or its B-subunit as a carrier bound to high to lower molecular weight Vi. *Infect Immun* 1989;57:3823–7
40. Schneerson R, Robbins JB, Chu C, et al. Serum antibody responses of juvenile and infant rhesus monkeys injected with *Haemophilus influenzae* type b and pneumococcus type 6A polysaccharide-protein conjugates. *Infect Immun* 1984;45:582–91
41. Granoff DM, Holmes SJ, Beishe RB et al. Effect of carrier protein priming on antibody response to *Haemophilus influenzae* type b conjugate vaccines in infants. *JAMA* 1994;272:1116–21
42. Barington T, Gyhrs A, Kristensne K, Heilmann C. Opposite effects of actively and passively acquired immunity to the carrier of responses of human infants to a *Haemophilus influenzae* type b conjugate vaccine. *Infect Immun* 1994;62:0–14
43. Greenberg DP, Leiberman JM, Marcy SM et al. Enhanced antibody responses in infants given different sequences of heterogeneous *Haemophilus influenzae* type b conjugate vaccines. *J Pediatr* 1994;126:206–11
44. Robbins JB, Schneerson R, Szu, SC. Perspective. Hypothesis: serum IgG antibody is sufficient to confer protection against infection disases in inactivating the inoculum. *J Infect Dis* 1995;171:1387–98
45. MacLeod CM, Hodges RG, Heidleberger M et al. Prevention of pneumococcal pneumonia by immunization with specific capsular polysaccharides. *J Exp Med* 1945;82:445–65
46. Gotschlich EC, Goldschnieder I, Artnstein MS. Human immunity to the meningococcus. V. The effect of immunization with meningococcal group C polysaccharide on the carrier state. *J Exp Med* 1969;129:1385–95
47. Takala Ak, Eskola J, Leinonen M et al. Reduction of oropharyngeal carriage of *Haemophilus influenzae* type b (Hib) in children immunized with Hib conjugate vaccine. *J Infect Dis* 1991;164;982–6
48. Jónsdóttir KE, Steingrímsson O, Ólafsson O. Immunization of infants in Iceland against *Haemophilus influenzae* type b. *Lancet* 1992;340:252–3
49. Vadheim CM, Greenberg DP, Eriksen E et al. Eradication of *Haemophilus influenzae* type b disease in southern California. *Arch Pediatr Adolesc Med* 1994;148:51–6
50. Van Alphen L, Spanjaard L, van der Ende A, Dankert J. Predicted disappearance of *Haemophilus influenzae* type b meningitis in the Netherlands. *Lancet* 1994;344:195
51. CDC. FDA approval of use of a new *Haemophilus* b conjugate vaccine and a combined diphtheria-tetanus-pertussis and *Haemophilus* b conjugate vaccine for infants and children. *MMWR* 1993;42:296–8
52. CDC. Progress toward elimination of *Haemophilus influenzae* type b disease among infants and children – United States, 1993–1994. *MMWR* 1995;44:545–50
53. Baker CJ, Kasper DL. Correlation of maternal antibody deficiency and susceptibility to neonatal group B streptococcal infection. *N Egl J Med* 1976;294:753–6
54. Robbins JB, Schneerson R, Vann WF, Bryla DA, Fattom A. Prevention of systemic infections caused by group B *Streptococcus* and *Staphylococcus aureus* by multivalent polysaccharide-protein conjugate vaccines. In: *Combined Vaccines and Simultaneous Administration: Current Issues and Perspectives*. Williams JC, Goldenthal KL, Burns DL, Lewis BP Jr, eds, New York Academy of Sciences: 1995;754:68–82
55. Robbins JB, Chu CY, Schneerson R. Hypothesis for vaccine development: protective immunity to enteric diseases caused by nontyphodial Salmonellae and Shigellae may be conferred by serum IgG antibodies to the O-specific polysaccharide of their lipopolysaccharides. *Clin Infect Dis* 1982;15:346–61
56. Szu SC, Gupta RK, Robbins JB. Induction of serum vibriocidal antibodies by O-specific polysaccharide-protein conjugate vaccines for prevention of cholera. In: Wachsmuth IK, Blake PA, Olsvik O, eds. *Vibrio cholerae and Cholera. Molecular to Global Perspectives.* Washington DC: American Society for Microbiology, 1994;381–94

57. World Health Organization Expert Committee on Biologic Standardization. *Requirements on Vi Polysaccharide for Typhoid. Tech Rep Ser* 840, 43rd ed. Geneva, Switzerland, 1993;14–32
58. Szu SC, Taylor DN, Trofa AC et al. Laboratory and preliminary clinical characterization of Vi capsular polysaccharide-protein conjugate vaccines. *Infect Immun* 1944;62:4440–4
59. Taylor DN, Trofa AC, Sadoff J et al. Synthesis, characterization and clinical evaluation of conjugate vaccines composed of the O-specific polysaccharides of *Shingella dysentieriae* type 1, *Shingella flexneri* type 2a and *Shigella sonnei (Plesiomonas shingelloides)* bound to bacterial toxoids. *Infect Immun* 1993;61:3678–87
60. Cohen D, Ashkenazi S, Green M et al. Safety, immunogenicity of preliminary efficacy of Shingella conjugate vaccine in Israeli soldiers (manuscript in preparation)
61. Svenson SB, Nurminen M, Lindberg AA. Artificial *Salmonella* vaccines: O-antigen oligosaccharide-protein cinjugates induce protection against infection with *Salmonella typhimurium*. *Infect Immun* 1979;25:863–70
62. Watson DC, Robbins JB, Szu SC et al. Protection of mice against *Salmonella typhimurium* with an O-specific polysaccharide-protein vaccine. *Infect Immun* 1992; 60:4679–86
63. Peeters CAM, Evenberg D, Hoogerhout P et al. Synthetic trimer and teramer of 3-β-D-ribose-(1,1)-D-ribitol-5-phosphate conjugated to protein induce antibody responses to *Haemophilus influenzae* type b capsular polysaccharide in mice and monkeys. *Infect Immun* 1992;60:1862–33
64. Pavliak V, Nashed J, Pozsgay V et al. The binding of the O-antigen of *Shingella dysenteriae* type 1 and 26 related synthetic fragments to a monoclonal IgM antibody. *J Biol Chem* 1993;268:25797–802

After Pasteur: History of New Rabies Vaccines

Hilary Koprowski

There are questions which we could never get over if we were not delivered from them by the operation of nature.

Kafka, *Reflection on Sin, Pain, Hope and the True Way*

One of the major problems presented by the Pasteur rabies vaccine and its numerous varieties was the presence of animal nervous tissue in the final product and the potential allergic responses to that tissue, leading to encephalitis in humans and animals. As the vaccine was also occasionally applied indiscriminately to subjects not exposed to rabies, the price of vaccination was indeed high.

A solution to this problem necessitated a change in the way the virus was grown to ensure replication in a vehicle free from nervous tissue elements, in which the absolute amount of viral protein would be increased, but the proportion of allergenic cellular antigen drastically reduced. With this in mind, in 1946, I developed a technique for growing rabies virus in chick embryos (tissue culture methods were

H. Koprowski (✉)
Jefferson Cancer Institute, Thomas Jefferson University,
1020 Locus Street, Philadelphia, PA 19107-6799, USA
e-mail: hilary.koprowski@jefferson.edu

S.A. Plotkin (ed.), *History of Vaccine Development*,
DOI 10.1007/978-1-4419-1339-5_13, © Springer Science+Business Media, LLC 2011

not available at that time). I chose a strain of virus adapted to a nonmammalian host, the so-called Flury strain, which was passaged serially by Harold Johnson in the brain of a 1-day-old chick.

Harold Johnson was a member of the staff of the Rockefeller Foundation in New York. He had worked with rabies for a decade and was not infrequently exposed to street rabies virus from animals used in his laboratory experiments. Once, in Mexico, he developed encephalitis accompanied by general paralysis. He miraculously recovered from the disease, but with sequelae involving difficulty in walking and weakness in his arms. Johnson maintained that he had had an abortive attack of rabies, even though he had previously received on several occasions the Pasteur vaccine. He was a very good pianist, and even after his illness he was still able to play the piano.

After I met Johnson in 1945, we decided to take lessons in two-piano playing at the Turtle Bay Conservatory, in New York. During our sessions, we frequently discussed rabies and Johnson offered for my experimentation a strain of street virus originally isolated from a girl who died of rabies infection that he had adapted to 1-day-old chicks [1]. This virus had undergone 136 passages in a 1-day-old chick brain. There were no difficulties in adapting it to chick embryo through inoculation into the yolk sac of 7-day-old embryos and harvesting the embryonic tissue, 10 days later [1]. After the third passage, pathogenicity testing of the virus revealed a loss of infectious capability in rabbits or dogs injected parenterally, but a virulence, albeit greatly reduced, for mice and hamsters injected intracerebrally [1, 2].

However, we continued to passage the virus many more times in chick embryo and observed a sudden change in its pathogenic properties at the level of the 180th passage in chick embryo. At that passage level, the Flury strain lost its virulence for mice and hamsters injected intracerebrally [3]. To differentiate this strain of virus from one at low embryo-passage levels, we called it the HEP (high-egg passage), in contrast to LEP (low-egg passage). The LEP-Flury was found to be nonvirulent for canine species and was used as a live-virus vaccine for dogs. After vaccination with Flury, the dogs remained immune for at least 3–4 years (the longest period tested). At that time, the law did not require revaccination during the dog's lifetime.

Then, there occurred an interesting incident which permitted safety testing of LEP-Flury for humans. Cornelius (Dusty) Rhoads, Director of the Memorial Cancer Center in New York, phoned me one day and asked for rabies vaccine to be used in cancer patients. He described to me an unusual case of a woman who, 3 years before, had been found to have inoperable uterine cancer with abdominal metastases and was sent home to die. Three years later, she appeared at the Memorial Cancer center for a check-up and was found to be free of cancer. There was a great commotion in the Institute over this case and the patient was queried in detail about any unusual treatment she may have received after returning home. She replied that, after returning home, she had been bitten by a dog (rabid?) and received 14 injections of Semple vaccine (modified Pasteur method). Dusty Rhoads hypothesized that it was the rabies virus that caused regression of the cancer and, in order to obtain a preparation better than inactivated virus in the Semple vaccine, he requested the LEP-Flury vaccine. Despite my warnings that we knew only about the safety of LEP for dogs, and not humans, Rhoads was insistent and 33 cancer patients were subsequently given LEP intramuscularly. As expected, the vaccine had no effect on the course of cancer, but none of the recipients showed any untoward reaction following vaccination. Notwithstanding these encouraging results, I remained hesitant to use LEP for routine immunization of humans. On the other hand, HEP had been extensively tested in clinical trials for prophylactic immunization of humans against rabies [4], but had not been accepted for general use because it induced relatively low levels of antibodies in monkeys, as compared to other vaccines [5].

In an alternative approach to the development of a potent human rabies vaccine to use as a vector for production of vaccine material innocuous for human subjects, we chose cultures of human fibroblasts that had been recently developed at The Wistar Institute [6] and were collectively named human diploid cell strain (HDCS). These were cells originally obtained from human embryos, preserved by freezing at early passage levels and reconstituted at will. After serial passages in vitro, they died because of senescence [6]. The absence of microbial contaminants and the fact that the chromosome complement of the cells remained unchanged during their lifetime assured us of the safety of products of microorganisms grown in these cells.

We chose several rabies strains for propagation in HDCS. Most of the initial studies involved adaptation of HEP to HDCS. We found that at the 47th HDCS passage level, HEP completely lost its virulence for monkeys injected intracerebrally, but retained virulence for newborn mice injected by the same route [7]. Although the HEP-HDCS-vaccinated monkeys showed a high titer of antirabies antibodies, it was difficult to overcome resistance from those reluctant to use live virus for vaccination of humans, and we turned to inactivated virus.

After the adaptation of several virus strains to the WI-38 strain of HDCS by the late Tad Wiktor, Mario Fernandes, and me [8], we decided to use the PM strain to produce the vaccine. The virus was inactivated by β-propiolactone and found to be immunogenic in monkeys [9, 10]. In 1971, Wiktor, Stanley, Plotkin, and I decided one day to give each other the first vaccinations with the rabies HDCV [11]. Plotkin then went on to organize the first clinical trials of the vaccine [12, 13], which eventually led to extensive field trials [14–17]. Today, HDCV is the gold standard of rabies vaccines, both for preexposure and postexposure treatment. It is remarkably well tolerated, free of paralytogenic properties, and because of its immunogenicity, can be given in four to five injections postexposure, rather than the 14–21 required for earlier vaccines [18]. There are few vaccination failures after HDCV, and the only reason it has not completely replaced other vaccines is its cost. A less expensive cell culture vaccine based on PM has been produced in the Vero continuous cell line [19].

The history of rabies vaccination is incomplete without mentioning two more recent developments. One is the production of recombinant vaccine by insertion of the rabies glycoprotein gene into live vaccinia virus [20]. This resulted in the first successful oral immunization against rabies in wildlife [21]. The vaccinia-rabies recombinant is extensively used countrywide in a bait for immunization of wildlife such as foxes and raccoons [21]. Mass vaccination has led to complete control of rabies spread among wildlife in several countries [21]. The second development arose from the need to produce an economically feasible vaccine for human use. We thus searched for a vehicle to produce the vaccine in plants as the cheapest source of material. We have succeeded in the expression of rabies glycoprotein in transgenic tomatoes which was found to be immunogenic by injection or by the oral route [22, 23].

References

1. Koporowski H, Cox HR. Studies on chick embryo adapted rabies virus. I. Culture characteristics and pathogenicity *J Immunol* 1948;60:533–54
2. Koprowski H, Black J. Studies on chick embryo adapted rabies virus. II. Pathogenicity for dogs and use of egg-adapted strains for vaccination purposes *J Immunol* 1950;64:185–96
3. Koprowski H, Black J, Nelseen DJ. Studies on chick embryo adapted rabies virus. VU. Further changes in pathogenic properties following prolonged cultivation in the developing chick embryo *J Immunol* 1954;72:94–106
4. Ruegsegger JM, Black J, Sharpless GR. Primary antirabies immunization of man with HEP Flury virus vaccine Am J Public Health 1961;51:706–16
5. Wiktor TJ, Koprowski H. Successful immunization of primates with rabies vaccine prepared in human diploid cell strain WI-38 *Proc Soc Exp Biol Med* 1965;118:1069–73
6. Hayflick L, Moorhead PS. The serial cultivation of human diploid cell strains. *Exp Cell Res* 1961;25:595
7. Wiktor TJ, Fernandez MV, Koprowski H. Potential use of human diploid cell strains for rabies vaccine. In: *Proc Symp on Characterization and Uses of Human Diploid Cell Strains.* Yugoslavia: Opatija, 1963
8. Wiktor TJ, Fernandez MV, Koprowski H. Cultivation of rabies virus in human diploid cell strain WI-38. *J Immunol* 1964;93:353–66
9. Koprowski H. In vitro production of antirabies virus vaccine. *Int Symp on Rabies, Talloires* 1965; *Symp Series Immunobiol Standard* 1966;1:357–66
10. Wiktor TJ, Sokol F, Kuwert E, Koprowski H. Immunogenicity of concentrated and purified rabies vaccine of tissue culture origin. *Proc Soc Exp Biol Med* 1969;131:799–805
11. Wiktor TJ, Plotkin SA, Grella DW. Human cell culture rabies vaccine. *JAMA* 1973;224:1170–1
12. Plotkin SA, Wiktor TJ, Koprowski H, Rosanoff EI, Tint H. Immunization schedule for the new human diploid cell vaccine against rabies. *Am J Epidemiol* 1976;103:75–80
13. Plotkin SA, Wiktor TJ. Rabies vaccination. *Annu Rev Med* 1978;29:583–91
14. Aoki FY, Tyrrell DAH, Hill LE. Immunogenicity and acceptability of a human diploid cell culture rabies vaccine in volunteers. *Lancet* 1975;1:660–2
15. Kuwert EK, Marcus I, Werner J, Scheiermann N, Thraenhart O. Some experiences with human diploid cell strain (HDCS) rabies vaccine in pre-and post-exposure vaccinated humans. *Dev Biol Stand* 1978;40:79–88
16. Cox JH, Kleitmann W, Schneider LG. Human rabies immunoprophylaxis using HDC (MRC-5) vaccine. *Dev Biol Stand* 1978;40:105–8
17. Ajjan N, Soulebot JP, Triau R, Biron G. Intradermal immunization with rabies vaccine: inactivated Wistar strain cultivated in human diploid cells. *JAMA* 1980;244:2528–31
18. Immunization Practices Advisory Committee (ACIP). Rabies prevention-United States, 1991. MMWR 1991;40(RR-3):1–19
19. Montagnon BJ. Polio and rabies vaccines produced in continuous cell lines: a reality for Vero cell line. *Dev Biol Stand* 1989;70:27–47
20. Wiktor TJ, MacFarlan RI, Reagan KJ et al. Protection from rabies by a vaccinia virus recombinant containing the rabies virus glycoprotein gene. *Proc Natl Acad Sci USA* 1984;81:7194–8

21. Rupprecht CE, Wiktor TJ, Johnston DH et al. Oral immunization and protection of raccoons *(Procyon lotor)* with a vaccinia-rabies glycoprotein recombinant virus vaccine. *Proc Natl Acad Sci USA* 1986;83:7947–50
22. McGarvey P, Hammond J, Dienelt MM et al. Expression of the rabies virus glycoprotein in transgenic tomatoes. *Biotechnology* 1995;13:1484–7
23. Yuslbov V, Hooper DC, Spitsin SV et al. Expression in plants and immunogenicity of plant virus-based experimental rabies vaccine. *Vaccine* 2002;20:3155–64

Yellow Fever Vaccines: The Success of Empiricism, Pitfalls of Application, and Transition to Molecular Vaccinology

Thomas P. Monath

In 1951, Max Theiler was awarded the Nobel Prize in Medicine and Physiology for the development of yellow fever vaccine. The discovery phase of Theiler's research preceded the prize by only about 20 years, during which time his vaccine against yellow fever had been put into wide-scale use in Africa and South America, and tens of thousands of yellow fever deaths had been averted. After smallpox vaccine, which had been discovered some 135 years before, yellow fever was the first human vaccine to be used at a population level for control of a major epidemic disease. Yellow fever vaccines were developed during an early stage in the history of virology, using empirical methods. Approximately 50 years later, research was initiated on the molecular basis of attenuation; progress is briefly reviewed here. The history of yellow fever vaccines provided a number of important paradigms for vaccine development in general (Table 1).

T.P. Monath (✉)
Kleiner Perkins Caufield & Byers, Cambridge, MA 02139, USA
e-mail: tmonath@kpcb.com

S.A. Plotkin (ed.), *History of Vaccine Development*,
DOI 10.1007/978-1-4419-1339-5_14,

Table 1 Yellow fever vaccine as a paradigm for vaccine development in general

Empirical development
Uncontrolled passage, leading to overattenuation or reactogenicity
Two competing vaccines
Developed during period when ethics of clinical research were not clearly established
No formal tests of efficacy
Problems with adventitious agents
Problems with thermostability
Requirement for combined immunization
Difficulties in the implementation of effective vaccination at a population level

Early Work: Immunization by "Mosquito Inoculation"

Yellow fever was recognized as a major cause of human morbidity in the eighteenth and nineteenth century, but without knowledge of the causative agent or mode of transmission, no effective strategy was available to prevent the disease. The first attempts at immunization against yellow fever were made by Carlos J. Finlay, a Cuban physician who first proposed, in 1881, that the disease caused by an unidentified "germ" was transmitted by mosquitoes [1]. Finlay sought to prove this theory by allowing mosquitoes that had previously bitten yellow fever patients to feed on nonimmune recent immigrants to Cuba. He proposed that controlled inoculation by the bite of a single mosquito could confer immunity against subsequent "severe" exposure to multiple bites of infected mosquitoes [2, 3]. Between 1883 and 1890, Finlay compared the incidence of fatal yellow fever in a group of 65 Spanish priests, 33 of whom had received such "mosquito inoculations." None of the inoculated priests died of yellow fever during their service in Cuba, leading Finlay to conclude that they had been effectively immunized, whereas five (15.6%) of the "controls" died of the disease. In retrospect, Finlay's experiments were inconclusive, principally because, believing the "germ" of yellow fever was mechanically transmitted on the proboscis of the mosquito, he used mosquitoes that had only recently fed on a yellow fever patient and thus could not have undergone an extrinsic incubation period sufficiently long for disseminated mosquito infection and virus transmission to have occurred.

In 1900, Walter Reed and his colleagues definitively proved in a series of human volunteer studies that *Aedes aegypti* mosquitoes transmitted yellow fever and showed that the agent was a filterable virus [4]. None of the 14 intentionally inoculated subjects died, and only four had severe disease. This result, which seemed to corroborate Finlay's earlier suggestions, stimulated Juan Guiteras, Professor of Pathology and Tropical Medicine, University of Havana), to further induce infections by mosquito bite "… with the hope of propagating the disease in a controllable form, and securing among the recently arrived immigrants, immunization, with the minimum amount of danger to themselves and to the community" [5]. From Reed's studies, Guiteras knew that it was necessary to allow an interval of time to elapse between initial mosquito feeding on a yellow fever patient and refeeding on the experimental subject.

Guiteras was able to elicit clinical yellow fever in eight volunteers, three of whom died. His results stood in stark contrast to the total experience of the Reed Commission, in which none of the 14 persons who developed the disease died. Guiteras' unfortunate experience put a halt to the mosquito immunization approach. Guiteras was apparently unlucky enough to have used a virus strain (Alvarez) that produced a low infection-to-case ratio (of only 1.3:1) and a high case-fatality ratio (43%). In studies conducted shortly thereafter in Rio de Janeiro by Marchoux and Simond, there were no fatalities among volunteers experimentally infected by mosquito bite [6].

1927: Isolation of Yellow Fever Virus, a Requisite Step Towards a Defined Vaccine

The development of a defined vaccine against yellow fever obviously depended upon the isolation, characterization, and ability to grow the etiologic agent. In 1925, the Rockefeller Foundation established the West Africa Yellow Fever Commission, with headquarters at Yaba (near Lagos), Nigeria. An important objective of the Commission was to isolate and characterize the etiologic agent [7]. As luck would have it, a widespread epidemic of yellow fever occurred along the West African coast in 1926–1927. In July 1927, blood obtained from a nonfatal case named Asibi yielded a pathogenic agent when inoculated into Indian crown monkeys. The agent was subsequently passed to rhesus monkeys, shown to be a filterable virus [8], maintained in the laboratory by continuous passage in monkeys or mosquitoes, and ultimately stabilized by lyophilization [9]. The Asibi strain became the focus of attempts to develop a vaccine at the Rockefeller Institute laboratories in New York.

Contemporary efforts to isolate the etiological agent of yellow fever were also made by French workers at the Pasteur Institute in Dakar, Senegal. In 1927, a virus was recovered from a Syrian patient with mild yellow fever and was named the French strain [10]. The virus was graciously distributed to interested researchers in the United States and Europe, a number of whom initiated work towards the development of a vaccine.

1928–1930: Early Attempts at Vaccination

The earliest attempt to use the newly isolated yellow fever virus for immunization was made by Edward Hindle at the Wellcome Research Laboratories in London [11]. Hindle was aware of the recent reports of inactivated vaccine preparations against veterinary pathogens, including foot-and-mouth disease, fowl plague, and canine distemper viruses. Since yellow fever virus could, at the time, be propagated only by passage in monkeys, he used liver and spleen tissue from a monkey infected with the French strain as a vaccine substrate. The tissue emulsion was inactivated with formalin or phenol. Rhesus monkeys inoculated subcutaneously with putatively inactivated virus survived when challenged a week later with a large dose of infected liver suspension, whereas unvaccinated animals succumbed. Hindle's work was followed by other attempts to produce chemically inactivated vaccines. For example, Petit and Stefanopoulo reported protection of monkeys after two injections of inactivated vaccine prepared from monkey liver tissue [12]. In Brazil, Arago administered an inactivated vaccine to 25,000 residents of Rio de Janeiro during an epidemic. The results were inconclusive [12]. The preparation in inactivated vaccines was hampered by the lack of efficient methods for propagation and titration of the virus, and difficulty in controlling the inactivation process. Some vaccine preparations, including those of Hindle [11] and Davis [13], probably contained residual live virus, whereas others were degraded during the inactivation process or during storage [14]. In 1929, Wilbur Sawyer and his colleagues at the Rockefeller Institute in New York produced similar vaccine preparations from infectious monkey serum. Vaccines exposed to inactivating agents for short periods of time cause fatal disease in inoculated monkeys, and preparations exposed for long periods failed to immunize. These workers concluded that the chemical treatment of virulent virus was unlikely to produce a safe and dependable vaccine.

At the time, there was no convenient way of measuring the immune responses to vaccine preparations, other than by challenging monkeys with infectious liver suspensions. Most workers of the era also assumed that a single inoculation of inactivated vaccine was sufficient for immunization. The rudimentary virological methods, lack of an understanding of immunological responses, and the requirement for priming and boosting were obstacles to the development of a successful inactivated vaccine.

In 1928, Petit, Stefanopoulo, and Frasey conducted a series of studies employing immune sera prepared in monkeys and horses. They showed that monkeys could be actively immunized by inoculation of mixtures of virulent yellow fever virus and immune serum. In 1928, Theiler and Sellards also reported that monkeys given simultaneous inoculation of convalescent serum and virulent virus could raise active immunity to subsequent challenge [15]. Workers in the veterinary field had previously taken this approach for immunization of animals against rinderpest and hog cholera [16]. In the case of yellow fever, the potential dangers were appreciated, and the method was not used for human immunization. However, a modification of this approach was later used with partially attenuated vaccine strains (sero-immunization).

1931–1934: Partially Attenuated Live Vaccines and "Sero-Immunization"

Research on the French virus became a collaborative effort between Jean Laigret and his colleagues at the Pasteur Institute and Andrew Watson Sellards of the Harvard Medical School, Boston. In 1928, Sellards had visited Dakar and returned to Harvard with infected rhesus liver containing the French virus. Max Theiler, who had been recruited by Sellards in 1922 and held the academic rank of Instructor at

Harvard, began to work with this material [17]. Theiler was cognizant of the achievements of Pasteur and Roux, 50 years before, in the isolation of rabies virus by intracerebral inoculation, first of dogs and subsequently of monkeys, guinea pigs and rabbits. He also credited the work of Andervont, who in 1929 had shown that mice were susceptible to herpes simplex encephalitis after intracerebral inoculation [18].[1] Theiler inoculated a small number of adult Swiss mice by intracranial route with liver suspension from a monkey infected with the French strain. All the animals died of encephalitis, without signs of liver damage. Brain tissue from the mice passed to a rhesus monkey caused typical fatal yellow fever. In 1930, Theiler reported in *Science* the development of a convenient small animal model for the isolation and study of yellow fever virus [19].

Theiler knew that Pasteur had modified the virulence of rabies virus by serial passage in rabbit brains, and he believed that smallpox virus had been attenuated naturally, probably by passage through an unusual (bovine) host. He therefore undertook a series of sequential brain-to-brain passages of yellow fever virus in mice in an attempt to attenuate its pathogenicity. At the 29th and 42nd passage levels, the monkeys inoculated with the mouse brain material survived infection without developing yellow fever hepatitis. Moreover, they had developed immunity, as illustrated by resistance to challenge with virulent virus [20]. Theiler also noted that sequential passage of the virus in mice led to an increase in its neurotropism, suggesting that, with further passage, the virus might eventually become "fixed" in its neurovirulence for mice, in the same way that rabies virus had after sequential passage in rabbit brain. The implications for preparation of a vaccine were clear! The French virus was therefore passed in mice over 100 times, eventually reaching a "fixed" interval of 4 days between virus inoculation and death [21].

Theiler's discovery ushered in an intensive effort to develop a satisfactory means of immunization based on the virus attenuated by passage in mouse brain. This was spurred not only by the medical impact of the disease in affected populations, but also importantly by the devastating effects of laboratory-acquired yellow fever. In 1932, Wilbur Sawyer, Director of the International Health Division Laboratories at the Rockefeller Institute in New York, remarked that "In the four and a half years which have elapsed since rhesus monkeys first came into use as experimental animals in yellow fever studies, there have been reported 32 infections with yellow fever in laboratories devoted to research in this disease … and five scientists have lost their lives." [22]. The vaccine development efforts would then proceed as two largely independent efforts by Theiler who joined Sawyer at the Rockefeller Institute, and by Lairget and his colleagues in French West Africa, in collaboration with Andrew Sellards at Harvard.

Sawyer, Kitchen, and Lloyd in New York were the first to exploit Theiler's mouse brain virus for human immunization [23]. The French strain – fixed by intracerebral passage up to 176 times – was used to prepare a mouse brain vaccine.

---

[1] In fact the first studies showing mice to be susceptible to herpes virus were done earlier by G. Blanc and J. Camio-Petros (Recherches expérimentales sur l'herpès. *CR Soc Biol* 1921; 84:859).

Human immune serum was added to the mouse brain-virus suspension, since the virus was thought to be insufficiently attenuated for direct application. Indeed, laboratory-acquired cases of yellow fever due to “fixed” virus had occurred [24], and monkeys inoculated intracerebrally [25] or even parenterally with the fixed virus developed encephalitis.

The approach taken by Rockefeller investigators was thorough and meticulous, and represented the best methodology of the era. Eight lots of vaccine were prepared, differing principally in mouse passage levels. The vaccine was tested for bacterial sterility and for the presence of infectious vaccine virus following intracerebral inoculation of mice. A quantitative potency assay was not performed, however, and the dose of vaccine for inoculation was defined by the weight of mouse brain represented in the vaccine rather than by a virus titer. The human sera – recognized to be potentially hazardous – were added to the vaccine or administered at separate sites simultaneously with the vaccine, but were inactivated with tricresolether, tested for sterility, and evaluated for potency by monkey challenge test prior to use. The determinant of effective immunity was the neutralization test performed in mice, which had been developed by Theiler just before he left Harvard [26]. The preliminary studies in monkeys demonstrated that vaccine and serum produced no untoward reactions and conferred active immunity.

This satisfactory result led to the first human inoculations, conducted on an in-patient basis at the Hospital of the Rockefeller Institute for Medical Research. As these were no significant adverse reactions to the vaccine, 15 other volunteers underwent vaccination and were observed as out-patients. Reactions (attributed to the brain tissue component of the vaccine, including local tenderness and redness, fever and arthralgia) were considered benign. Attempts were made to isolate infectious virus from the patients’ blood, and no virus was recoverable. However, the presence of leucopenia and the development of immunity indicated that vaccination resulted in a true infectious process.

Sero-vaccination came into standard use for the immunization of laboratory workers. By 1934, 56 persons had been immunized in New York [27], and similar studies had been conducted at the Wellcome Bureau of Scientific Research in London by Findlay [28]. However, the requirement for human immune serum and the difficulty in establishing standard conditions for preparation of vaccine for passive–active immunization were major obstacles for neutralization of virus-serum mixtures was extremely problematic. At best, the procedure was applicable to the protection of laboratory workers at the highest risk.

Various attempts were made to replace human serum with heterologous antisera. In Paris, Petit and Stefanopoulo used equine serum [29], and at the Rockefeller laboratories, hyperimmune goat serum was investigated. The goat serum, which was tested in New York and in Brazil proved reactogenic, possibly because it was rapidly cleared and did not check replication of the neurotropic virus [30]. Theiler and Hugh Smith therefore prepared hyperimmune monkey serum, which was evaluated in Brazil with promising results [31]. However, there were concerns about allergic reactions, transfer of monkey diseases, and the high costs of preparing monkey serum for wide-scale use.

Clearly, a better approach was required. To address this need, French workers pursued the development of a neurotropic vaccine *without* the addition of serum. However, believing the neurotropic virus to be too dangerous as a stand-alone vaccine, the Rockefeller group initiated a search for a less pathogenic variant and for improved methods for propagation of the virus.

1932–1941: Development of the French Neurotropic Virus and Initial Field Trials

Andrew Sellards at Harvard collaborating with Jean Laigret at the Pasteur Institute in Tunis were the first to inoculate humans with the French strain in the absence of immune serum [32]. A 10% suspension of mouse brain infected with the French strain at the 134th passage level was prepared in water, and normal rabbit serum was added. Five nonimmune subjects were inoculated. The authors reported that these regimens were well-tolerated and resulted in the development of neutralizing antibodies. Laigret then conducted a second study of seven subjects with virus passed 143 times in mice [33]. It is interesting to note that the subjects were patients with syphilis, a number of whom had central nervous system (CNS) disease, reflecting the ethic of the day with respect to human experimentation. Several vaccines developed nausea, vomiting, abdominal pain, hyperactivity, abnormal reflexes, hemoptysis, and albuminuria. Laigret considered a number of possibilities for these symptoms, including superinfection with another agent, contamination of the vaccine with an adventitious virus, and the contribution of the underlying disease (syphilis). He concluded that the neuroadapted virus was responsible for the adverse events and that it was therefore insufficiently attenuated for humans. Laigret redirected his efforts to develop a nonreactogenic regimen. He believed that the severe reactions to the neuroadapted virus were dose-related, and could be overcome by using an appropriate formulation and dosing regimen. He developed a method for "attenuating" the mouse brains by "aging" the vaccine for varying lengths of time, a process probably based on Pasteur's original approach to rabies vaccine. Humans were given the most "attenuated" virus (brains incubated for 4 days), followed at intervals of 20 days by the virus treated for 2 days and 1 day, respectively. In 1934, Mathis, Laigret, and Durieux reported vaccination by this method of over 3,000 persons, mainly expatriates living in French West Africa [34]. Approximately one-third of the vaccines reported febrile reactions, and there were two cases with meningitis and myelitis. The next year, Nicolle and Laigret modified the formulation to a *single infection* of mouse brain-virus coated with egg yolk or olive oil (to retard diffusion of the virus from the inoculation site) and "attenuated" for 24 h at 20°C [35]. By 1939, more than 20,000 persons in West Africa had received either the three-injection series [36].

Thus, over a period of approximately 7 years, the neurotropic vaccine had been transitioned from initial testing in humans to large scale trials. Initial concerns about reactogenicity were dissipated as larger numbers were vaccinated. Although painstaking follow-up studies were not generally conducted, the evidence suggested the occurrence of some severe CNS reactions [37, 38]. Nevertheless, the risks associated with the vaccination were considered much lower than the risks of acquiring yellow fever.

Around 1939, efforts were made to simplify the vaccination by conversion from subcutaneous injection to scarification of the skin. Earlier work by Beeukes at Yaba had demonstrated that yellow fever virus applied to scarified skin could infect monkeys. Peltier and his colleagues at the Pasteur Institute, Dakar [39] used mouse brain vaccine was also administered by scarification, and Peltier showed that both vaccines could be given as a mixture. In 1939, approximately 100,000 persons in Senegal were vaccinated against yellow fever and smallpox without incident. The method was effective, since 95.6% of 1,630 vaccines developed neutralizing antibodies [40]. In 1940–1941, trials were extended to populations in the Ivory Coast and Sudan [41]. The results, which appeared to be favorable, paved the way for a much-expanded program of immunization and for refinements in the production of mouse brain vaccine at Dakar.

1931–1938: Development and Initial Field Trials of Yellow Fever 17D Vaccine

Research efforts at the Rockefeller Institute in New York were directed towards the development of an attenuated vaccine that had no neurovirulence properties. The French strain "fixed" by passage in mouse brain was considered too dangerous for use in humans, since it was capable of producing yellow fever encephalitis in monkeys and was associated with neurological accidents in humans. There was also a concern that the virus had residual viscerotropism, since Findlay and Clarke had shown reversion on repeated direct liver passage in monkeys [42].

Theiler and his colleagues decided to utilize recently described methods for cultivation of tissue in attempts to induce attenuation of wild-type yellow fever virus. Virus growth in minced tissue cultures prepared from mouse and chicken embryos had been developed at the Rockefeller Institute by Alexid Carrell and Thomas M. Rivers [43, 44]. In 1932, Theiler and Eugen Haagen demonstrated that the neurotropic French virus could also be propagated in chick-embryo tissue cultures [45], but attempts to grow unadapted virus strains failed. To avoid the danger of the neurotropic virus, it was important to derive a method adaptation in mouse brain. The operating principal was that the virus propagated under conditions that were generally unfavorable or restrictive, would lead to the selections of variants with altered phenotypic characteristics. This principle appeared to explain Findlay and Stern's observation that viserotropic yellow fever virus passaged in a transplantable mouse carcinoma had partially lost its virulence [46].

In 1936, Wray Lloyd (Fig. 1), Theiler, and technician Nelda Ricci reported the first successful in vitro cultivation of the Asibi strain [47]. After 240 passages, the virus became progressively less viscerotropic for rhesus monkey, although it retained its capacity to produce encephalitis after inoculation. The virus (designated

Fig. 1 Dr Wray Lloyd played a key role in the adaptation of yellow fever virus to tissue culture during studies in the early 1930s at the Rockefeller Foundation in New York. He investigated the use of the 17E virus (grown in mouse embryo tissue culture) with immune serum from human immunization. Lloyd took the 17E virus to Brazil in 1935, where he conducted further clinical studies. He died of an accidental fall in 1936. (Photo courtesy of the Rockefeller Archive Center, North Tarrytown, New York)

17E) was deemed too virulent for human inoculation without coadminstration of immune serum. The 17E virus replaced the mouse brain virus for sero-immunization of laboratory workers [48]. In November, 1935, speaking at the Annual Meeting of the American Society of Tropical Medicine in Baltimore, Sawyer noted that " ... a safer strain has supplanted the original neurotropic strain for use with immune serum in vaccination and it is confidently expected that a strain of virus safe for use without protective immune serum will finally be achieved." [49].

Between 1934 and 1936, multiple attempts were made to cultivate the Asibi virus in other substrates that might favor selection of attenuated variants, including minced tissue culture of mouse and guinea pig testicle, and of chick embryo [49]. After initial propagation in whole mouse embryo tissue culture, passage to these alternative substrates was achieved, in each case with an attendant decrease in viscertotropism of the virus. Since neurotropism was not markedly diminished it was decided to attempt sequential passages in chicken embryo tissues from which brain and spinal cord had been removed before mincing. The most important experiment, designated 17D, was initiated after 18 subcultures in whole mouse embryo cultures, as which point the virus as passed to whole minced chick-embryo cultures. After 59 subcultures in the latter tissue, the virus was then passed in minced chick embryo devoid of nervous tissue, each subpassage being checked for virus intracerebral inoculation of mice. Hugh H. Smith took responsibility for the oversight of these subculture experiments.

After 100 passages in chick embryo without nervous tissue (i.e., at the 176th passage since initiating *in vitro* culture), Smith noted a decrease in the neurovirulence of the virus, with mice surviving or developing nonlethal paralysis. Theiler and Smith confirmed that this virus induced minimal viremia and no hepatitis in rhesus monkeys after subcutaneous inoculation and, most importantly, that it had a markedly diminished neurovirulence for mice [50] (Table 2). The loss of neurovirulence for monkeys diminished between the 89th and 114th passage subcultures, and

Table 2 Biological characteristics of wild-type yellow fever virus and attenuated vaccine viruses derived empirically by serial passage (after Theiler and Smith [54])

Virus	Virulence for		
	Mice AST (days)	ic	Monkeys sc or ip
Wild-type	8–10	Fatal hepatitis	Fatal hepatitis
French neurotropic	4–10	Fatal encephalitis (100%)	Viremia, fatal encephalitis (30%)
17D	8–20	Encephalitis (<10%)	Minimal viermia, no illness

the neurovirulence in mice diminished between the 114th and 176th passage. Monkeys inoculated with the attenuated virus were protected against lethal peripheral challenge with the Asibi virus. At last, a strain was at hand that could be tested in humans without the addition of protective immune serum! In March 1937, Theiler and Smith submitted landmark papers to the *Journal of Experimental Medicine*, describing the development of and the first clinical trials with 17D viruses [51].

The vaccine was prepared from infected chick embryos ground with normal human serum to stabilize the virus, centrifuged, sterilized by filtration, and lyophilized. Bacterial sterility was checked and a potency test was performed by intracerebral inoculation of mice. The first two subjects to take the vaccine were Theiler and Smith themselves, both being immune (Theiler by virtue of an accidental infection while at Harvard in 1929, Smith by immunization with the French neurotropic virus plus immune goat serum) [31]. Two other immune subjects received minor febrile reactions were recorded, but all subjects had increases in serum protective antibody levels. Theiler and Smith had reached a critical milestone in vaccine research.

Theiler concluded that the secret to success had been the elimination of nervous tissue from minced chick embryos used to propagate the virus, since parallel passage series carried out over several hundred subcultures in minced whole chick embryos (experiment 17D WC) and in chick-embryo brain tissue (17D CEB) had not led to a decrease in viscerotropism or neurotropism. Theiler attempted to confirm that the absence of neural tissue had been responsible for attenuation. Starting with the virus at the 212th subculture, he established a new series of passages in which 17D WC and CEB viruses were passed in chick-embryo brains only. No modification occurred in the pathogenic properties of any of the viruses. Thus he was neither able to reproduce the level of the attenuation of 17D nor revert 17D to neurovirulence.[2] The reasons for the rapid change in 17D between the 89th and

[2]Later, Theiler passes the 17D virus sequentially by intracerebral inoculation of mice. Starting with the attenuated virus (176th subculture), 195 mouse brain passages were made, with period checks for monkey neurovirulence. After 106 passages, the virus causes encephalitis in monkeys. A parallel since 120 passages in chick embryo showed no phenotypic change. The virus thus appeared to be stable when maintained in chick embryo tissue, but with selective pressure could revert to neurovirulence.

114th subcultures in the original series remained unexplained [52]. The mutational events (or selection of preexisting variants) responsible for attenuation of 17D had occurred by chance during the course of systematic experiment that could not be readily duplicated. Virus attenuation by serial passage was an unpredictable procedure. Theiler and Smith's achievements were the result of a systematic and meticulous application of empirical process and keen, continuous observation by prepared minds, but they also had been extraordinarily lucky!

Field Trials in Brazil

The success of 17D virus in its initial tests in New York was followed by a rapid transition into practical use where yellow fever threatened human life. In early 1937, Hugh Smith left New York destined for Rio de Janeiro with a supply of 17D vaccine. A group of 24 mosquito control workers, shown first to be seronegative, were immunized in February–March 1937 [53], demonstrating that the vaccine was well-tolerated and immunogenic. Between January and March, Henrique Penna and Smith established yellow fever 17D vaccine production in the yellow fever laboratory in Rio de Janeiro. By June, over 100 persons had been immunized under controlled conditions in the laboratory, and the decision was made to broaden the trials under field conditions [32]. The site selected was at Varginha in Mineas Gerais state, and the target population was composed of coffee plantation worked at high risk of yellow fever. By August, Smith and Penna had inoculated over 2,800 people (Fig. 2), without observing any untoward reactions, and by October, this number had increased to 17,500. The trials established confidence in the safety of the 17D vaccine, and seroconversion rates exceeded 95%.

During the latter months of 1937 and early 1938, the methods for vaccine manufacture in eggs were refined, and production of the vaccine was scaled up in Rio de Janeiro.

In 1938, Fred L. Soper of the Rockefeller Foundation and the Brazilian government developed and implemented a plan for the first mass immunization campaign, and by the end of the year, nearly one million inhabitants of Brazil had received the 17D vaccine (Fig. 2).

A true efficacy study in Brazil was not undertaken at the time nor has one been conducted since. The complete absence of cases among vaccinated laboratory workers (who had suffered a high incidence of disease before the advent of a vaccine), as well as observations from Brazil and Colombia that yellow fever cases rarely occurred among vaccinated individuals, constituted "proof" that the vaccine was effective. The fact that the vaccine elicited neutralizing antibodies in nearly all recipients provided additional evidence. A series of studies subsequently demonstrated that the vaccine immunity was also remarkably durable.

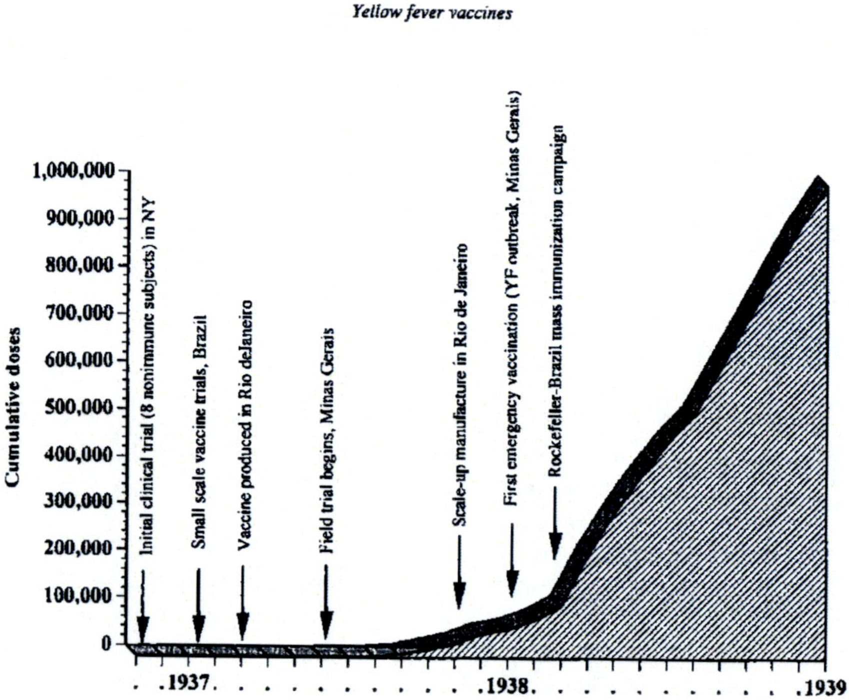

Fig. 2 Milestones in the clinical development of yellow fever 17D vaccine: transition from experimental studies in New York to the first mass immunization in Brazil

1941: Problems with Vaccine Manufacturing and Control: Neurological Accidents Caused by 17D Virus

By 1941, approximately two million persons had received 17D vaccine in Brazil, Colombia, Bolivia, the United States, England, and West Africa, without reports of serious vaccine-related adverse events. However, in July of that year, a Brazilian physician in Minas Gerais noted more than 20 cases of encephalitis occurring after yellow fever immunization. Investigations were undertaken by the Yellow Fever Research Service (jointly maintained by the Brazilian Ministry of Health at the Rockefeller Foundation), resulting in a published report by John Fox and his colleagues the following year [54]. The epidemiological investigation which ultimately encompassed 55,073 vaccines, revealed the occurrence of 273 (0.5%) unusually severe reactions, 199 (0.36%) with central nervous system disease signs, and 1 death with encephalitis. Similar disease in the unvaccinated population was significantly less frequent. The highest incidence of encephalitis occurred in individuals who had received vaccine prepared from the NY104 substrain of 17D (especially Lot E718).

It was apparent that the 17D vaccine, and particularly certain lots of vaccine, had the potential to cause postvaccinal encephalitis, and that this represented a change in the characteristic of the vaccine. Fox and Penna reviewed the manufacturing procedures

[55]. Vaccines were prepared from a number of independent virus substrains with no controls on passage level. The possibility of reversion to virulence was considered and guarded against by safety test in monkeys. Careful retrospective inspection revealed that certain substrains were more neurovirulent that others, but that these difference has been disregarded because the dose inoculated intracerebrally was high in relation to the dose given to humans by the subcutaneous route, and because no neurological accidents had been noted during the field trials conducted prior to 1941.

It was clear that uncontrolled passage of yellow fever vaccine could select for variants with increased virulence characteristics or with over attenuation and decreased immunogenicity [56, 57]. In 1941, workers at the Oswaldo Cruz Institute devised the "seed lot system" as a method for reducing the number of passages during 17D vaccine production, thereby stabilizing its biological behavior [58]. This was accomplished by preparing a primary seed lot (for "Master Seed" in today's parlance), which was used to prepare a secondary seed (working seed) for manufacturing. The seed lot requirements were set forth in a document published in 1945 [59], and were implemented by most manufacturers at the time. Implementation by vaccine manufacturers was not universal, however, until the mid-1950s, and further cases of neurological accidents occurred elsewhere in the world in conjunction with uncontrolled passage of the 17D virus [58].

The biological standards for vaccine production were refined by an expert group convened by the World Health Organization (WHO), resulting in the publication in 1957 of the WHO Requirements for Yellow Fever Vaccine [60]. These standards have been updated from time to time with increasingly stringent requirements for reduced neurovirulence, based on monkey safety tests. Since application of the seed lot system, the incidence of neurological accidents has been remarkably low. Only 21 cases of postvaccinal encephalitis and one death have been officially reported, despite the administration of over 250 million doses – a truly remarkable record of safety. Eighteen (86%) of the encephalitis cases were in children, of whom 16 were infants under the age of 7 months. The recognitions of postvaccinal encephalitis as a complication of immunization of young infants occurred during the 1950s, and in the 1960s regulations were changed, setting the lower age limit for immunization at 9 or 12 months of age. Since these limits have been established, there have been only six reported encephalitis cases (four of them in teenagers or adults).

1941–1953: Prevention of Yellow Fever in French West Africa Using the French Neurotropic Vaccine

The initial success of field trials with the French neurotropic vaccine between 1937 and 1940 was followed by a massive campaign to immunize the entire population of francophone colonial West Africa. Yellow fever vaccination was made compulsory in 1941, and over the next six years, nearly all inhabitants of the vast territory – approximately 14 million persons – received the vaccine [61]. By 1953, 56 million vaccinations had been performed, a number twice that of the population of the region [62]. This remarkable feat was accomplishable because of the simplicity and low cost of manufacturing the mouse brain vaccine, its relative stability under field conditions without refrigeration, and its simple method of administrations via scarification.

The vaccine was prepared in Dakar by the intracerebral inoculation of weaned mice. When the mice showed signs of paralysis (day 4–5), brains were aseptically removed, frozen, and desiccated, and a powder prepared by grinding a mortar and pestle with kaolin. After verification of bacterial sterility, the powder was distributed to ampoules contained 1/10 of a mouse brain (0.4 g) and equivalents to 100 human doses. The vaccine powder was found to be quite stable, allowing its use under field conditions without refrigeration. At the time of use, it was reconstituted in a mortar by addition of a gum arabic solution, the suspension being placed on the skin prior to scarification with bifurcated needle as for smallpox vaccine. For combined vaccinations, dried smallpox vaccine was mixed with yellow fever, prior to the addition of the gum solution. Of the 14 million doses administered up to 1945, 12 million were given with smallpox vaccine. The vaccine appeared to be well-tolerated, although Peltier acknowledged the occurrence of rare cases of CNS reactions [63], some of which were fatal [64]. The coadministration of smallpox vaccine was a confounding factor in the etiology of some encephalitis cases.

With the wide-scale use of the French vaccine, the incidence of yellow fever disease abated in francophone countries where the vaccine was used (Fig. 3), but

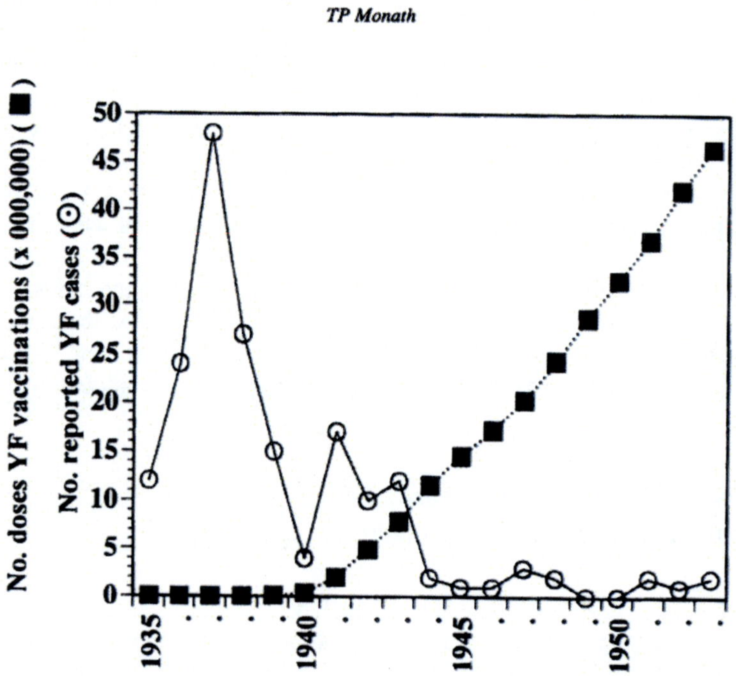

Fig. 3 Cumulative number of vaccination and number of reported yellow fever cases, French West Africa, 1935. Fifty three (data from Durieux [62]), showing decline in the disease with implementation of compulsory immunization and wide-scale coverage

not in other areas of Africa. The effectiveness of the immunization campaign was also affirmed by serological surveys, which showed the overall prevalence of yellow fever immunity in French West Africa to be approximately 20% in collections made between 1931 and 1940, but 86% in collections made in 1952–1953 [63]. Follow-up studies in selected villages indicated that the vaccine resulted in durable immunity. Peltier concluded in 1947 that "the results of long duration (of immunity) obtained in the bush, far from the conditions realized in the laboratory, are excellent and indicate the confidence one may place upon the wide use to the method, carried out with the periodicity of vaccinations every four years." The French vaccine remained in production until 1982, although its use was modified in the 1960s by the recognition of safety problems and the dismantling of colonial administrations and infrastructure for compulsory vaccination.

Neurologic Accidents Caused by the French Neurotropic Vaccine

Neurologic reactions noted by French workers during the wide-scale use of the mouse brain vaccine were attributed to the inherent neurotropism of the vaccine to low doses and a delayed immune response when the vaccine administered by poor scarification technique or after deterioration during field use. The evidence suggested that the accidents were the result of invasion of the brain by the vaccine virus, and not an allergic demyelinating process of the presence of an adventitious agent in the vaccine.

American and English workers were reluctant to utilize the Dakar vaccine, but the occurrence of severe epidemics in eastern Nigeria in 1951–1952 and in Panama and Central America in 1950–1952 presented a formidable problem, as it was not practical to deploy the thermolabile 17D virus under prevailing conditions in remote areas. To combat the outbreaks, emergency use of the French vaccine appeared justified [65, 66]. Because of the concern about the safety of the vaccine, a careful follow-up study was conducted in Nigeria [67]. The incidence of postvaccinal encephalitis was 3–4% in children, with a few cases noted in adults. A high case-fatality rate (40%) was observed, indicating that many cases of milder encephalitis were probably missed.

The unfortunate experience reinforced the conviction of the public health authorities in Anglophone countries of West Africa that the French vaccine was unsafe, and left to a major research effort to adapt the 17D vaccine to use in the tropics.

Increasing recognition of the problem of age-related CNS reactions in francophone Africa was followed in about 1959–1960 by a change in the routine immunization practices, such that the French vaccine was no longer administered to children 0–9 years of age [68]. The result was a rapid accumulation of a young, nonimmune population. In 1965, 5 years after suspending immunization of children, Senegal, which had not suffered a yellow fever outbreak since 1937, experienced one of the largest epidemics on record [67]. A mass vaccination campaign was therefore instituted with the French vaccine. Because of the high incidence of the disease of children, the age limit for use of the French vaccine was lowered from 10 to 2 years, and the 17D vaccine, available in a very limited supply, was used for immunization of children less that 2 years of age [69]. Among 498,887 persons vaccinated with the French vaccine, there were 231 documented cases of encephalitis and 23 deaths [70]. Over 90% of the cases occurred in children 2–11 years of age, and there was an inverse relationship between the age and the encephalitis incidence. In children, the incidence of encephalitis was 1–2%, and the case fatality rate was 22%.

This episode provided the mandate for a safer approach to immunization. It was clear that the risk of epidemic yellow fever was tied to the presence of an unprotected childhood population, but that the neurotropic vaccine was unacceptable for use in this age group. Thus, in 1966, with assistance from the WHO, the Pasteur Institute in Dakar established an expanded facility of the 17D vaccine in eggs. In 1970, a policy was established for the use of the 17D vaccine in all persons under the age of 15 years [71]. By 1970, the distribution of the 17D vaccine by the Pasteur Institute in Dakar exceeded that of the French neurotropic vaccine, and by 1982, production of the French vaccine was discontinued altogether (Fig. 4).

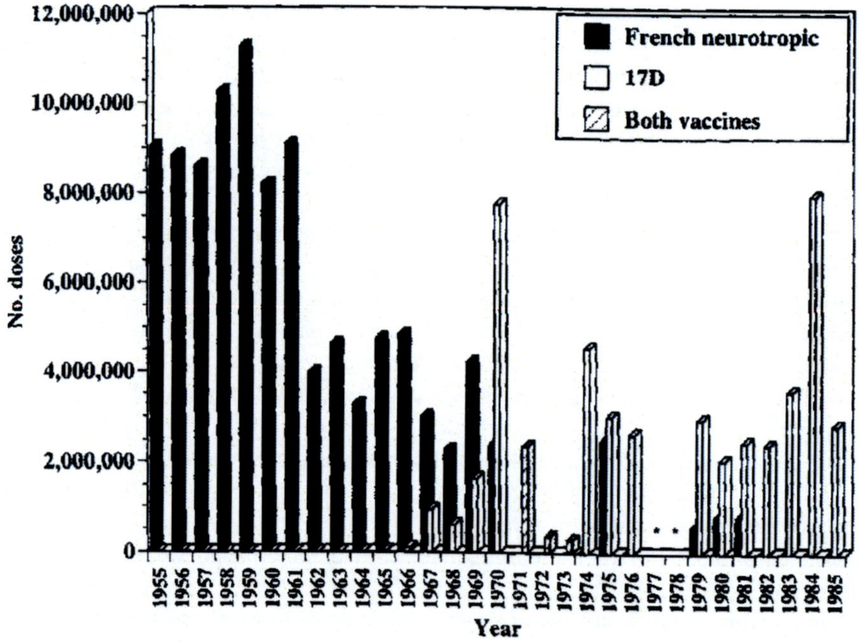

Fig. 4 Distribution of the French neurotropic and 17D vaccine from the Pasteur Institute, Dakar, the leading supplier of vaccine in Africa. Prior to the yellow fever epidemic in 1965, the French neurotropic mouse brain vaccine was widely used in francophone Africa. The occurrence of post-vaccinal encephalitis during mass immunization campaigns for emergency control of the outbreak led to a sea change towards 17D vaccine production in Dakar, facilitating the incorporation of the safer vaccine into childhood immunization programs. Manufacturing of the mouse brain vaccine ceased in 1982. ((*Asterisk*) Data unavailable (data from Annual Reports, Institut Pasteur, Dakar))

An important chapter in yellow fever vaccinology was closed. The French vaccine had saved many lives, but it had taken a number too, and the events of 1965–1966 provided the mandate for change. The most important target for immunization was children for whom the neurotropic vaccine was clearly unsafe. New methods of rapid immunization with 17D had been developed using the Ped-o-Jet, which reduced the relative value of scarification method, and improvements were made in the cold chain required for transport of the 17D virus. Taken together these changes led to the abandonment of mouse brain in favor of 17D.

Adventitious Agents: Hepatitis B and Avian Leucosis Viruses in 17D Chick-Embryo Vaccine

Early manufacturing procedures for yellow fever 17D vaccine included the addition of nonimmune human serum to the tissue culture fluid and to the final vaccine as a stabilizer. As early as 1937, Findlay and McCallum noted the occurrence of acute hepatitis occurring 2–7 months after 17D yellow fever vaccination [72]. Similar cases were reported in Brazil by Soper and Smith [73]. There were lessons from history: transmission of hepatitis by human blood had been described as early as 1885, caused by smallpox vaccine prepared from human lymph [74], and in the 1930s, there was an outbreak of jaundice among recipients of human measles and mumps-convalescent plasma [75].

The occurrence of jaundice associated with specific lots of 17D vaccine were investigated in Brazil in 1939 and 1940 [76]. At first it was suspected that such cases could be due to reversion of yellow fever virus, but this was excluded based on the occurrence of cases in vaccines known to be immune to yellow fever prior to immunization. Blame fell on an agent introduced with the human serum used in preparations of the vaccine, and in 1940, human serum was eliminated from 17D vaccine produced in Brazil [77].

However, with the advent of World War II, large numbers of military personnel were immunized with vaccine prepared in New York with pooled human serum. In 1942, a massive outbreak of jaundice occurred, with approximately 28,000 cases and 62 deaths from fulminant hepatitis [78]. Investigation into this outbreak was coordinated by the Commission on Tropical Diseases and the Armed Forces Epidemiological Board, under the direction of Wilbur Sawyer, and involved studies at the Rockefeller laboratories, several universities, and a number of Army bases. As a result of these investigations and of earlier studies in Brazil which clearly implicated the human serum component of the vaccine, 17D was henceforth manufactured without serum by the Rockefeller Foundation and the US Public Health Service [79] and no further cases were recorded. In a serological study conducted many years later, Seeff et al. confirmed that the hepatitis B virus was responsible for the jaundice epidemic in yellow fever vaccines during World War II [80].

In 1966, a thornier, if hypothetical problems arose when it was reported that a secondary seed of lot of chick-embryo yellow fever 17D vaccine (and measles vaccine) produced in the United Kingdom contained Rous sarcoma virus (RSV) (avian leukosis-sarcoma group) [81]. This was not the first concern about potentially oncogenic adventitious agents in viral vaccines, as SV40 virus had previously been discovered in polio vaccine. Avian leukosis viruses were known to cause a variety of tumors in chickens, and RSV had also been reported to cause tumors in experimentally inoculated monkeys. It was not surprising that yellow fever vaccines would be contaminated, as surveys showed a high prevalence of infection in commercial chicken flocks used to supply eggs [82]. 17D vaccines produced in the United States [83] and elsewhere were confirmed to be contaminated, with virus titers of leukosis virus in the range of 5–6 $\log_{10}$/mL. The contaminated vaccines were shown to cause malignant lymphomatosis in chickens.

The presence of an oncogenic virus in the 17D vaccine raised the concern of producers of the vaccines, the Food and Drug Administration (FAD), the National Institutes of Health (NIH), and the American Cancer Society. The susceptibility of humans to avian leukosis virus was unknown, and the mechanism of oncogenesis was not understood. The problem was addressed in several ways. On the one hand, studies were performed to determine whether persons who received the vaccine developed an immune response to the avian leukosis virus or had a higher incidence of malignancies. On the other hand, efforts were initiated to free the vaccine from the contaminating leukosis virus and to establish methods for producing leukosis-free vaccines.

Several serological surveys were conducted showing that persons who had received egg-based vaccines, including yellow fever vaccine as well as multiple doses of influenza vaccine did not develop detectable serological responses to avian leukosis virus [83, 84]. A retrospective survey of World War II veterans who had received yellow fever vaccine failed to show an increased risk of leukemia or other malignancies [85]. However, these approaches lacked sensitivity and specificity, and it was clear that 17D vaccines should be produced free from contaminant leukosis viruses. This presented certain problems, because it would have to be shown that passages in the presence of leukosis antiserum required to clear the vaccine substrain of leukosis virus would not be introduce biological changes in 17D.

The first successful leukosis-free vaccine was prepared in 1967 and tested in humans at the Wellcome Research Laboratories [86]. Similar results were subsequently reported in the United States [87]. All yellow seed stocks worldwide are now free of leukosis virus; however, neither the WHO nor some national regulatory authorities require that eggs used for yellow fever vaccine production be free of leukosis virus, and some yellow fever vaccines are still highly contaminated. Recent concern about leukosis viruses has arisen because use of more sensitive tests has revealed the presence of avian retrovirus gene sequences in egg-based vaccines.

Vaccine Thermostability: The Chink in the Armor

A major limitation for use of yellow fever 17D vaccine in tropical countries has been its thermolability, both in the lyophilized state and after reconstruction. The early vaccines were produced without stabilizers and were found to deteriorate very rapidly when exposed to temperatures greater than 20°C [88, 89]. Between 1940 and 1970, this problem was a major obstacle in the distribution and use of the 17D vaccine, especially in Africa, and one of the principal reasons why immunization was not widely implemented in Anglophone countries. The difficulty in using a thermolabile product in areas with limited cold chain capability stimulated research on vaccine stabilizers.

In 1970s, researchers in England [90] and France [91] systematically investigated stabilizing agents and devised successful formulations, now used by a number of manufacturers. One widely used stabilizer employs sugars (lactose, sorbitol), amino acids, and divalent cations. Addition of stabilizer reduces losses of virus titer both during lyophilization and storage of dried vaccine. However, it took a number of years to reformulate 17D vaccines to bring them to a point of acceptable stability. In 1986, a working group of the WHO conducted a study of the thermostability of yellow fever 17D vaccines produced by 12 approved manufacturers [92] and found that only a few vaccines were stable. In the following year, the WHO published a guideline specifying that each lot of vaccine should be tested and meet a stability standard [93]. To meet this standard, a lyophilized vaccine must fulfill two criteria after being held for 2 weeks at 37°C: (1) maintenance of potency (>1,000 mouse 4ic LD_{50}/human dose) and (2) mean loss of titer <1.0 $\log_{10}LD_{50}$. A recent survey indicates that most vaccines meet these criteria. The improved vaccine is well-suited for use in routine and emergency cold chain operations in the tropics.

Combined Immunization

Strategies for combining vaccines have been in place for many years, inspired by the need to simplify the logistics of immunization, reducing to a minimum the number of contacts with the at-risk population. More recently, the practical problems of incorporating a growing number of vaccines into childhood immunizations schemes has stimulated intense competition among vaccine producers to develop combined and multivalent vaccines.

In the case of yellow fever vaccines, it has already been mentioned that early work centered on the combination of yellow fever and smallpox vaccines delivered by scarification, and that this strategy was widely used in the case of the French neurotropic vaccine. Similar studies of the 17D virus, however, indicated that combined intradermal inoculation gave poor results, possible due to interference events.

Important factors in the quest for combined vaccination strategies for developing countries occurred in the 1960s, with the advent of jet inoculation and the appearance of new vaccines, especially the live measles vaccine. In 1964, Meyer and colleagues conducted a study in West Africa, in which infants received combined measles, smallpox, and yellow fever vaccines by jet inoculation [87]. The results indicated that slight interference with seoconversion to 17D virus may have occurred. A few years later, during the smallpox eradication era in West Africa, a second study of simultaneous smallpox–measles–yellow fever immunization by jet injector was conducted, in which the vaccines were given at separate sites [94], providing further evidence for interference of coadminstration of yellow fever vaccine with other live vaccines at the same site. Current strategies in the Expanded Program on Immunization (EPI) call for the administration of measles and yellow fever 17D vaccine at 9 months of age at separate inoculation sites.

A few other studies have been conducted on the simultaneous administration of other bacterial and viral vaccines with yellow fever 17D. In the early 1970s, several workers reported the unexplained antagonism of whole cell cholera and yellow fever vaccines [95, 96]. In 1986 Yvonnet et al. studied infants given yellow fever simultaneously, but at different sites, together with DPT, hepatitis B, and measles or with DTP and measles [97]. Both groups had a satisfactory and similar serconversion rate to yellow fever (>91%), but the mean neutralizing antibody titer to tallow fever was significantly lower in infants that had received hepatitis B vaccine. In a recent study, yellow fever and a purified typhoid vaccine were injected simultaneously at the same or in separate site(s) [98]. In this case, the bacterial antigen (the VI capsular polysaccharide of *Salmonella typhi*) appeared to have an adjuvant effect, so that higher neutralizing titers were found in subjects receiving both vaccines than in those inoculated with 17D alone. Taken together, these results indicate that the outcome of combined administration of viral and bacterial vaccines may be accompanied by both interference or enhancement phenomena, which are unpredictable. Future efforts to develop advantageous vaccine combination will require specific studies to elucidate such events on a case-by-case basis.

The Molecular Basis of Attenuation

Yellow fever vaccine had been the subject of a considerable body of research aimed at unraveling the molecular basis of attenuation. As the biological characteristics of yellow fever vaccine viruses and their virulent parent strains have been extensively studied in animals and vector mosquitoes, it was of obvious interest to compare these strains at the genome level. The identification of virulence genes of the prototype flavivirus might provide useful information for the rational design of other vaccines or antiviral drugs.

A major obstacle to the genetic evaluation, however, was the long series of empirical passages made in the derivation of the French neurotropic and 17D vaccines, and the fact that neither viruses were derived by biological cloning. The high rates of mutation of RNA viruses assured that multiple genetic changes would occur during the course of the development of these vaccines, many of which might not be responsible for alterations in virulence. Employing a relatively insensitive method (T_1 oligonucleotide fingerprinting), Monath et al. found minor differences between the genomes or several substrains of 17D virus used for vaccine production [99] indicating the presence of genetic changes that were not linked to attenuation. In one case, a difference in the T_1 oligonucleotide map was noted between seed and vaccine viruses, indicating that mutation (or variant selection) could occur within one or two passages in chick embryos.

The complete nucleotide sequence of the 17D virus genome was first reported in 1985 by Rice and his colleagues [100]. The genome was found to contain 10,862 nucleotides with a single, long open reading frame (10,233 nucleotides) encoding all of the structural and nonstructural proteins of the virus, and short 5′ and 3′ non-coding regions. In 1987, the nucleotide sequence of the parental Asibi strain was described [101]. Comparison of the Asibi and 17D viruses revealed 68 nucleotide changes resulting in 32 amino acid differences scattered across the genome. A high rate of amino acid substitutions was found in the envelope glycoprotein E gene. One or more of the five tropism for hepatic and neural cells is specified by this protein.

A somewhat clearer picture of the genetic basis for attenuation emerged with the sequencing of further substrains of 17D virus [102–104] (reviewed by Monath and Heinz [105] and Barrett [106]). This analysis narrowed the list of changes that might be responsible for attenuation from 32 to 20 (Fig. 5), including four nonconservative changes in the E gene (shaded boxes, Fig. 5) that might significantly alter protein function.

An additional and interesting piece of evidence resulted from an analysis of the 17D virus recovered from the brain of the sole fatal case of postvaccinal encephalitis by Jennings et al. [107]. Compared to the 17D vaccine from the same manufacturer, the brain isolate had increased neurovirulence for mice and monkeys, and differed at three amino acids (two in the E gene and one in NS4b). Of these changes, the substitution at E-303 may be considered the most likely to have been implicated in increased neurovirulence, since it is close to the E-305 mutation in the 17D virus, whereas the other mutation (at E-155) is present in other vaccine strains and thus could not be responsible for neurovirulence.

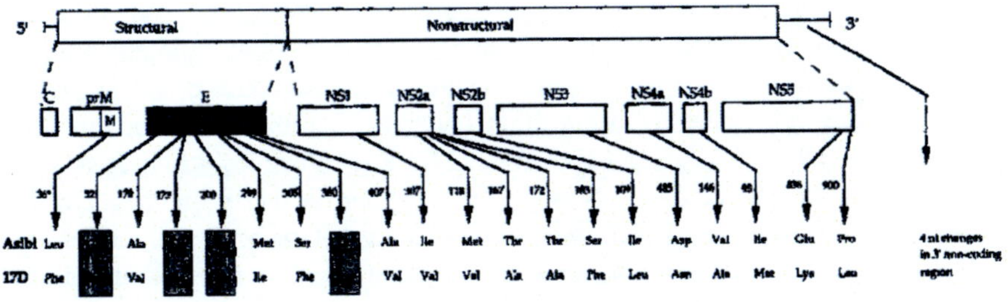

Fig. 5 Genomic organization of the yellow fever virus showing untranslated and translated portions of the genome, and the encoded structural (C-prM-E) and nonstructural (NS). The location of 20 mutations in the structural genes and four in the 3′ untranslated region are shown. These sites have been identified by comparison between the parental Abisi sequences and the sequence of various 17D substrain viruses (reviewed in Barrett [106]). Four nonconservative changes in the E glycoprotein (*shaded boxes*) are suspected to play a role in viscera- or neurotropism, but other mutations may also be important, especially those at E170 and E305 (see text). Mutations in the NS genes may influence virus replication in hepatic or neural cells. The complex, multigenic factors responsible for virulence/attenuation may be elucidated in the future by site-specific mutagenesis of a full-length cDNA clone. The amino acid number within each protein is indicated)

A quantum leap in our understanding of flaviviruses came in 1995, when Rey et al. published the three-dimensional crystallographic structure of the E polypeptide dimer [108]. The mutations suspected to have caused attenuation of the Asibi virus in the evolution of the 17D vaccine are located at sites exposed on the surface of the E glycopeptides. Two of the mutated amino acids (at positions E-52 and E-200) are present in the fusion sequence in domain II. Two other changes (at amino acids E-305 and E-380) are present in domain III (C-terminus in the crystallized membrane anchor-free fragment). The importance of domain III in neurovirulence has been demonstrated by the analysis of single amino acid mutants of other flaviviruses, and amino acid substitutions in the yellow fever 17D vaccine strain also cluster within this domain. Another mutation in the E protein suspected to play a role in virulence occurs at position E-173 in domain I. The mutation corresponds to a neurovirulence determinant defined by a monoclonal antibody escape of tick-borne encephalitis virus.

Although these observations provide a sharper image of the genetic basis of yellow fever virulence, particularly with respect to the E glycoprotein, it is still impossible to identify at a functional level the precise changes responsible for the attenuation of the 17D virus. A potential breakthrough came in 1989 when Rice et al. succeeded in generating a full-length cDNA clone of 17D virus that yielded infectious RNA transcripts [109] (It thus became theoretically possible to introduce site-specific wild-type (Asibi) mutations and to determine whether the progeny viruses are reconstituted to virulence.). Data are awaited with interest.

The application of infectious clone technology provides a potential means of reducing the neurovirulence of the 17D vaccine, both by decreasing the potential for selection of heterogeneous virus populations in the vaccinated host and by the individual and alterations of specific neurovirulence determinants. A vaccine with lower neurovirulence might be useful in the protection of infants during yellow fever epidemics (the current vaccine cannot be given to infants less than 6 months of age, and there is probably an increased risk of encephalitis is infants 6–12 months of age). In addition, as the optimal time for measles immunization in developing countries may be at 6 months of age, there would be a practical advantage to have a yellow fever vaccine that could be coadministered in the EPI. Recently, the cDNA clone of the 17D virus (17D-204 substrain) was used to generate a seed virus and vaccine lots in Brazil [110] and further refinements of the cDNA clone are underway. It will be instructive to learn whether a vaccine derived in this way and composed of a homogeneous, clonal population of virions will be safe and immunogenic.

Acknowledgements Dr A. Barrett kindly shared unpublished data on yellow fever vaccine and parental strain molecular comparisons and insights as to their relevance. Dr V. Deubel assisted in finding, photographs of Drs Mathis and Laigret in the Pasteur Institute archives. Dr F. Rey, Harvard Medical School, kindly provided the figure showing the location of yellow fever 17D virus mutations in the crystal structure of the flavivirus E protein dimer. The author is indebted to the Rockefeller Archive Center, North Tarrytown, NY and the Francis A. Countway Library of Medicine, Boston, MA for photographs of yellow fever researchers.

References

1. Finlay C. El mosquito hipoteticamente considerado como agente de transmission de la fiebre amarilla. *Anal Real Acad Ciencias Med Fisicas Naturales* 1881;18:147–69
2. Finlay C. Yellow fever: its transmission by means of the Culex mosquito. *Am J Med Sci* 1886;92:395–409
3. Finlay C. Inoculations for yellow fever by means of contaminated mosquitoes. *Am J Med Sci* 1891;102:264–8
4. Yellow fever. A compilation of various publications. Results of the work of Maj Walter Reed, Medical Corps, United States Army, and the Yellow Fever Commission. 61st Congress Doc No 822, Washington DC; Govt Printing Office, 1911
5. Guiteras J. Experimental yellow fever at the inoculation station of the sanitary department of Havana, with a view to producing immunization. *Amer Med* 1901;28:809–17
6. Löwy I. Yellow fever in Rio de Janeiro and the Pasteur Institute mission (1901–1905): the transfer of science to the periphery. *Med Hist* 1990;34:144–63
7. Warren AJ. Landmarks in the conquest of yellow fever. In: Strode G, ed. *Yellow Fever.* New York: McGraw Hill, 1951;5–37
8. Stokes A, Bauer JH, Hudson NP. Transmission of yellow fever virus to *Macacus rhesus,* preliminary note. *JAMA* 1928;90:604–6
9. Sawyer WA, Lloyd WDM, Kitchen SF. Preservation of yellow fever virus. *J Exp Med* 1929;50:1–13
10. Mathis C, Sellards AW, Laigret J. Sensibilité du *Macacus rhesus* au virus de la fièvre juane. *C R Acad Sci* 1928;186:604–6
11. Hindle E. A yellow fever vaccine. *Br Med J* 1928;1:976–7
12. Petit MA. Rapport sur la valeur immunisante des vaccins employés contre le fièvre juane et la valeur thérapeutique du sérum antiamaril. *Bull Acad Natl Méd* 1931;105:522–6
13. Davis NC. Attempts to determine the amount of yellow fever virus injected by the bite of a single infected *Stegomyia* mosquito. *Am J Trop Med* 1934;14:343–54
14. Okell CC. Experiments with yellow fever vaccine in monkeys. *Trans Roy Soc Trop Med Hyg* 1930;24:251–4
15. Theiler M, Sellards AW. The immunological relationship of yellow fever as it occurs in West Africa and in South America. *Ann Trop Med Parasitol* 1928;22:449–60
16. Duval CW. Observations upon the nature of the virus of hog cholera. *Proc Soc Exp Biol Med* 1929;27:87–9
17. Bendiner E. Max Theiler: Yellow Jack and the jackpot. *Hosp Pract* 1988;23:211–44
18. Andervont HB. Activity of herpetic virus in mice. *J Infect Dis* 1929;44:383–93
19. Theiler M. Susceptibility of white mice to yellow fever virus. *Science* 1930;71:367
20. Theiler M. Studies on action of yellow fever in mice. *Ann Top Med Parasitol* 1930; 24:249–72
21. Lloyd W, Penna HA, Mahaffy AF. Yellow fever in rodents. *Am J Hyg* 1933;18:323–44
22. Sawyer WA, Kitchen SF, Lloyd W. Vaccination against yellow fever with immune serum and virus fixed for mice. *J Exp Med* 1932;55:945–69
23. Sawyer WA, Kitchen SF, Lloyd W. Vaccination of humans against yellow fever with immune serum and virus for mice. *Proc Soc Biol Med* 1931;29:62–4
24. Berry GP, Kitchen SF. Yellow fever accidentally contracted in the laboratory: study of seven cases. *Am J Trop Med* 1931;11:365–434
25. Sellards AW. Behavior of virus of yellow fever in monkeys and mice. *Proc Natl Acad Sci USA* 1931;17:339–43
26. Theiler M. Neutralization tests with immune yellow fever sera and a strain of yellow fever virus adapted to mice. *Ann Trop Med Parasitol* 1931;25:69–77
27. Sawyer WA. Enquête sur l'immunité vis-à-vis la fièvre juane au moyen de l'épreuve de protection de la souris. *Bull Off Intl Hyg Pub* 1934;26:1057–9

28. Findlay GM. Immunization against yellow fever with attenuated neurotropic virus. *Lancet* 1934;2:983–5
29. Petit A, Stefanopoulo GJ. Utilisation du sérum antiamaril d'origine animale pour la vaccination de l'homme. *Bull Acad Natl MédI* 1933;110:67–76
30. Smith HH. *Life's a Pleasant Institution: the Peregrination of a Rockefeller Doctor*, Tucson AZ: Hugh Smith, 1978
31. Theiler M, Smith HH. Use of Hyperimmune monkey serum in human vaccination against yellow fever. *Bull Off Intl Hyg Pub* 1936;28:2354–7
32. Sellards AW, Laigret J. Vaccination de l'homme contre la fièvre juane. *C R Acad Sci* 1932;194:1609–11
33. Laigret J. Recherches expérimentales sur la fièvre jaune. *Arch Inst Pasteur Tunis* 1933;21:412–30
34. Mathic C, Laigret J, Durieux C. Troi mille vaccinations contra la fièvre juane en Afrique Occidentale Française au moyen du virus vivant de souris, atténué par vieillissement. *C R Acad Sci (Paris)* 1934;199:742–4
35. Nicolle C, Laigret J. La vaccination contre la fièvre jaune par le virus amaril desséché et enrobé. *Compt Rend Acad Sci (Paris)* 1935;201:312–4
36. Durieux C. Preparation of yellow fever vaccine at the Institute Pasteur, Dakar. In: Smithburn K et al., eds. *Yellow Fever Vaccination.* Geneva: World Health Organization, 1956;31–43
37. Sorel F. La vaccination anti-amarile en Afrique occidentale française, mise en application du procédé de vaccination Sellards-Laigret. *Bull Off Intl Hyg Pub* 1936;28:1325–56
38. Laigret J. De l'interprétation des troubles consécutifs aux vaccinations par les virus vivants en particulier à la vaccination de la fièvre jaune. *Bull Soc Path Exot* 1936;29:230–4
39. Peltier M, Durieux C, Jonchère H, Arquié E. Pénétration du virus amaril neurotrope par voie cuntanée: vaccination contra la fièvre jaune et la variole, note préliminaire. *Bull Acad Méd (Paris)* 1939;121:657–60
40. Peltier M. Yellow fever vaccination, simple or associated with vaccination against smallpox, of the population of French West Africa by the method of the Pasteur Institute of Dakar. *Am J Pub Health* 1947;37:1026–32
41. Peltier M. Vaccination antiamarile simple et associée à la vaccination antivariolique par scarification. *Méd Trop* 1941;1–49
42. Findlay GM, Clarke LP. Reconversion of neurotropic into viscerotropic strain of yellow fever virus in rhesus monkeys. *Trans Roy Soc Trop Med Hyg* 1935;28:579–600
43. Carrell A, Rivers TM. La fabrication du vaccin in vitro. *C R Soc Biol* 1927;96:848–50
44. Rivers TM. Cultivation of vaccine virus for Jennerian prophylaxis in man. *J Exp Med* 1931;54:453–61
45. Haagen E, Theiler M. Untersuchungen über das Verhalten des Gelbfiebervirus in der Gewebekultur. *Zentralbl Bakt* 1932;125:145–58
46. Findlay GM, Stern RO. Essential neurotropism of yellow fever virus. *J Pathol Bact* 1935;41:431–8
47. Lloyd W, Theiler M, Ricci N. Modification of the virulence of yellow fever virus by cultivation of tissues in vitro. *Trans Roy Soc Trop Med Hyg* 1937;24:481–529
48. Lloyd W. Use of cultivated virus together with immune serum in vaccination against yellow fever. *Bull Off Intl d'Hyg Pub* 1935;27:2365–8
49. Sawyer WA. A history of the activities of the Rockefeller Foundation in the investigation and control of yellow fever. *Am J Top Med* 1937;17:35–50
50. Theiler M, Smith HH. The effect of prolonged cultivation in vitro upon the pathogenicity of yellow fever virus. *J Exp Med* 1934;65;767–86
51. Theiler M, Smith HH. The use of yellow fever virus modified by in vitro cultivation for human immunization. *J Exp Med* 1937;65:787–800
52. Theiler M. The virus. In: Strode G, ed. *Yellow Fever.* New York: McGraw Hill, 1951; 43–136
53. Smith HH, Penna HA, Paoliello A. Yellow fever vaccination with cultured virus (17D) without immune serum. *Am J Trop Med* 1938;18:437–68

54. Fox JP, Lennette EH, Manso C, Aguiar JRS. Encephalitis in man following vaccination with 17D yellow fever virus. *Am J Hyg* 1942;36:117–42
55. Fox JP, Penna HA. Behavior of 17D yellow fever virus in rhesus monkeys. Relation to sub-strain, dose and neural or extraneural inoculation. *Am J Hyg* 1943;38:152–72
56. Fox JP, Kossobudzki SL, Fonseca da Cunha J. Field studies on immune response to 17D yellow fever virus: relation to virus substrain, dose, and route of inoculation. *Am J Hyg* 1943;38:113–38
57. Soper FL, Smith HH. Vaccination with virus 17D in the control of jungle yellow fever in Brazil. *Trans 3rd Intl Congr Trop Med & Malaria* 1938;1:295–313
58. Panthier R. A propos de quelques cas de reactions nerveuses tardives observées chez de nourissons après vaccination antiamarile (17D). *Bull Soc Pathol Exot* 1956;49:477–94
59. Standards for the manufacture and control of yellow fever vaccine. *Epidem Inform Bull* 1945;1:365–8 (original standards for the vaccine developed by the United Nations Relief and Rehabilitation Administration (UNRRA), according to Article XI of the International Sanitary Convention for Aerial Navigation)
60. Requirements for Biological Substances No 3. *WHO Tech Rep* Ser No 1957;136:Annex 1
61. Peltier M. Yellow fever vaccination, simple or associated with vaccination against smallpox, of the population of French West Africa by the method of the Pasteur Institute of Dakar. *Am J Pub Health* 1947;37:1026–32
62. Durieux C. Mass yellow fever vaccination in French West Africa south of the Sahara. In: Smithburn K et al., eds. *Yellow Fever Vaccination*, Geneva: WHO, 1956:115–21
63. Peltier M. Vaccin antiamarile et vaccinvations antiamariles et antivariolo-amarile par la méthode dakaroise en Afrique occidentale française. *Proc 4th Intl Cong Trop Med & Malaria* 1948;1:489–97
64. Jadin J, Arnaldi E. Considérations au sujet de l'épidémie intérigène de Zongo et vaccination anti-amarile. *Ann Soc belge Méd Trop* 1939;19:377–91
65. MacNamara FN. Reactions following neurotropic yellow fever vaccine given by scarification in Nigeria. *Trans Roy Soc Trop Med Hyg* 1953;47:199–208
66. Elton NW. Public health aspects of the campaign against yellow fever in Central America. *Am J Public Health* 1952;42:170–4
67. Chambon L, Wone I. Une épidémie de fièvre jaune au Sénégal en 1965. *Bull WHO* 1957; 11:325–50
68. Brés P, Lacan A, Diop I, Michel R, Peretti P, Vidal C. Des campagnes de vaccination antiamarile en République du Sénégal. *Bull Soc Pathol Exot* 1963;64:1038–43
69. Sankalé M, Bourgeade A, Wade F. Bèye B. Contribution à l'étude de reactions vaccinales observée en dehors de Dakar. *Bull Soc Méd Afr Noire Lgue Fr* 1966;11:617–24
70. Rey M, Satge P, Collomb H et al. Aspects épidémiologiques et cliniques des encephalites consécutives à la vaccination antiamarile. *Bull Soc Méd Afr Noire Lgue Fr* 1966;11:560–74
71. Ricossé JH, Albert JP. La vaccination antiamarile dans les états de l'OCCGE. Conférence sur l'épidémiologie et le contrôle de la fièvre jaune en Afrique de l'Ouest, Bobo Dioulasso, March 20–23, 1971 (unpublished document No 266, OCCGE, Centre Muraz, Bobo Dioulasso, Haute Volta)
72. Findlay GM, MacCallum FO. Note on acute hepatitis and yellow fever immunization. *Trans Roy Soc Trop Med Hyg* 1937;31:297–309
73. Soper FL, Smith HH. Yellow fever vaccination with cultivated virus and immune and hyper-immune serum. *Am J Trop Med* 1938;18:111–34
74. Lurman A. Eine Ictrusepidemie. *Berl Klin Wochenschr* 1885;22:20–3
75. Anon. Acute infectious jaundice and the administration of measles serum. *In*: McNalby A, ed. On the state of public health. Annual Report of the Chief medical officer, Ministry of Health, for the year 1937. London: HMSO 1937;4:235
76. Fox JP, Manso C, Penna HA, Para M. Observations on the occurrence of icterus in Brazil following vaccination against yellow fever. *Am J Hyg* 1942;36:68–116
77. Smithburn KC. Immunology of yellow fever. In: Smithburn et al., eds. *Yellow Fever Vaccination*. Geneva: WHO, 1956;11–30

78. Sawyer WA, Meyer KF, Eaton MD, Bauer JH, Putnam P, Schuentker FF. Jaundice in Armey personnel in the western region United States and its relation to vaccination against yellow fever. *Am J Hyg* 1944;39:337–430
79. Hargett MV, Burrus HW, Donovan A. Aqueous-base yellow fever vaccine. *Public Health Rep* 1943;58:505–12
80. Seeff LB, Beebe GW, Hootnagle JH. A serologic follow-up of the 1942 epidemic of post-vaccination hepatitis in the US Army. *N Engl J Med* 1987;316:965–70
81. Harris RJC, Dougherty RM, Biggs PM et al. Containment viruses in two live virus vaccines. *J Hyg (Cambr)* 1966;64:1–7
82. Rubin H, Fanshier L, Cornelius A, Hilaghes WF. Tolerance and immunity in chicken after congenital and contact infection with a avian leukosis virus. *Virology* 1962;17:143–56
83. Piraino F, Krumbiegel ER, Wisniewski HJ. Serologic survey of man for avian leukosis virus infection. *J Immunol* 1967;98:702–6
84. Richman AV, Alusio CG, Jahnes WG, Tauraso NM. Avian Leukosis antibody response in individuals given chicken embryo derived vaccines, *Proc Soc Exp Biol Med* 1972;139:235–7
85. Melnick JL. Latent virus infections in donor tissues and in recipients of vaccines. *Natl Cancer Inst Mongr* 1968;29:337–49
86. Draper CC. A yellow fever vaccine free from avian leukosis viruses. *J Hyg (Cambr)* 1967;65:505–13
87. Tauraso NM, Spector SL, Jahnes WG, Shelokov A. Yellow fever vaccine: I. Development of a vaccine seed free from contaminating avian leukosis viruses. *Proc Soc Exp Biol Med* 1968;127:1116–20
88. Burrus HW, Hargett MV. Yellow fever vaccine inactivation studies. *Public Health Rep* 1947;62;940–56
89. Robin Y, Saenz AC, Outschoorn AS et al. Etude de la thermostabilité du vaccin anti-amaril sur des échantillons de huit lots provenant de divers pays. *Bull WHO* 1971;44:729–37
90. Burfoot C, Yound PA, Finter NB. Thermal stability of stabilized 17D yellow fever virus vaccine. *J Biol Stand* 1977;5:173–9
91. Barme M, Bronnert C. Thermostabilisation du vaccin antiamaril 17D lyophilize. I. Essai des substances protectrices. *J Biol Stand* 1984;12:435–42
92. World Health Organization. Yellow fever vaccines: thermostability of freeze-dried vaccines. *WHO Wkly Epidemiol Rep* 1988;62:181–83
93. World Health Organization. Requirements for yellow fever vaccine, Addendum 1987. *WHO Tech Rep Ser No. 71,* Annex 9, 1988;208–9
94. Ruben FL, Smith EA, Foster SO et al. Simultaneous administration of smallpox, measles, yellow fever, and diphtheria-pertussis-tetanus antigens to Nigerian Children. *Bull WHO* 1973;48:175–81
95. Gateff C, Le Gonidec G, Boche R et al. Influence de la vaccination anti-cholérique sur la vaccination anti-amarile associée. *Bull Soc Path Exot* 1973;66:266–75
96. Felsenfeld O, Wolf RH, Gyr K et al. Simultaneous vaccination against cholera and yellow fever. *Lancet* 1973;i:457–8
97. Yvonnet B, Coursaget P, Deubel V, Diop Mar I, Digoutte JP, Chiron JP. Simultaneous administration of hepatitis B and yellow fever vaccines. *J Med Virol* 1986;19:307–11
98. Ambrosch F, Fritzell B, Gregor J et al. Combined vaccination against yellow fever and typhoid fever, a comparative trial. *Vaccine* 1994;12:625–8
99. Monath TP, Kinney RM, Schlesinger JJ, Brandriss MW, Bres P. Ontogeny of yellow fever 17D vaccine: RNA oligonucleotide fingerprint and monoclonal antibody analyses of vaccines produced world-wide. *J Gen Virol* 1983;64:627–37
100. Rice CM, Lenches EM, Eddy SR. Nucleotide sequence of yellow fever: implications for flavoring gene expression and evolution. *Science* 1985;229:726–33
101. Hahn CS, Dalrymple JM, Strauss JH, Rice CM. Comparison of the virulent Asibi strain of yellow fever virus with the 17D strain derived from it. *Proc Natl Acad Sci USA* 1987;84:2019–23
102. Duarte dos Santos CN, Post CR, Carvalho R, Ferreira H, Rice CM, Galler R. Complete nucleotide sequence of yellow fever virus vaccine strains 17DD and 17D-213. *Virus Res* 1995;35:35–41

103. Jennings AD, Whitby JE, Minor PD, Barrett ADT. Comparison of the nucleotide and deduced amino acid sequences of the structural protein genes of the yellow fever 17D vaccine strain from Senegal with those of other yellow fever vaccine viruses. *Vaccine* 1993;11:679–81
104. Dupuy A, Despres P, Cahour A et al. Nucleotide sequence comparison of the genome of two 17D-204 yellow fever vaccines. *Nucl Acid Res* 1989;17:39–89
105. Burke DS, Monath TP, Flaviviruses. In Knipe OM, Howley PM (Eds) Fourth Eds Lippincott, Williams and Williams, Philadelphia, vol 1, 2001;1043–1126
106. Barrett ADT. Yellow fever vaccines. *Biologicals* 1997;25:17–25
107. Jennings AD, Gibson CA, Miller BR et al. Analysis of a yellow fever virus isolated from a fatal case of vaccine-associated human encephalitis. *J Infect Dis* 1994;169:512–8
108. Rey FA, Heinz FX, Mandl C, Kunz C, Harrison SC. The envelope glycoprotein from the tick-borne encephalitis virus at 2 Å resolution. *Nature* 1995;375:291–8
109. Rice CM, Grakoui A, Galler R, Chambers TJ. Transcription of infectious yellow fever RNA from full-length templates produced by in vitro ligation. *New Biol* 1989;1:285–96
110. Marchevsky RS, Mariano J, Ferreira VS et al. Phenotypic analysis of yellow fever virus derived from complementary DNA. *Am J Trop Med Hyg* 1995;52:70–80

A Race with Evolution: A History of Influenza Vaccines

Edwin D. Kilbourne†

Nothing endures but change.

Heraclitus, 540–480 BC

I would like to begin by showing a graph depicting the dramatic impact of influenza vaccination on the morbidity and mortality of the disease during the decades of its use. This, unfortunately, I cannot do; not because the vaccine does not work, but because it is an underutilized vaccine for an unreportable disease. The data, therefore, are not available. I hope the following discussion will resolve this paradox. I hope, also, that I will be forgiven if I have focused narrowly on the complicated history of influenza vaccine development in a way which may appear to slight the scientific basis for its success. I am well aware of this basis, and in another recent

†Deceased

S.A. Plotkin (ed.), *History of Vaccine Development*,
DOI 10.1007/978-1-4419-1339-5_15,

essay on "A History of Influenza Virology" [1], I emphasize not only basic discoveries, but the contribution of influenza virology to virology as a whole.

The Pioneers of the 1930s

After 75 years, the first demonstrably effective vaccine for influenza was administered to 800 retarded male subjects in a state colony in Pennsylvania [2]; it is sobering to consider how little the basic principles governing influenza vaccination have changed in the interim. Although, Chenoweth and his colleagues [2] used filtered suspensions of mouse lung containing live virus, administration of the preparation by the parenteral route insured that it was functioning as a nonreplicating agent analogous to present inactivated vaccines. Before we condemn the use of retarded subjects, it should be pointed out that the concept of informed consent had not yet evolved, that similar closed populations still are used in such studies because they have high attack rates, and that, in fact, protection occurred. I also recall "volunteering" as a medical student at Cornell in 1942 for a study of chick embryo-derived vaccine in that pre-consent, pre-zonal centrifugation era and experiencing most of the side effects now listed in package inserts as uncommon. The vaccine was given in the student microbiology lab by the investigators themselves, Thomas Magill and John Sugg, the former having collaborated with Thomas Francis, Jr in the discovery of influenza B virus 2 years before. (A cryptic effect of this inoculation may have been my unwitting recruitment to the pursuit of the virus, a priming dose (FM-1 virus), boosted in 1947 by experience as an army medical officer at Fort Monmouth, NJ where I witnessed a massive failure of vaccination) [3].

In retrospect, it is somewhat ironical that Chenoweth et al. [2] used mouse lung vaccine, because the virus had already been successfully cultivated in chick embryo tissue culture (independently, by Wilson Smith [4] and by Francis et al. [5]) to which fact reference is made in their publication. Virus yields were higher in mouse lung. Then as now, yields influenced policy. A little impatience may also have prompted this decision. We have always been in an evolutionary race with the virus, but early students of the virus were in a bit of a priority race, as well, with the usual trans-oceanic competition. Both races, I think, continue.

Although I have ascribed priority for the first *demonstrably protective* vaccine to the Philadelphia group of Chenoweth and colleagues, the basis for vaccine development was laid by those who first isolated or worked with the virus in the early 1930s. It was recognized by Wilson Smith, a co-discoverer of the virus, that the principal experimental animal, the ferret, became refractory to challenge after a single infection, but this immunity waned with time so that reinfection could then occur [6]. These observations simultaneously demonstrated that (1) immunity to the infection could be established and (2) this immunity was not sustained – observations abundantly confirmed since in humans. To quote Smith's remarkable insights, "...most human beings already possess some degree of influenzal immunity...(and are) comparable not with normal ferrets, but with...partially resistant ferrets when the immunity following infection has declined to a low level." In this early report of progress which is buried in the archives of *St Mary's Hospital Gazette* [6], and which reviews prior investigations with Christopher Andrewes and Charles Stuart-Harris, Smith also describes the successful use of a "formol vaccine" for the subcutaneous immunization of ferrets which were challenged not only by artificial, but also by contact infection. It is reassuring that in introducing the subsequent use of a similar vaccine in man, Smith cautions that "with human volunteers one must tread warily." Indeed, he did, getting "volunteers" (if such there be) from the military. In the epidemic of 1936–1937, the vaccine appeared to fail – presaging all the things that can go wrong with a vaccine trial: (1) the epidemic struck only 3 days after the vaccination, (2) the attack rate was low, and (3) the epidemic probably comprised more than one virus, because ferret inoculation was often negative. Smith also recognized the possibility of antigenic variation from the vaccine strain. Taylor and Dreguss [7] later showed for the first time a vaccine failure due to antigenic variation of the epidemic virus.

Coincident with the English studies, Francis and Magill in the United States used minced chick embryo tissue cultures as the source of the virus vaccine given in non-inactivated form to medical students, demonstrating antibody response by mouse protection tests and no evidence of infection by the parenteral routes of injection used [8].

The first attempt to infect humans with influenza virus by Andrews, Laidlaw, and Smith [9] were unsuccessful. However, a paper published from the Soviet Union by Smorodintsev et al. in 1937 [10] – frequently cited as the first paper on live virus vaccine – described the administration of a mouse-lethal strain of the original Wilson and Smith (WS) virus by protracted inhalation of atomized virus. Typical febrile influenza developed in 20% of volunteers, hardly an acceptable vaccine by present standards, and certainly not attenuated, as claimed by the authors. Remarkably, they claimed, as well, that the virus appeared not to multiply in men, but the study was a landmark in establishing unequivocally the role of the virus in the development of the disease and in demonstrating antibody response to the virus during convalescence. Subsequently, the Russians took the lead in the early development of live virus vaccines and used them widely in the population.

If I dwell in disproportionate detail on the accomplishments of this early period, it is because they set the stage for most of the later developments in a quite remarkable way at a time when intuition seemed to overcome great technological limitations. Table 1 summarizes these accomplishments.

Table 1 The fundamental accomplishments of the 1930s

First human influenza isolated – (ferrets, mice, and tissue culture)
Animal models developed
Experimental infection of humans
Protection with inactivated virus
Resistance related to serum antibody levels
Antigenic variation described

Emergence of the Perception of Influenza as a Unique Problem Vaccinology

If the 1930s initiated vaccine development, the 1940s were a period during which it became clear that no easy victory was in sight. It was evident that influenza was a disease caused by more than one virus, that antigenic variation of each of these occurred frequently and some times compromised immunity, and that immunity from vaccination was only briefly sustained. The direst antecedents of today's vaccine were first introduced at this time with the disappearance of mouse lung vaccines in favor of relatively pure virus from chick embryo allantoic fluid, purified by red blood cells elution and formalin inactivation.

Trials of a trivalent vaccine (A/PR8, A/Weiss and B) at multiple sites by the US Armed Air Forces Commission on Influenza repeatedly demonstrated a reduction in illness, and by 1945 the entire US Army had been immunized [11].

Against this background, the almost total, repeatedly confirmed failure of this vaccine in US military [11] and in a variety of British populations [12] in 1947 was a sobering indication that influenza was not yet under control.

From the present day perspective, antigenic variation of influenza virus represents a continuous, unidirectional acquisition of point mutations in the virus hemagglutinin molecule in response to immunologic selection by population of antibodies, the response of our predecessors to what I can only call the progressive revelations by the virus of its nature and potential response prevailed as I have outlined in Table 2. If we have returned to the trivalent vaccine of the early 1940s, it is because we currently face 2 co-circulating A (and one B) strain(s) – not because we are unaware, as was the case in 1942, that antigenic drift is progressive and that the addition of a *replaced* strain would not be helpful.

Both American [13] and British [14] virologists viewed potential antigenic changes as limited in number [1] – a concept which led Francis to develop the polyvalent A vaccines (which included even the swine virus) used in the 1950s.

Table 2 Antigenic variation – revelations and response

Date	Revelation	Response
1936	Antigenic change – vaccine failure	*Add* new variant to vaccine
1940	More than one causative agent	*Add* newly revealed agent (influenza B)
1957–1968	New A (pandemic) subtype	*Replace* polyvalent A, B with New A+B
1972 to present	Sequential antigenic change pattern	*Replace* annually
1977	Co-circulation of two type A subtypes	*Add* additional subtype – replace variants annually

Vaccine Reactogenicity: Technical Solutions

In part related to the increasing antigenic mass of polyvalent vaccines, and in part to poor standardization of production, early influenza vaccines gained a bad reputation as inducers of local and systemic symptoms. These problems have been largely solved by improved methods of viral purification (zonal centrifugation and chromatography) [15, 16] and by splitting of the viral lipid membrane with ether or detergents [15, 16, 18]. "Split" and subunit vaccines – the latter further purified by removal of most internal viral proteins – now dominate the market.

The Practical Application of Influenza Virus Genetics

The first genetically engineered, licensed vaccine of any kind was produced in my laboratory by reassortment of the standard A/PR/8/34 virus with a 1968 "Hong Kong" (H3N2) pandemic isolate [19]. The high yield properties of PR8 [20] were transferred to the reassortant (x-31) [21], facilitating vaccine production. Since 1971, most influenza vaccines have been produced from similar reassortants, and several experimental live virus vaccines have used reassortment for attenuation [22]. Other applications of genetics to vaccine development are discussed below in the Live Virus Vaccine section.

Adjuvants

Adjuvants have been used experimentally in humans as immunopotentiators of influenza vaccines for 50 years. There is no doubt that they enhance the magnitude and duration of the short-lived response to influenza vaccines [23–26], but concern about toxicity has made them unacceptable, thus far, for licensure.

Live Virus Vaccines

In this brief review, I have chosen to focus principally on the history of non-replicating vaccines. They represent, therefore, an historical if not scientific endpoint of vaccine development. But from the beginning, live virus vaccines have been the subject of intensive study by talented investigators, including Smorodintsev [27] and Slepushkin [28] in Russia, Burnet and Bull [29] in Australia, and more recently, Chanock [30], Murphy [31], and Maassab [32] in the United States. The American scientists were the first to employ genetic manipulation of the virus (conditional lethal mutants and reassortment) in the attenuation of vaccine strains. Earlier vaccines comprised host range mutants empirically selected by passage in laboratory hosts (see reviews by Beare [33], Stuart-Harris [34] and Kilbourne [22]. The licensure of reassortants developed using Maasab's attenuation strains has given us a new vaccine against influenza.

Surveillance as an Essential Part of Influenza Vaccine Strategy

As a continually reemerging threat [35, 36], influenza has been under systematic surveillance by the World Health Organization [37] in an effort to detect early virus mutations and this to anticipate prevalence in the coming year. In recent years, improved monitoring of China, the apparent source of most new epidemic strains, has made it possible to predict their seasonal invasion of the Western Hemisphere and to prepare appropriate vaccines in advance. Prognostic hints have also been provided by detection of so-called herald-strains [38], which may appear at the end of epidemics.

Postlude and Summary (see also Table 3)

This brief history of influenza vaccine development has not stressed the obvious – that the present day relatively pure, effective, and nontoxic vaccines are grossly underutilized and therefore have had little effect on the overall prevalence of the disease. Are we losing the evolutionary race to the virus – or at least, not catching up – because the mutation rate of RNA exceeds that of our DNA by six orders of magnitude? There seems to be an enormous discrepancy between the rapidity of advances in the basic understanding of the virus and their successful application to vaccine development. Granted that more than just better vaccines are needed to control this complex disease, they are still badly needed. An historical review written 10 years hence will describe better vaccines than we now have, but influenza will still be with us. It may be true that "Those who do not remember the past are condemned to repeat it" [39]. But, uniquely with influenza, remembrance of things past may be of little avail if that next pandemic reveals yet another unsuspected face of this wily foe.

Table 3 Summary

Effective inactivated and live virus vaccines have been used for >50 years
Vaccine-induced immunity is not durable because of:
(a) Intrinsic brevity of inactivated vaccine effect
(b) Progressive evolutionary drift of the virus
(c) Punctuated evolutionary – pandemic shift of major antigen(s)
Annual vaccine changes depend on global surveillance
Control of influenza is not imminent

References

1. Kilbourne ED. A history of influenza virology. Presented May 23, 1995 at a meeting in Chavoires, France on "*Microbe Hunters: Then and Now*". In press
2. Chenoweth A, Waltz AD, Stokes Jr J, Gladen RG. Active immunization with the viruses of human and swine influenza. *Am J Dis Children* 1936;52:757
3. Kilbourne ED. Influenza A prime: a clinical study of an epidemic caused by a new strain of virus. *Ann Int Med* 1950;33:371–9
4. Smith W. The complement-fixation reaction in influenza. *Lancet* 1936;2:1256–9
5. Francis T Jr, Magil TP. The antibody response of human subjects vaccinated with the virus of human influenze. *J Exp Med* 1937;65:251–9
6. Smith W. The influenza problem. *St. Mary's Hosp Gazette* 1937;43:112–20
7. Taylor RM, Dreguss M. *Amer J Hyg* 1940–31:31
8. Francis T Jr, Magil TP. Vaccination of human subjects with virus of human influenza. *Proc Sco* 1936;33:604–6
9. Andrewes CH, Laidlaw PP, Smith W. Influenza observations on recovery of virus from man and on antibody content of human sera. *Brit J Exp Path* 1935;16:566
10. Smorodintsev AA, Tushinsky MD, Drobyshevskaya AI, Korovin AA. Investigation on volunteers infected with the influenza virus. *Am J Med Sci* 1937;194:159–70
11. Francis T JR. Vaccination against influenza. *Bull World Hlth Org* 1953;8:725–41
12. Mellanby H, Dudgeon JA, Andrewes CH, Mackay DG. *Lancet* 1948;1:978–82
13. Magill TP, Francis T Jr. Antigenic differences in strains of human influenza virus. *Proc Soc Exp Biol Med* 1936;35:463–8
14. Smith W, Andrewes CH. Serological races of influenza virus. *Br J Exp Pathol* 1938;19: 293–314
15. Williams MS, Wood JM. A brief history of inactivated influenza virus vaccines. In: Hannoun C, eds. *Options for the Control of Influenza II.* Amsterdam: Elsevier Science Publishers, 1993;169–70
16. Reimer CB, Baker RS, van Frank RM, Newlin TE, Cline GB, Anderson NG. Purification of large quantities of influenza virus by density gradient configuration. *J Virol* 1967;1:1207–16
17. Hoyle L. Structure of the influenza virus. *J Hyg* 1952;50:229–45
18. Cromwell HA, Brandon FB, McLean IW Jr., Sadusk JF, Jr. Influenza immunization: a new vaccine. *JAMA* 1969;210:1438–42
19. Kilbourne ED. Influenza 1970: unquestioned answers and unanswered questions. *Arch Environ Health* 1970;21:284–92
20. Kilbourne ED, Murphy JS. Genetic studies of influenza viruses. I. Viral morphology and growth capacity as exchangeable genetic traits. Rapid *in ovo* adaptation of early passage Asian strain isolated by combination with PR8. *J Exp Med* 1960;111:387–406
21. Kilbourne ED, Schulma JL, Schild GC, Schloer G, Swanson J, Bucher D. Correlated studies of a recombinant influenza-virus vaccine. I. Derivation and characterization of virus and vaccine. *J Infect Dis* 1971;124:449–62
22. Kilbourne ED. *Influenza.* Plenum Publishing, 1987
23. Henle W, Henle G. Effects of adjuvants on vaccination of human beings against influenza. *Proc Soc Exp Biol Med* 1945;59:179–81
24. Salk JE, Bailey ML, Lavrante AM. The use of adjuvants in studies on influenza immunizations. II. Increased antibody formation in human subjects inoculated with influenza virus vaccine in water-in-oil emulsion. *Am J Hyg* 1952;55:439
25. Davenport FM. Applied immunology of mineral oil adjuvants. *J Allergy* 1961;32:177–89
26. Hilleman MR. The roles of early alert and of adjuvant in the control of Hong Kong influenza by vaccines. *Bull Wld Hlth Org* 1969;41:623–8
27. Zhilova GP, Alexandrova GI, Zykov, Smorodintsev AA. Some problems of modern influenza prophylaxis with live vaccine. *J Infect Dis* 1977;135:681–6

28. Slepushskin AN, Ellegorn NS, Russina AE, Vitkina BS, Bobyleva TK. Assessment of the efficacy of mass influenza vaccinations. Assessment of efficacy of mass influenza vaccinations. Report II. *J Microbiol Epidemiol Immunobiol* 1968;10:26–32
29. Bull DR, Burnet FM. Experimental immunization of volunteers against influenza virus B. *Med J Australia* 1943;1:389–94
30. Mills J, Van Kirk J, Hill DA, Chanock RM. Evaluation of influenza virus mutants for possible use in a live virus vaccine. *Bull WHO* 1969;41:599–606
31. Murphy BR, Buckler-White AJ, London AJ et al. Avian-human reassortant influenza A viruses derived from mating avian and human influenza A viruses. *J Infect Dis* 1984;150:841–50
32. Reeve P, Gerandas B, Moritz A, Liehl E, Kunz C, Hofmann H, Maassab HF. Studies in man with cold-recombinant influenza virus (H1N1) live viruses. *J Med Virol* 1980;6:75–83
33. Beare AS. Live viruses for immunization against influenza. *Progr Med Virol* 1975;20:49–83
34. Stuart-Harris C. The present status of live influenza virus vaccine. *J Infect Dis* 1980;142: 784–93
35. Kilbourne ED. New viral diseases. A real and potential problem without boundaries. *J Amer Med Assoc* 1990;162:800–9
36. Kilbourne ED. New viruses and new diseases: mutation, evolution and ecology. *Curr Opin Immunol* 1991;3:518–24
37. Assaad F, Bres P, Delon P. Surveillance internationale de la grippe. *Rev Epidemiol Sante Publ* 1977;25:441–5
38. Glezen WP, Couch RB, Six HR. The influenza herald wave. *Am J Epidem* 1982;116:589–98
39. Santayana G. *The Life of Reason: Phases in Human Progress*. New York: Charles Scribner's sons, 1953

The Role of Tissue Culture in Vaccine Development

Samuel L. Katz, Catherine M. Wilfert, and Frederick C. Robbins†

Jhon Enders, Frederick Robbins, Thomas Weller
1954

In 1954, John F Enders, Frederick C Robbins and Thomas H Weller received the Nobel Prize in Physiology and Medicine for their successful propagation of poliovirus in tissue culture. Like many great scientific advances, this was the culmination of a lengthy series of laboratory investigations that had begun more that 20 years previously and had included the work of a great number of scientists. The publication by the Enders' group in 1949 of their seminal work provided the

S.L. Katz (✉)
Duke University School of Medicine, Box 2925, Durham, NC 27710, USA
e-mail: katz0004@mc.duke.edu

S.A. Plotkin (ed.), *History of Vaccine Development*,
DOI 10.1007/978-1-4419-1339-5_16,

"Rosetta Stone" that integrated the results of previous research and added those techniques necessary to its final completion and widespread applicability [1].

Enders' leadership of this group was exemplary of his role for more than 20 years as a basic microbiologist. He had come late to the field, following an unsuccessful career as a real-estate salesman and 4 years in the study of Celtic and Teutonic languages as a doctoral candidate at Harvard. In one of those serendipitous events that markedly alter life patterns, he had obtained a room in a boarding house shared with several Harvard medical students, including Hugh Ward, an Australian Rhodes Scholar who was working in the laboratories of the famed Hans Zinsser. Captivated by the work of Ward and his associates, Enders abandoned his nearly completed PhD in English and joined the Department of Bacteriology and Immunology at Harvard Medical School where he finally received a doctorate and his first faculty appointment at the age of 32. He was a Renaissance scholar who knew literature, music and philosophy as well as microbiology. When, after 15 years with Zinsser, he established his own laboratory at the Children's Hospital in Boston, Massachusetts, he was already fascinated by viruses, having previously investigated feline panleucopenia with his colleague, Bill Hammon, and prepared a vaccine which was quickly adopted by veterinary medicine. He had also collaborated in studies of mumps virus during World War II, a cause of serious morbidity among military recruits in the past.

As early as 1928, Maitland and Maitland had shown that vaccinia could be grown in a system consisting of kidney tissue fragments nurtured by a mixture of serum and organic salts. Their observations led to a proliferation of attempts to exploit tissue culture techniques for virus growth. Until then, nearly all work had been conducted in laboratory animals (rodents or nonhuman primates). Over the succeeding years, a large roster of investigators manipulated tissue culture systems in a variety of ways to grow viruses as varied as ectromelia, equine encephalitis, herpes simplex, influenza, Japanese encephalitis, mumps, rabies, vaccinia, yellow fever and others. Although individual successes were reported, they were rarely sufficiently simple to be repeated easily in the laboratories of other investigators, or to be maintained for lengthy periods of time. It was also difficult to prove that virus had multiplied rather than merely persisted in the varied culture systems. To avoid the requisite use of animal inoculation to demonstrate presence and multiplication of virus, a number of indirect methods were developed, depending on attributes of individual viruses. For example, both mumps and influenza caused agglutination of red blood cells, and the presence and amount of hemagglutinin produced was an indirect measure of viral replication, or at least the multiplication of viral antigens [2].

Diminished cellular metabolism of cells injured by virus provided another indirect indicator of probable viral replication. Huang had reported the decreased acid metabolism of chick embryo tissues that had been infected with equine encephalomyelitis virus [3]. Similar pH change had been observed in infected tissues by Plotz using fowl plague virus. In 1936 Sabin and Olitsky were successful in culturing poliovirus in human embryonic brain tissue, but in no cells of nonnervous system origin [4]. This had lent further credence to the dogma prevalent that poliovirus was strictly a neurotropic agent, but conflicted with the observations that in infected patients and monkeys, virus was present in both the nasopharynx and the gastrointestinal tract for long periods of time. It was not until quite recently that the specific cell receptors for poliovirus were isolated and identified on many human and other primate cells [5].

At his new laboratories at the Boston Children's Hospital in 1946, Enders was joined by two former medical school roommates, Tom Weller and Fred Robbins. Both had returned from World War II armed forces service and were developing their careers in pediatrics and infectious diseases. This then was the team that went on to capture polioviruses successfully in cell culture systems. Their poliovirus studies were an interesting, almost tangential result of Robbin's concern with infantile diarrhea, Weller's with varicella and Enders' with vaccinia. Because of some laboratory findings from the National Foundation for Infantile Paralysis, polio became almost as a "fellow traveler" in the laboratory. At Enders' suggestion, they inoculated some of the extra tubes from the varicella and diarrheal cultures with poliovirus that was stored in the laboratory freezer. Employing fibroblasts from human foreskin (Weller) and embryonic tissues (Robbins), they inoculated a mouse-adapted strain of type 2 polio in roller tube cultures not used for the primary studies. Once again with Enders' encouragement, they maintained these cultures for long periods of time changing their nutrient media several times each week, providing an optimal environment for the cells and therefore the virus. Because this particular poliovirus strain (Lansing) was mouse-adapted, they were able to test for its presence and multiplication by inoculation of culture fluids intracerebrally into susceptible mice. Much to their amazement, the inoculated mice became paralyzed, indicating that successful growth in their cell culture systems. Their results were duplicated not only with cultures of central nervous tissue but with skin, muscle, kidney and intestine. Their first publication of these observations appeared in January 1949 [1] and soon thereafter they found that types 1 and 3 polio grew similarly in cell culture systems [6, 7]. Because cells in the inoculated cultures appeared to die more rapidly than in the controls, the capacity of virus to kills cells was a convenient indication of successful replication, obviating the need for inoculation of experimental animals (monkeys for types 1 and 3, mice as mentioned for type 2). In addition to the long time maintenance of their cultures and the periodic change of nutrient medium, another contributory feature to success with the availability in the late 1940s of antibiotics to protect the cultures from contaminating bacteria. In addition to the altered metabolism of the infected cells, they later found that morphologic changes could be observed ("cytopathology") which reflected specific cellular damage by replicating virus. From these observations then came the ability to culture and observe the replication of viruses totally in a cell culture system, without the need to check for replication by animal inoculation. Additionally, it was then possible to assess the presence or absence of virus-specific serum antibody in tissue culture neutralization procedures, once again avoiding the tedium of time-consuming, expensive, and demanding animal inoculations.

When in 1954 the Nobel Prize was awarded for this work, it was characteristic of John Enders, who could easily have been the sole recipient, that the award was shared with Weller and Robbins. He always made it clear that his associates in his laboratory were full participants and recognition for a resultant work should be shared [8]. This was not only true of the polio studies, but later of measles, interferon, SV 40 and all other output of his fruitful laboratory. Indeed he was a scientist with a green thumb and a great heart. He maintained unwaveringly impeccable standards of personal and scientific honesty and an openness that was striking. Following the publication of the polio results (and later similarly with measles), the laboratory was visited by dozens of investigators from around the world who never failed to leave without packages of virus, cell cultures, sera, reagents, or whatever might have been requested to enhance their own studies when they returned home [8].

Within only 4 years of the initial publication from the Enders laboratory, both Jonas Salk and Albert Sabin were able to report success with two differing approached to immunization against poliomyelitis, a formalin-inactivated preparation of the three virus types (Salk's IPV) and three attenuated live variants of poliovirus (Sabin's OPV) [9, 10]. Drs Beale and Melnick will discuss these at greater length in their articles. Moreover, in the succeeding years a host of new virus vaccines were developed exploiting the cell culture techniques that stemmed from the work of the Enders group. These have included measles, mumps, rubella, adenoviruses, varicella-zoster, rabies, hepatitis A, rotavirus, cytomegalovirus and others still under research and development (Table 1). None of these was "easy," but the barriers has been eliminated and the pathway illustrated by their work [11]. No longer was it necessary to prepare vaccines on calf skin, in the brain or spinal cord of various species, or in the fertile hen's egg (Table 2).

Cell culture of viruses was seized upon as a convenient technology for a great variety of research including quantitative assay, diagnostic tests, and cell transformation in vitro by tumor viruses. The ability to maintain cells in culture and to observe their response to a variety of manipulations opened the way for expansion to many fields beyond virology including cellular genetics, pharmacology, immunology, cellular biology, and indeed every area of basic biomedical science. John Enders and his colleagues left a legacy to which all modern biomedical science remains indebted.

Table 1 Tissue culture virus vaccines

Polio	Rabies
Measles	Varicella
Mumps	Hepatitis A
Rubella	Rotavirus
Adenovirus	Cytomegalovirus

Table 2 Nontissue culture virus vaccines

Smallpox (vaccinia): bovine lymph or skin
Rabies: various brain or spinal cords (rabbit, sheep, goat, mouse), duck embryo
Japanese encephalitis: mouse brain
Yellow fever: mouse brain, hen's egg
Influenza: chick embryo allantoic fluid

References

1. Enders JF, Wellers TH, Robbins FC. Cultivation of the Lansing strain of poliomyelitis virus in cultures of various human embryonic tissues. *Science* 1949;109:85–7
2. Robbins FC, Enders JF. Tissue culture techniques in the study of animal viruses. *Am J Med Sci* 1950;220:316–38
3. Huang CH. Further studies on the titration of the Western strain of equine encephalomyelitis virus in tissue culture. *J Exp Med* 1943;78:111
4. Sabin AB, Olitsky PK. Cultivation of poliomyelitis virus in vitro human embryonic nervous tissue. *Proc Soc Exp Biol Med* 1936;31:357–9
5. Mendelsohn CL, Wimmer E, Raconiello VR. Cellular receptor for poliovirus: molecular cloning, nucleotide sequence, and expression of new member of the immunoglobulin superfamily. *Cell* 1989;56:855–65
6. Weller TH, Robbins FC, Enders JF. Cultivation of poliomyelitis virus in cultures of human foreskin and embryonic tissues. *Proc Soc Exp Biol Med* 1949;72:153–5
7. Robbins FC, Weller TH, Enders JF. Studies on the cultivation of poliomyelitis virus in tissue culture. II. The propagation of poliomyelitis viruses in roller-tube cultures of various human tissues. *J Immunol* 1952;69:673–94
8. Katz SL, John Franklin Enders. In: Bentinck Smith W, Stauffer E, eds. *More Lives of Harvard Scholars*. Cambridge: Harvard University Press, 1986;166–73
9. Salk J, Bennet BL, Lewis LJ et al. Studies in human subjects on active immunization against poliomyelitis. 1. A preliminary report of experiments in progress. *JAMA* 1953;151:1081–98
10. Sabin AB, Hennessen WA, Winsser J. Studies of variants of poliomyelitis virus. I. Experimental segregation and properties of avirulent variants of three immunologic types. *J Exp Med* 1954;99:551–76
11. Robbins FC. Polio-historical. In: Plotkin SA, Mortimer EA Jr, eds. *Vaccines*. Philadelphia: WB Saunders Co,1994;137–54

Viral Vaccines and Cell Substrate: A "Historical" Debate

Florian Horaud[†]

The article by S. Plotkin concerning the history of rubella vaccines and the cell substrate is an excellent illustration of the "ballet" performed around the problem of the acceptability criteria of cell substrate used in the development and preparation of viral vaccines. This problem came to my attention many years ago when I worked in the World Health Organization (WHO) experts groups, but in the last decade this topic had been an important concern for me since I act as an expert in virology operating in French and European regulatory systems.

The importance of the cell substrate in the development of viral vaccines is illustrated in Table 1 in which it is shown that the outcome of a new generation of viral vaccines depended on an improvement of the technology used in the preparation of cell substrates destined for virus propaganda.

When viral vaccine development is correlated with the appearance of adverse reactions after vaccine administration, it can be observed that each generation of viral vaccines has been responsible for nontargeted effects [1]. The turning point in the new technology of viral vaccine preparation was the discovery of Enders, Weller, and Robbins [2] that poliovirus replicates in nonnervous tissue culture of

[†] Deceased

S.A. Plotkin (ed.), *History of Vaccine Development*,
DOI 10.1007/978-1-4419-1339-5_17,

Table 1 A historical account of viral vaccine development

Year	Vaccine	Cell substrate
1795	Smallpox (Jenner)	Calf lymph
1885	Rabies (Pasteur)	Rabbit CNS
1937–1940	Influenza Yellow fever	Embryonated eggs (Woodroof and Goodpasture
1953	Polio	Nonneural cell culture (Enders, Weller, and Robbins)
1963–1965	Measles	Chick embryo fibroblast
1967	FMDV	BHK 21[a] (McPhearson and Stocker)
1968	Rubella	Human diploid cells (W138, Hayflick, and Morhead)
1981	Polio (killed)	Vero cell line[a]
1985	Live polio	Vero cell line[a]

CNS central nervous system
[a]Continuous cell line

human and nonhuman primates. Since then, tissue culture has become a major technology in the preparation of viral vaccines. However, the real debate about the criteria for the acceptability of cell substrate in the preparation of viral vaccine originated with the discovery of the simian virus 40 (SV40) as a common contaminant of polio and other viral vaccines.

SV40 was present, as adventitious virus, in primary cell culture of rhesus monkey kidney used as a substrate for the preparation of killed and live polio vaccines [3]. Another event that addressed safety questions in connection with the cell substrate used for viral vaccine was the development by Hayflick of human diploid cell lines [4].

During the 1960s and 1970s, the dominant ideology concerning cell substrate was the acceptability of "normal cells" exclusively, i.e., primary culture or diploid cell lines having a finite life span. This view was, for instance, clearly stated in the US Regulation for the Manufacture of Biological Products [5], which forbade the use of heteroploid continuous cell lines (CCL) derived from in vitro cell transformation or from neoplastic tissue as substrates for the preparation of biologicals.

Although cell substrates used for the production of viral vaccines are defined by regulations established by international and/or national control authorities, the philosophy and the requirements concerning tests used for the quality control are periodically revised as a result of the progress made in the fields of virology, cell biology and molecular separation, and protein purification

A new approach toward the acceptability of CCL for the preparation of biologicals in general, and of viral vaccines in particular, was emphasized in an international meeting held in Lake Placid, New York, in 1978 [6]. The meeting concluded that compared to the classical cell substrates, the use of CCL in the manufacture of viral vaccines had four main advantages: (1) the cell culture substrate is more consistent and "clean" than primary culture, since the production is based on the cell bank system that allows characterization of the cell source and detection of microbial contaminants; (2) it allows more efficient and more reproducible cell growths by using new large-scale tissue culture procedures; (3) it provides higher yields of virus; and (4) it could reduce or preclude the use of animals (monkey, in the case of polio vaccine).

Despite these advantages, the acceptability of CCL as substrate in the production of viral vaccines was controversial in the late 1960s and early 1970s, since the oncogenic potential of heteroploid cells that have an infinite life span was already well known. However, at this early time, the molecular biology of oncogenic transformation was little understood, and the scientific community was not prepared to accept the use of transformed cells in the production of biologicals destined for humans. The hesitation with regard to the approval of CCL in the production of biologicals mainly came from the idea that the final preparations might contain enough cellular nucleic acid to "induce cancer" in humans receiving the product.

In 1978, a real breakthrough in the production of biologicals on CCL was realized by the preparation of interferon from Nemalva cells, a lymphoblastoid cell line derived from a Burkitt lymphoma [7]. The powerful purification method used in this case (i.e., immunoaffinity chromatography) was able to reduce to practically undetectable levels the potential contaminants of the product. The result of this study, together with the development by van Wezel of a new technology for large-scale cultivation of anchorage-dependent cells on microcarriers [8], encouraged and justified further studies of the use of CCL in the manufacture of viral vaccines.

In the early 1980s, the first inactivated polio vaccine prepared in Vero cells – a CCL – and manufactured by Institut Mérieux was licensed. This approach was then successfully followed by the same producer for rabies and live polio vaccine [9].

The risk raised by the use of CCL in the manufacture of vaccines and other biologicals was evaluated in 1986 by a WHO study group [10]. The experts examined various facets of this topic and strongly encouraged further development of this promising field.

In recent years, the acceptability of CCL in the manufacture of biologicals made considerable progress. Rodent CCL able to induce tumors in animals and harboring endogenous retroviruses such as hybridoma of Chinese hamster ovary (CHO) cells are widely used in the production of biologicals. Products derived from these cells (e.g., monoclonal antibodies, cytokines, hormones, etc.) are purified to eliminate cell DNA [11, 12]. In any case, the safety of this category of products is presently well established and the debate about their use in the production of biologicals of cells having tumorigenic properties is now over.

The issue of using CCL in the production of viral vaccines was greatly facilitated by the existence of Vero cells. This cell line has a negligible oncogenic potential, is sensitive to numerous viruses, and does not contain known adventitious or endogenous viruses [13].

During recent years, the question of using CCL in the production of live vaccines was raised. Despite advances in the understanding of oncogenic process, the introduction of CCL in the production of live vaccines had not developed. Live polio vaccine prepared on Vero cells is licensed and available on the market of some countries, but surprisingly, in other areas live polio vaccine is still prepared on primary monkey kidney cell culture.

It is now well established that only intact cells and nucleic acids extracted from oncogenic viruses are capable of inducing tumors in animals, while DNA obtained from highly tumorigenic cells had no detectable activity in vivo [10]. Taking into consideration the great advantages of CCL, such as Vero cells, in the production of biologicals, it is therefore justified to continue the effort to replace primary cell culture by CCL, without being haunted by the ghost of the cancer risk represented by residual cell DNA present in the vaccines.

The new developments in recombinant DNA technology for the obtention of vaccines, such as DNA vaccines or immunogenic peptides expressed by plant viruses, might make this debate obsolete. In this perspective, it will be possible to obtain immunogenic viral peptides without previous amplification of the infectious virions in animal cells in culture. This perhaps opens the way to a new generation of revolutionary vaccines.

References

1. Horaud F. Viral safety of biologicals. *Dev Biol Stand* 1991;75:3–7
2. Enders JF, Weller TH, Robbins FC. Cultivation of Lansing strain of poliomyelitis virus in culture in various human embryonic tissues. *Science* 1949;109:86–7
3. Keereti S, Nathanson N. Human exposure to SV40: review and comment. *Am J Epidmiol* 1976;103:1–12
4. Hayflick L. History of the cell substrates used for human biologicals. *Dev Biol Stand* 1989;70:11–26
5. US Department of Health, Education and Welfare, Public Health Service. Regualtion for the Manufacture of Biological Products, title 42, part 73. DHEW publication no (NIH) 71–161, formerly PHS publication no 437, revised 1971–1976
6. Petricciani J, Hopps H, Chapple PJ eds. Cell substrates: their use in the production of vaccines and other biologicals. In: *Advances in Experimental Medicine and Biology* 118:9–21, New York: Plenum Press, 1979
7. Finter NB, Fantes KH, Lockier MJ, Lewis GD, Ball GD. The DNA content of crude and purified human interferon prepared by Wellcome Biotechnology Limited. In: Hopps HE, Petricciani JC, eds. *Abnormal Cells, New Product and Risk. In vitro Cellular & Development Biology*, Tissue Culture Association USA. 1985; monograph no 8:125–8
8. Van Wezel AL. Growth of cell strain and primary cells on micro carriers in homogenous culture. *Nature* 1967;216:65–5
9. Montagnon BJ. Polio and rabies vaccines produced in continuous cell line: a reality for Vero cell line. *Dev Biol Stand* 1988;70:27–47
10. Petriccianni J. Cell, products, safety: background papers from the WHO Study Group on biologicals. *Dev Biol Stand* 1987;68:43–9
11. Horaud F. Viral vaccines and residual cellular DNA. *Biologicals* 1995;23:225–8
12. Petricciani J, Horaud F. DNA, dragons and sanity. *Biologicalsl* 1995;23:233–8
13. Horaud F. Absence of viral sequnces in the WHO-Vero cell bank: a collaboration study. *Dev Biol Stand* 1992;76:43–6

History of Koprowski Vaccine Against Poliomyelitis

Hilary Koprowski in collaboration with Stanley Plotkin

The Past Was Not a Prolog

It all started, like much in science, with prior failure and frustration. Poliomyelitis had become prominent late in the nineteenth century, and by the middle of the twentieth was an epidemic disease that fascinated laymen and scientists alike. In 1946, we knew from the work of Landsteiner and Popper in 1909 [1] that polio was caused by a virus, and from the work of Burnet and McNamara [2] that there was more than one type of poliovirus, although we did not know how many until 1951 [3]. Nevertheless, it had been established that prior infection protected primates against subsequent homologous infection, which suggested that a vaccine was possible. But how? The only possible substrate at that time was monkey spinal cord. Two separate attempts were made in 1935 to inactivate such suspensions, one with formalin and one with sodium ricinoleate: both failed, and many cases of poliomyelitis were caused by the vaccines, some fatal [4, 5]. Thus, it was back to the drawing board for virologists.

"Genius is one percent inspiration and ninety-nine percent perspiration" – Edison

There is a continuity of ideas in science which depends on contact between generations. During World War II, I was working at Rockefeller Institute Yellow Fever Research Laboratory in Brazil, where I met Max Theiler, who recently developed a live yellow fever vaccine. This friendship led to other conversations held later in New York, after I had immigrated to the USA in 1945, and took a job with Lederle Laboratories. By 1941, chimpanzees had been infected by poliovirus given orally, and it seemed likely that the disease was transmitted by the oral route [6].

Accordingly, we decided that a live attenuated poliovirus given orally was the best way to immunize, but we needed a means of attenuation. Fortunately, it had just been shown that the cotton rat (which is a mouse) and the mouse could be

H. Koprowski (✉)
Jefferson Cancer Institute, Thomas Jefferson University,
1020 Locus Street, Philadelphia, PA 19107-6799, USA
e-mail: hilary.koprowski@jefferson.edu

S.A. Plotkin (ed.), *History of Vaccine Development*,
DOI 10.1007/978-1-4419-1339-5_18, © Springer Science+Business Media, LLC 2011

Fig. 1 Members of the Koprowski laboratory at Lederle, including Koprowski (*upper center*) and Tom Norton (*upper right*). Photo taken in 1949

Table 1 History of TN (type 2) strain

Brockman strain of polio (supposedly type 1 strain)
Monkey cord
↓
Intraspinally: 8 mouse brain passages
↓
Intracerebrally: 1–3 cotton rat passages

infected by human poliovirus [7]. Theiler helped not only with his ideas, but also by allowing us to employ Tom Norton, his chief technician, an accomplished laboratory worker (Fig. 1). The polio project started in 1948, but even 2 years earlier we had helped Walsh McDermott, a professor of public health at New York Hospital, to isolate a poliovirus from the blood of a 29-year-old bank employee who was acutely ill with polio [8]. The first virus we attenuated, however, was the Brockman strain obtained from Kessel in California (Table 1). A suspension of monkey spinal cord infected with the Brockman virus was adapted to Swiss albino mice by means of serial intracerebral transfers. At level 7 mouse passage, the virus was harmless for intracerebrally injected monkeys. Cotton rats were injected with the mouse-adapted virus, and at the level of 1–3 passages, the vaccine was prepared [9]. Later on, it was also found that in contrast to virulent virus, this attenuated strain failed to cause any cytopathic effect on monkey kidney tissue culture cells [10]. Although Brockman was reported to be a type 1 virus, the virus adapted to grow in mice and cotton rats proved later to be a type 2 virus. We named this virus TN in honor of Tom Norton.

TN virus at various levels of passage in mouse, cotton rat or baby hamster brain was tested in monkeys and chimpanzees. The results in monkeys did not appear to reflect the natural history of polio, whereas chimpanzees seemed more useful, as they excreted virus and developed antibodies to polio [10–12]. The results of those studies suggested that the poliovirus adapted to cotton rats, which showed a low pathogenicity for rhesus monkeys inoculated intracerebrally, was a candidate for trials in humans.

Crossing the Rubicon

Lederle Laboratories is in Pearl River, New York, a village located in a rural area. Not far is a home for retarded children, Letchworth Village. Although even in 1950s, Letchworth was run on advanced principles of human care for the mentally retarded, the behavior of the inmates was such that infectious diseases were a constant preoccupation of the staff. The Director of Research at Letchworth was a pathologist named George Jervis, who came to see me one day about studies of allergic encephalitis. George was a tall, lanky, shy, and soft-spoken sort, but an independent thinker, and when he heard about the work we had started on polio, he proposed that we try to protect his patients against the possible introduction of polio virus into the wards of the institution. Permission was obtained from the institutional authorities as there were no human subject committees in those days!

On 27 February 1950, we took the first step on the path that ultimately led to the control of polio, with the administration of TN orally as a suspension of infected cotton rat brains to an 8-year-old with no antibodies to type 2 poliovirus. This child was kept under careful observation until it was clear that there were no ill effects, and gradually 19 other children were enrolled in the trial, of whom 16 would prove to be antibody negative at enrollment. The results of these studies are shown in Table 2 [13], which indicates that seroconversion was induced in all 17 seronegative subjects. In addition, the same subjects excreted TN virus in their stools, but when TN was readministered, 10 of 12 subjects resisted a second infection. Thus, we established early on that orally administered attenuated virus that could induce a state of intestinal resistance to subsequent challenge.

However, it had been found in the first study that the neutralizing antibody response induced to TN did not cross-protect against Brunhilde, which later was identified as type 1 strain. Thus, we attempted to adapt a type 1 strain to cotton rats. The starter material was a mixture of two known laboratory strains, Sickles and Mahoney, which were inoculated together into the brains of Swiss mice and

Table 2 Responses to the attenuated TN

Subjects	Seroconversion[a]	Virus excretion	Past dose number		
			1	2[b]	3[c]
Seronegative	17/17	0	5	10	5
		1+	3	0	0
		2+	6	2	1
		3+	3	0	0
Seropositive	0/3	0+	1	–	–
		1+	2	–	–

[a] After the first dose
[b] 12 Subjects were given a second dose
[c] 9 Subjects were given a third dose

passaged together until adapted to the Swiss mouse, and then subsequently to the PRI mouse [14, 15]. The resulting virus, the SM strain, was given orally to humans at Letchworth in 1953 [16]. Seroconversion was again shown to the homotypic virus. In that same study, we showed that the SM and TN strain could be administered together, each infecting the intestine and each producing a seroconversion. This was the first demonstration of a combined oral polio vaccine containing more than one serotype.

At that point, we chanced to meet a fascinating character, Karl P. Meyer, a Swiss veterinarian who had become a prominent virologist in California. He was a magisterial individual both in appearance and in manner, who by virtue of his eminence in chlamydial disease and plague was an expert in public health. He obtained permission for us to vaccinate at an institution for retarded children in Sonoma, CA, where we expanded our studies of immunization, using the TN and SM stains.

At about that time, two important advances in virology came into widespread application, both of which were to be later recognized by the Nobel Prize: the discovery of virus growth in cell culture by Enders, Wellers and Robbins, and the invention of plaquing in cell culture by Renato Dulbecco and Peter Vogt. Their effect on our work was profound, as we could now isolate viruses more easily in mice, and clone these viruses in order to study individual strains for their immunizing properties.

The SM strain of type 1 poliovirus was further modified through serial monkey kidney plaque passages originating from human feces from a child fed the SM virus. This strain was renamed CHAT strain after the initial of the patient [15] (Table 3). This type1 isolate became the extremely stable, attenuated, and immunogenic type 1 strain that we used in all later studies. Moreover, other workers, using the same cell culture techniques, could now more easily isolate viruses from subjects in epidemiologic studies. One such worker who contributed immensely to our understanding of the natural history of poliovirus was John Fox, then at Tulane University in New Orleans, where he was studying the circulation of poliovirus in families. A virus belonging to the third serotype of polio was isolated from a child with asymptomatic infection and furnished to us. We called the isolate Fox [15] in his honor, and we had little difficulty adapting it to monkey kidney cell culture, and

Table 3 History of the attenuation of CHAT virus (type 1)

Adaptation to mice from monkey kidney (MK)
↓
1 Swiss+27 PRI mice
↓
14 Successive chick embryo (CE) tissue culture passages
↓
5 Alternative MK-CE tissue culture passages
↓
4 Serial human passages
↓
4 Serial MK plaque purifications
↓
MK-CHAT strain

in showing that it was attenuated for the central nervous system of monkeys and immunogenic by the oral route in humans.

As the number of vaccinees increased, our confidence grew, and in February 1955 at a meeting of the New York Academy of Science [17] I said, "It is time now to attempt breaking through the fear barrier to apply live immunization to the normal population."

In addition, we made a decision to abandon the type 2 TN strain, as it was no longer practical to consider viruses grown in mouse brain. Instead, we obtained a type 2 strain from Fox, called P-712, which was shown to be highly attenuated in the monkey neurovirulence test.

Hither, Thither, and Yon

We now entered a period of expanding studies in many different parts on the world. An important American collaborator was Joseph Stokes Jr., a Quaker pediatrician, who was one of the first great pediatric infectious diseases pediatricians. Despite his peaceful Quaker background, Joe would brook no difficulties, and his reputation for probity opened doors all around Pennsylvania and New Jersey. In his vintage Plymouth, which he drove like a racing car, Joe accompanied us to various homes for children [18], and also to a women's prison in New Jersey, called Clinton Farms [19, 20] to help us perform studies. Although it was nominally a closed institution, the liberal policies of this prison without walls resulted in a constant incidence of pregnancy.

At Clinton Farms [19, 20], we performed the first studies of vaccination of infants, in which it was shown that maternal antibodies did not prevent successful vaccination, even in the newborn period. This ultimately led to the current regimens of early poliovirus vaccination used in developing countries. Also through Joe Stokes, we were able to perform a family study, largely among Quaker families in Moorestown, NJ [21]. In those studies, we showed that excreted attenuated virus could spread from a vaccinated child to another family member, thus propagating the effect of the vaccination beyond the vaccinated.

A less successful collaboration at this stage was with George Dick, a microbiologist from Belfast, who came to me one day to suggest a clinical study in Belfast, with the microbiology to be performed by him. Dick tested the viruses excreted by the vaccinees and found that in some cases monkey neurovirulence had increased [22]. Although today that would evoke no surprise, Dick considered it the signal to campaign in the newspaper against live attenuated virus vaccination. Fortunately, Dick's diatribes left no lasting effect.

On the other hand, K.F. Meyer introduced me to a young, athletic Swiss virologist named Meinrad Schar, who was working at the central public health laboratory in Zurich, Switzerland. Schar was quite interested in oral polio vaccination, but had no access to a clinical population. He contacted various physicians, who were not willing to take any chances. Finally, he came across a private pediatrician named Fritz Buser who had a practice in Bern. Buser, an intense, short, and wiry man with a strong interest in vaccines, decided that attenuated polio vaccine had merit, and accordingly set up vaccine studies in children with Schar, using the CHAT and Fox strains. The studies started in 1958–1959 with small groups of children, but by 1960, 40,000 children were vaccinated in the cantons of Aargau and Basel, followed by over 300,000 in Bern, Lucerne, and Aargau [23, 24].

Another person who appeared on the scene at this time was Drago Ikic, head of the Immunology Institute of Zagreb in the former Yugoslavia. Ikic is tall, dark, and laconic. Once decided, he carries out projects meticulously. So it was when he decided to vaccinate Croatia with our types 1 and 3 stains. In early spring of 1961, over 1,300,000 children in Croatia were given a mixture of the two strains [25] and were controlled clinically and serologically. No postvaccination polio was seen, and the serologic studies suggested that there were at least 100,000 triple seronegatives in the vaccinated population. The data showed a seroconversion of 91.5% for type 1 and 93.5% for type 3 [25, 26].

Late, Ikic produced attenuated polio vaccine locally using human diploid cells, and became well-known for his use of these cells for vaccine production [27].

Although it is justly said that no one is a prophet in his own country, we also had success in introducing live attenuated polio vaccine in Poland, where I was born. As Poland was then a communist state, numerous visits and correspondence were necessary to overcome the political opposition that sprang up immediately. The man who made the vaccination campaign possible was a courtly gentleman by the name of Felix Przesmycki, who was the Head of the National Hygiene Institute in Warsaw. Przesmycki knew his way around the labyrinthine communist bureaucracy, and also was quite persistent. Perhaps part of his success was that he was quite deaf, and therefore could pretend not to take no for an answer.

On one occasion, however, this got him into trouble. On a visit to the USA, we introduced Przesmycki to an American dowager of excellent family connections. Just as we completed the introduction, and the good lady extended her hand, Przesmycki noted her very large wolfhound. Ignoring the lady's proffered hand, he seized the paw of the dog and shook it vigorously. Needless today, urgent action on our part was required to restore the social amenities.

The vaccine trial organized in Poland with live attenuated oral vaccine by Przesmycki and his associate, Dr. Dobrowolska, was the largest trial conducted with this vaccine. Between 20 October 1959 and 30 March 1960, 7,239,000 children and adolescents were vaccinated with the CHAT type 1 virus and between 20 October 1959 and 15 April 1960, 6,818,500 persons were vaccinated with type 3 virus [28]. The incidence of polio dropped from 1,112 cases in 1959 to 28 cases in 1963 [28, 29].

What mighty contests arise from trivial things! – Pope

At this point, I would like to clarify my relationship with Herald Cox, Director of Viral and Rickettsial Diseases at the Lederle Laboratories. Cox was supportive of my beginning poliovirus attenuation studies and participated in some of the original conversations with Theiler. However, he was not directly involved with the project, and indeed the truth is that he did not even know of our first vaccination experiments at Letchworth until they were completed. As a young man, I might be forgiven for taking the bit in my mouth, although I still regret that rupture caused between us, which separated him from our project in 1952. Afterward, he developed his own set of strains based on our earlier work. These strains later came to grief in clinical trials conducted in Florida [30] and in Berlin during the later 1950s [31]. A high incidence of vaccine-associated polio led to the abandonment of the Cox strains by Lederle.

By 1956, it was evident that the situation at Lederle was untenable, and in any case I wanted more freedom to choose my own projects. Thus, when an offer came from The Wistar Institute to become director of that already venerable institute, I seized the chance. The move to Philadelphia in 1957 enabled me to deal more freely with other investigators, and to gather around me a new team of scientists interested in the problems of immunologic prevention of diseases, polio in particular.

Heart of Darkness

Reading Joseph Conrad's famous novel did not sufficiently prepare me for the adventure of polio vaccination in what was then called the Belgian Congo, today's Democratic Republic of the Congo. At the time, the mid-1950s, the Belgian health administration was working hard to deal with epidemic diseases. Polio was one such disease, and was running rampant in both the major cities and in "the bush," as they called it. The dogma was that polio did not affect Africans because all were immune, but in fact there were thousands of cases concentrated in Congolese infants, which went largely unnoticed. There were also cases concentrated in Belgian adults who came to the Congo, and this caused much fear and consternation among the colonials.

In 1955, I participated in a WHO rabies course given in Kenya, where I met a jolly, vital, gravelly voice Belgian virologist named Ghislain Courtois. Courtois was interested in the scientific use of native chimpanzees, and at first our conversations concerned the establishment of a camp for chimpanzees near Stanleyville, now Kisangani. The idea was to perform pathogenesis experiments in chimpanzees, and by 1956 the camp had been established and work had begun on polio and hepatitis B.

The camp soon became an obligatory stop for visiting dignitaries, including Prince Baudoin. When Baudoin offered his hands to one of the chimpanzees who had been tamed, the animal refused to shake hands. Courtois promptly explained to Baudoin that the chimpanzee was a Republican.

Nevertheless, when experiments began with wild poliovirus, Courtois implored me to vaccinate the staff. This small vaccine study went well, and in view of the known epidemiological situation regarding polio in the Congo, Courtois and I began to discuss a mass vaccination campaign. Permission from the authorities came in early 1957, and George Jervis, together with Agnes Flack, the Head physician at Clinton Farms Women's Prison, left immediately for the Congo. They made quite a sight in their safari gear obtained from I do not know what clothing store in Manhattan.

Jervis and Flack moved up the Ruzizi River, vaccinating Congolese who were called by drums – the bush telegraph. Almost 250,000 people, mostly from the Ruzizi Valley itself (present day Rwanda), received the type 1 CHAT strain in the world's first mass vaccination with oral polio vaccine [32], of whom a small subgroup of about 2,500 also received the type 3 Fox strain (Fig. 2). Serology revealed that 12% had been seronegative for type 1. During the campaign, there was an

Fig. 2 Dr Agnes Flack vaccinating in the Ruzizi Valley

interesting occurrence. An epidemic of type 1 polio was reported among infants living in a village of about 4,000 inhabitants. Accordingly, Jervis, Flack, and their team rushed to the village, where they vaccinated every living soul. The epidemic promptly ended. Thus, by late 1957, we had demonstrated that oral polio vaccine could be delivered in mass campaigns and that application of the vaccine could terminate epidemic disease.

Soon thereafter, the Belgian authorities asked us for help in managing the constant outbreaks of polio in large cities, Leopoldville [33] (now Kinshasa) and Stanleyville. Mass vaccinations of children in Leopoldville began in August 1958 and continued until April 1959 by which time 46,000 had been vaccinated with type 1 virus. Efficacy against paralytic poliomyelitis in the tropical setting was 60%, and we speculated about the importance of interference by other enteric viruses. Nevertheless, efficacy of oral polio vaccine to protect against paralytic poliomyelitis had been established for the first time [34].

In the late 1950s, in many more trials undertaken in the USA, several aspects of vaccination with oral polio vaccine were investigated. We showed [33] that whereas premature and full-term infants less than 5 days old were highly susceptible to intestinal infection with the attenuated virus, infants 5–60 days old were more resistant to infection. Both groups, however, developed protective antibodies [35]. In a routine vaccine of 850 children in Philadelphia with the oral polio vaccine containing three types of attenuated virus strains, the percentage of infants with antibodies against all three types increased from 15 to 85% after vaccination [36]. It was also possible to return to the group of originally vaccinated children in 1950 and to find that their high titer polio antibody levels persisted for a long time [37].

Mes semblables, mes frères

The development of polio vaccines was highly competitive, something like the field of acquired immune deficiency syndrome (AIDS) in recent years. Polio was the number one disease for which a vaccine was needed, and the National Foundation for Infantile Paralysis with its large financial resources was organized for the purpose of producing such a vaccine.

In March 1951, the National Foundation organized a meeting in Hershey Pennsylvania [38] at which I was asked to speak about rabies. Jonas Salk was there to present results of his immunization experiments in monkeys involving inactivated polio viruses. He carried this work into human experiments in 1952 [39], which culminated in the 1954 field trial conducted by Thomas Francis and the licensure of inactivated polio vaccine in 1955. Also present was Howard Howe of Johns Hopkins, who vaccinated humans with an inactivated vaccine during the summer and fall of 1951 [40]. Albert Sabin was another attendee, although at the time he had not entered the field of polio.

The rabies talk was duly presented, but after lunch, in an atmosphere of postprandial somnolence, I rose and said that I would also speak about some polio experiments. I recounted the results obtained at Letchworth, but there were few questions. Tom Francis thought at first that I was talking about monkeys. Only later in the corridors did I realize from the comments made to me that many in the audience, including Albert Sabin, were disturbed by the audacity of the experiments and critical of the idea. However, later Sabin, recognizing our pioneer work with oral polio vaccine [41], decided to initiate his own studies on oral polio vaccination. He started laboratory work in 1953, and by 1954 Sabin tested candidate strains in inmate volunteers at the Federal prison in Cincinnati, OH [42]. He showed that the strains were attenuated both by the oral and intramuscular routes, and by 1957 he had developed a trivalent vaccine which was tested by Soviet scientists in enormous field trials starting in December 1958 [43].

On the strength of those trials, a US Public Health Service approved, over my protest [44], the Sabin strains for the USA Licensure in 1961 to the exclusion of other strains. In retrospect, we think it is fair to say that both our type 1 strain and Sabin's were highly attenuated and immunogenic, that our type 2 strains were identical and therefore equally acceptable, and that our type 3 strain called WM-3 [45], more stable genetically and therefore less likely to revert to virulence, but that strain was only used in Poland, Switzerland, and Croatia. As Sabin always insisted that his strains could not be mixed with any other, no mixture of WM-3 with his type 1 and type 2 strains was ever tested.

The passage of time has removed all rancor, however, Sabin and I exchanged reagents freely during the early days when we were both developing attenuated polio strains for practical use, and if we fell out during the later 1950s and 1960s owing to the pressures of competition, we resumed our friendship later and I mourned his passing both privately and in print [46].

L'envoi

The history of oral polio vaccine started in the mid-twentieth century and today, because of worldwide use of the oral polio vaccine, the number of paralytic cases reported in has fallen to 0 for the entire Western Hemisphere. Moreover, paralytic polio is being eradicated from the world. Although 50 years have involved all the elements of human drama, including inspiration, daring, disappointment, argument, intrigue, and above all the hard work of many, it has been a wonderful adventure.

References

1. Lansteiner K, Popper E. Ubertragung der Poliomyelitis acuta auf Affen. *Z Immun Forsch* 1909;2:377
2. Burnet FM, Macnamara J. Immunologic differences between strains of poliomyelitis virus. *Br J Exp Pathol* 1931;12:57–61

3. The Committee on Typing of the National Foundation for Infantile Paralysis: immunologic classification of poliomyelitis viruses. I. A cooperative program for typing of one hundred strains. *Am J Hyg* 1951;54:191–274
4. Brodie M, Park WH. Active immunization against poliomyelitis. *NY State J Med* 1935;35:815
5. Leake JP. Poliomyelitis following vaccination against this disease. *Am J Med Assoc* 1935;105:2152
6. Howe HA, Bodian D. Poliomyelitis in the chimpanzee: a clinical pathological study. *Bull Johns Hopk Hosp* 1941;69:149–81
7. Armstrong C. Successful transfer of the Lansing strain of poliomyelitis virus from the cotton rat to the white mouse. *Publ Health Rep* 1939;54:2302–5
8. Koprowski H, Norton TW, McDermott W. Isolation of poliomyelitis virus from human serum by direct inocu-lation into a laboratory mouse. *Publ Health Rep* 1947;62:1467–76
9. The Society of American Bacteriologists. Proceedings of the 51st General Meeting (Sponsored by the Illinois Branch). *Bacteriological Proceedings,* Chicago, Illinois, 1951;92–3
10. Koprowski H. Immunization of man with living poliomyelitis virus. *WHO Monogr Ser* 1955;26:335–56
11. Koprowski H, Jervis GA, Norton TW. Oral administration of poliomyelitis virus to man and ape – a comparative study. *Proc Natl Acad Sci USA* 1954;40:36–9
12. Koprowski H, Jervis GA, Norton TW. Oral administration of a rodent-adapted strain of poliomyelitis virus to chimpanzees. *Arch Gesamte Virusforsch* 1954;5:413–24
13. Koprowski H, Jervis GA, Norton TW. Immune responses in human volunteers upon oral administration of a rodent-adapted strain of poliomyelitis virus. *Am J Hyg* 1952;55:108–26
14. Koprowski H, Jervis GA, Norton TW, Pfeister K. Adaptation of type I strain of poliomyelitis virus to mice and cotton rats. *Proc Soc Exp Biol Med* 1954;86:238–449
15. Koprowski H. Vaccination with modified active viruses. *In: Poliomyelitis Papers and discussions presented at the Fourth International Poliomyelitis Conf (Sept 1957),* Philadelphia, PA: JB Lippincott Co, 1958;12–23
16. Koprowski H, Jervis GA, Norton TW. Administration of an attenuated type I poliomyelitis virus to human subjects. *Proc Soc Exp Biol Med* 1954;86:244–7
17. Koprowski H. Immunization of man against poliomyelitis with attenuated preparations of living virus. *Ann NY Acad Sci* 1955;61(4):1039–49
18. Koprowski H, Norton TW, Jervis GA, Stokes J Jr. McGee EL, Nelson DJ. Immunization of children by the feeding of living attenuated type I and type II poliomyelitis virus and the intramuscular injection of immune serum globulin. *Am J Med Sci* 1956;232:378–88
19. Koprowski H, Norton TW, Hummeler K, Stokes J Jr., Hunt AD Jr., Flack A, Jervis GA. Immunization of infants with living attenuated poliomyelitis virus. *JAMA* 1956;162:1281–8
20. Plotkin SA, Koprowski H, Stokes J Jr. Clinical trials in infants of orally administered attenuated poliomyelitis viruses. *Pediatrics* 1959;23:1041–61
21. Plotkin SA, Koprowski H, Richardson SN, Stokes J Jr. Vaccination of families against poliomyelitis by feeding and by contact spread of living attenuated virus including studies of virus properties after human intestinal passage. *Acta Paediatrica* 1960;49:551–71
22. Dane DS, Dick GWA, Connolly JH, Fisher OD, McKeown F. Vaccination against poliomyelitis with live virus vaccines I. A trial of TN type II vaccine. *Br Med J* 1957;1:59–64
23. Buser F, Schar M. Der gegenwartige Stand der Polimyelitisschutzimpfung mit lebenden avirulenten Polioviren. *Schweizz Med Wochenschr* 1958;88:1282–91
24. Schar M, Buser F. Immunisierung mit lebenden avirulenten Poliomyelitisviren. *Schweizz Med Wochenschr* 1960 90:618–23
25. Ikic D, Jacikic B, Lulic V, Cuk D, Juzbasic M, Manhalter T, Sindik A. Preliminary report on vaccination in Croatia against poliomyelitis with type 1 (CHAT) and type 3 (W-Fox) attenuated polioviruses of Koprowski. *Bull WHO* 1963;28:217–23
26. Ikic D, Jacikic B, Hrabar A. Epidemiological data on efficacy of live, oral poliovaccine Koprowski. In: *Proceedings – Symposium on the Characterization and Uses of Human Diploid Cells Strains.* 1963;541–4

27. Ikic D. Recent information on poliomyelitis vaccine, live, oral, prepared in human diploid cell strain system. In: *Proc 9th Int Cong Permanent Sect Microbiol Standard Int Ass Microbiol Soc, Lisbon: Prog Immunobiol Stand.* 1965;2:305–10
28. Plotkin SA. Recent results of mass immunization against poliomyelitis with Koprowski strains of attenuate live poliovirus. *Am J Public Health.* 1962;52:946–60
29. Przesmycki F. Paraliz Dzieciecy W Likwidacji. Na Terenic Polski. *Zycle Warszawy* 1964;236
30. Koprowski H, Jervis GA, Norton TW. Oral administration of a rodent-adapted strain of poliomyelitis virus to chimpanzees. *Archiv fur Virusforsch* 1954;5:413–24
31. Anonymous. Oral Poliomyelitis vaccine. *Lancet* 1962;2:647–8
32. Courtois G, Flack A, Jervis GA, Koprowski H, Ninane G. Preliminary report on mass vaccination of man with live attenuated poliomyelitis virus in the Belgian Congo and Ruanda-Urundi. *Br Med J* 1958;ii:187–90
33. Lebrun A, Cerf J, Gelfand HM, Courtois G, Koprowski H. Preliminary report on mass vaccination with live atte-nuated poliomyelitis virus in Leopoldville, Belgian Congo. *First Int Conf on Live Poliovirus Vaccines,* Washington DC, 1959;410–8
34. Plotkin SA, Koprowski H. Epidemiological studies of the safety and efficacy of vaccination with the CHAT strain of attenuated poliovirus in Leopoldville, Belgian Congo. *First Int Conf on Live Poliovirus Vaccines,* Washington DC, 1959;419–36
35. Pagano JS, Plotkin SA, Koprowski H. Variations in the responses on infants to living attenuated poliovirus vaccine. *N Engl J Med* 1961;264:155–63
36. Pagano JS, Plotkin SA, Janowski CC, Richardson SM, Koprowski H. Routine immunization with orally admi-nistered attenuated poliovirus: a study of 850 children in an American city. *JAMA* 1960;173:1883–9
37. Koprowski H, Jervis GA, Norton T. Persistence of neutralizing antibodies in human subjects three years after oral administration of a rodent-adapted strain of poliomyelitis virus. *Pediatrics* 1954;13;203–5
38. *Proceedings of the Round Table Conference on Immunization in Poliomyelitis.* The National Foundation for Infantile Paralysis, New York, New York, 1951
39. Salk JE. Studies in human subjects on active immunization against poliomyelitis. 1. A preliminary report of experiments in progress. *JAMA* 1953;151:1081–98
40. Howe H. Antibody response of chimpanzees and human beings to formalin inactivated trivalent poliomyelitis vaccine. *Am J Hyg* 1952;56:265–86
41. Sabin AB. Present status of attenuated live virus poliomyelitis vaccine. *Bull NY Acad Med* 1957;33:37–9
42. Sabin AB. Behavior of chimpanzee avirulent poliomyelitis viruses in experimentally infected human volunteers. *Am J Med Sci* 1955;230:1–8
43. Chumakov MP, Voroshilova MK, Vasilieva KA et al. Preliminary report of mass oral immunization of popula-tion against poliomyelitis with live virus vaccine from AB Sabin's attenuated strains. *First International Conference on Live Poliovirus Vaccines*, 1959;517–621
44. Koprowski H. Live virus vaccines against poliomyelitis, *JAMA* 1961;175:162–3
45. Plotkin SA, Norton TW, Cohen BJ, Koprowski H. A type 3 attenuated poliovirus genetically stable after human intestinal passage. *Proc Soc Exp Biol Med* 1962;829–34
46. Koprowski H, Albert B Sabin. *Nature* 1993;362:49

Oral Polio Vaccine and the Results of Its Use

Joseph Melnick[†] in collaboration with Stanley Plotkin

Reports on clinical and epidemiological aspects of poliomyelitis came in the late nineteenth and early twentieth centuries from von Heine, Medin, and Wickman in Europe, and Caverly in the USA, but basic studies on poliovirus only began in 1908 when Landsteiner, Popper, and Levaditi, again in Europe, transmitted the disease to monkeys, inducing typical histopathology lesions in the animals spinal cords. Shortly thereafter, Flexner and Wollstein in the USA achieved monkey-to-monkey passage.

By the late 1930s, investigators had begun turning toward the possibility of an oral alimentary route of infection [1]. Investigators at the Pasteur Institute in France had successfully infected *cynomolgus* monkeys by the oral route. Impetus was added when Howe and Bodian came across a report by Muller describing spontaneously acquired polio in chimpanzees in a children's zoo in Cologne, Germany. They followed up this clue with a series of experiments establishing that chimpanzees were far more susceptible to infection by the oral route than other experimental primates. Evidence from human cases also began to accumulate as – some 25 years after the almost-ignored reports of Kling and his associates in Sweden – investigators again

[†]Deceased

S.A. Plotkin (ed.), *History of Vaccine Development*,
DOI 10.1007/978-1-4419-1339-5_19, © Springer Science+Business Media, LLC 2011

began looking for poliovirus in the alimentary tract. At this point, it was not understood how poliomyelitis was spread. Although Wickman had postulated spread through inapparently infected persons, the circumstances under which the paralytic disease occurred, or where the virus was located in the infected or diseased individual, were unknown. The manner of transmission was puzzling: cases in epidemics seemed to move out from a focus, but in erratic patterns, for those afflicted usually had no direct contact with an earlier case. Even within a family where paralytic polio occurred, the disease was hardly ever seen in more than a single member.

At Yale University (Connecticut, USA), the search for viruses included specimens other than the spinal cord, and in 1938 John Paul reported to the American Epidemiological Society on some of the findings: not only was virus found in large quantities in feces, but it was also recovered repeatedly over a period of weeks, from both patients and healthy carriers. This report was met with skepticism by many in the audience, but it was soon confirmed and extended. The shift to the concept of poliomyelitis as an enteric infection was underway. Wickman's observations on the spread of virus by persons with inapparent infections could now be confirmed by laboratory documentation. The new concept also had direct implications for measures to control fecal contamination in hospitals and in households. It also had implications for immunization by means of a vaccine.

Still another series of questions was brought into focus: if polio was an enteric infection, could it be isolated from urban sewage? One of my first tasks upon joining the Yale Poliomyelitis Study Unit in 1940 was to devise methods for testing effluents from one other large sewage plants of New York City. During periods when paralytic polio was prevalent, I found polio virus to be present in New York sewage in huge quantities [2]. Knowing the quantity of virus in sewage came, I could establish a ratio of inapparent infection to paralytic cases. This ratio for the wild virus turned out to be 100 or more inapparent or subclinical infections for each paralytic case.

Despite the increasing knowledge about polio that was gained through investigations that had to depend on large and expensive laboratory animals (monkeys and chimpanzees), it was clear that simpler systems were much needed. Following original efforts by Levaditi in 1913, there had been two decades of repeated attempts to grow polioviruses in various kinds of culture substrates and conditions, with little success. Sabin and Olitsky in 1936 reported a study done with a strain of poliovirus that had gone through 20 years of brain-to-brain passage from monkey to monkey at the Rockefeller Institute. When these investigators attempted to grow this highly neurotropic strain in human embryonic nervous tissue or in non-nervous tissue from the viscera of the same embryos, the virus only grew in the nervous tissue. Thus, it became accepted doctrine that poliovirus could be grown only in human nervous system tissue. Such tissue is difficult to obtain, and furthermore its use as a substrate for growing antigenic material for a possible vaccine was precluded because of its known potential to cause brain damage when injected. Moreover, proof that poliovirus had indeed multiplied in these cultures still depended upon inoculating monkeys with tissue culture fluid and duplicating the disease from them. Another 14 years elapsed before the landmark report by Enders, Weller, and Robbins who used different strains of poliovirus and showed that poliovirus could indeed multiply in a variety of tissue cultures, particularly in cells that grow out from tissue fragments [3]. Moreover, the virus could be quantitatively assayed by observing its cytopathic effect in vitro.

About this time, the chimpanzee became more widely used as an experimental animal. Many new avenues of investigation were opened by the important findings that chimpanzees could readily infected by oral administration of poliovirus and become transient intestinal carrier of the virus, but seldom developed clinical signs [4]. These primates indeed show many parallels to the infection of human beings. Some of the unanswered questions that could now be pursued included the transmission of poliovirus in nature, the patterns of pathogenesis, and the site and course of the infection that usually results in the development of antibody in the human being. Thus, we could approach questions of how efforts toward the development of vaccines might best proceed.

In the 1930s, Burnet, looking at the studies of neutralization with epidemic poliovirus strains, and at the fact that antibody was already present in patients' blood at the onset of central nervous system (CNS)-related symptoms, suggested that if antibodies were already present at the "beginning" of the illness, then such antibodies were not able to protect against the disease. However, Herdi von Magnus and I, in studies with rhesus, cynomolgus and vervet monkeys, induced paralytic disease by feeding them virus. Paralysis occurred within 1–2 weeks but the monkey already had neutralizing antibodies in the blood by the first day of paralysis [5]. These findings reopened the question of whether, in human disease, antibodies found this early were indeed irrelevant to current disease – as Burnet thought – or were a response to a current pathogenic process that included growth of the virus in non-nervous tissue. A number of monkeys in our study had remained asymptomatic, but they also developed antibody; yet histologic examination showed them to be free of CNS lesions. Thus, not only the early antibody production in the paralyzed monkeys, but also in the development of antibodies in those without CNS lesions suggested that antibody development was an early response to infection of tissue outside the CNS, and that such antibody might actually protect the CNS from virus invasion.

In 1952, Dorothy Horstmann at Yale and David Bodian at Johns Hopkins reported the isolations of virus from the blood of *cynomolgus* monkeys a few days after oral administration of virus, but several days before the appearance of symptoms. When these findings were followed by observations of a similar pattern in humans, hope for immunization increased [6]. If viremia was the route by which the virus entered the CNS, then it might be blocked by circulating antibody and thus be prevented from invading the human CNS. One result of these developments was a trial to determine whether gammaglobulin could be effective in preventing paralytic polio. A large and well-conducted field trial under Hammon's direction in 1952 showed that passive antibody could indeed protect from paralytic polio, although only for about 2–5 weeks. This trial also showed that the quantity of antibody need not be large, and it favored the hope that an effective method of active immunization could be developed.

Let us return to epidemiological aspects. We had investigated a polio outbreak in North Carolina in 1944, but returned in 1948 when a much larger one occurred. Ultimately in 1948, there were more the 2,200 cases in the state, an attack rate of about 65 cases per 100,000 population. At the beginning of the summer "polio season," when we learned that four cases had been reported, we had thought that an epidemic might be in the making, and then in early June we obtained blood samples from healthy children in Winston-Salem [7]. These "pre-epidemic" specimens were then matched by second bleeding obtained in November for testing as "postepidemic" specimens. This was the first time that pre- and postepidemic sera taken from 248 normal children had been studied for polio antibodies – a measure of infection rather that disease caused by poliovirus. Nada Ledinko and I found that the antibody patterns in the spring of 1948 were similar for all three antigenic types of poliovirus; they indicated a low incidence of infection during the previous 4-year period. During the 1948 epidemic, antibodies against type 1 and type 2 were acquired; the antibody conversions indicated infections rates of 23% with type 1 and 17% with type 2. No child in the study produced type 3 antibodies that summer.

The age-specific rate of inapparent infection by wild type 1 was compared with age-specific attack rates for poliomyelitis in the epidemic. The number of sunclinical type I infections per case was found to be about 100, similar to the previous findings in the New York Sewage study.

The first efforts to develop a polio vaccine were rudimentary and ended in disaster. In 1936, two vaccines were prepared from spinal cords of monkeys. One was inactivated by formalin that was prepared by Brodie and Park, the other an "attenuated" vaccine made by Kolmer, who treated the spinal cords with ricinoleate [8, 9]. Trials of both vaccines were followed by cases of polio that probably resulted from residual live virus in the experimental vaccines [10].

The advent of cell culture technology allowed more precise attenuation of the polioviruses. The first successful attempts at vaccine development were by Hilary Koprowski in 1950 [11] and are described by Koprowski elsewhere in this book. Herald Cox at Lederle Labs also developed live virus strains, some related to strains of Koprowski, who had originally been at Lederle Labs [12]. However, the Cox strains proved to be too neurovirulent, and the Koprowski strains, although used for a time in Poland and Africa, were eventually discarded in favor of the strains developed by Albert Sabin.

Albert Sabin, who was born in Russia and who came with his parents to America as a boy, was already an accomplished virologist when he started work on polio in 1953, after hearing of Koprowski's work. He chose to grow polioviruses in cynomolgus monkey kidney culture, enabled by the prior work of the Enders group. For selection of the most attenuated clones of virus the plaquing technique developed by Dulbecco and Vogt [13] was critical. The measure of attenuation that Sabin used was intrathalamic and intraspinal injection of the virus variants in those monkeys. The objective of his research was to obtain strains that were immunogenic but that gave little viremia and low neurovirulence. Sabin also used chimpanzees to study the excretion of attenuated strains after oral administration, learning that the intestinal tract selected neurovirulent viruses. Between 1954 and 1956 Sabin performed studies involving 9,000 monkeys, 150 chimpanzees and 133 volunteers. The latter were prisoners in the Federal Penitentiary in Chillocothe, OH [14–16].

At that point, Merck produced large amounts of the candidate attenuated strains of polioviruses 1, 2 and 3 for further clinical trials. The first trial done by Sabin using that material was in his own triply negative wife and children. Although immunogenic and apparently safe, again it was shown that the viruses reverted to higher neurovirulence in the intestine. A crucial meeting of the WHO Expert Committee on Polio was held in July 1957, at which expanded clinical trials were urged. Accordingly, studies were conducted in the USA, Mexico and the Netherlands between 1957 and 1959. However, the truly crucial trials were undertaken in the former Soviet Union by Smorodintsev in St. Petersburg and Chumakov in Moscow in 1959, involving 15 millions of vaccinees [17–19]. The results they reported were excellent in terms of immunogenicity and effectiveness, as well as safety. Although there is some doubt as to the comprehensiveness of their safety observations, an inspection visit by the American Dr. Dorothy Horstmann led to a very positive report.

Meetings held at the Pan American Health Organization in Washington, DC in 1959 and 1960 convinced the majority of experts that the Sabin strains should be adopted as the prime weapon to control poliomyelitis. This decision was based on comparative neurovirulence studies that favored the Sabin strains over those of Koprowski and Cox [20, 21] as well as the large numbers of persons already vaccinated with the Sabin strains [22]. Moreover, Salk's injectable killed vaccine had been licensed in 1955 in the USA and Sabin's oral attenuated vaccine was licensed in 1961. How have the oral and injectable vaccines altered the situation? The world has two good vaccines against polio. Each has its advantages and disadvantages [23, 24].

Inactivated poliovirus vaccine (IPV), if properly prepared and administered, can confer humoral immunity if sufficient doses of the new vaccine of enhanced potency are given. IPV can be incorporated into a regular pediatric immunization schedule (along with other injectable vaccines, such as diphtheria–pertussis–tetanus). Because living virus is not present, the use of IPV excludes the possibility that the vaccine virus can revert toward virulence [25]. Also, for this reason it can be given to immunodeficient of immunosuppressed individuals and to the household contacts of these immunocompromised persons. Among the disadvantages of IPV, at least as originally prepared, was its low potency. The need for repeated booster injections added to higher costs, and in developing countries it also presented logistic problems of injecting several doses of vaccine into large numbers of infants and preschool children. The immunity conferred by IPV impedes pharyngeal and fecal shedding of virus to some extent, but it does not provide a high degree of intestinal resistance. When exposed to wild poliovirus, IPV vaccinees became infected, excreted the wild virus and thus because a source of infection to others. The new, more potent IPV prepared by the van Wezel procedure [26] is proving to be more effective with fewer doses than the original Salk vaccine. I believe the IPV has a role in disease eradication, especially when used in combination with the live oral vaccine.

Live trivalent poliovirus vaccine (OPV) has been widely used [23, 27]. By the oral route, it is able to induce not only serum antibodies but also intestinal resistance, and it rapidly induces an immunity similar to that induced by natural infection. OPV is also a good deal cheaper that IPV.

All living creatures undergo some degree of mutation, and live polioviruses are no exception [28]. The mutations that occur during replication of OPV have produced in very rare instances (about 1–500,000 in children receiving the first dose) viral progeny with neurovirulence sufficiently increased to cause paralysis in vaccine recipients and their susceptible contacts. From the study of poliovirus, we have learned a great deal about viral genetics [29–33]. Poliovirus replication is accompanied by an error frequency of about $10^{-3.5}$ per nucleotide incorporated into the nascent poliovirus RNA genome, which consists of about 7,400 nucleotides. This error rate suggests that most newly synthesized poliovirus RNA molecule differ in at least one nucleotide from the sequence of the parental template RNA. Thus, every batch of poliovirus is a population of viral genomic sequences that compete for dominance in the mixed population of sequences. Any change in growth conditions will alter the relative replicative efficiency of the competing viral sequences. This concept of a dynamic viral population explains the rapid changes of viral phenotype that can occur during manufacture or after application to a vaccinated person. A sensitive measure of reversions from uracil to cytosine at nucleotide 472 of type 3 OPV can be used to predict the results of the expensive and somewhat variable monkey neurovirulence test, the test being used by most manufacturers today.

Other types have similar reversions [34]. Thus, for type 1, a G to A reversion occurs at nucleotide position 480. Lots of type 1 in which the level of 480A was 2.7% of revertants have passed the monkey neurovirulence test. Thus, the in vitro genetic test which detects 480A is more sensitive than the monkey test for neurovirulence.

The insertion into mice of the human gene responsible for producing the cell receptor for poliovirus has yielded transgenic animals suitable for testing attenuated vaccine lots for neurovirulence [35, 36]. Such animals are replacing monkeys for safety testing of vaccine lots.

The proven risks of paralytic polio associated with OPV are exceedingly small. Three sequential 5-year studies of polio cases have been conducted among 12–15 nations, under the auspices of the World Health Organization (WHO). In these and other studies, live poliovirus vaccine have been repeatedly judged to be an extraordinarily safe vaccine [37].

The Global Advisory Group of WHO's Expanded Programme on Immunization (EPI) has concluded [38] that immunization of the newborn with OPV is a safe and effective means of protection, and that OPV may be administered simultaneously with BCG vaccine. Although the serological response to OPV in the first week of life is less than that observed after immunization of older infants, 70% or more of neonates benefit by developing local immunity in the intestinal tract. In addition, 30–50% of the infants develop serum antibodies to one or more poliovirus types. Many of the remaining infants have been immunologically primed, and they promptly respond to additional doses later in life. For those infants whose only encounter with preventive services is at the time of birth, this single dose of vaccine will offer some protection, and they will be less likely to be a source of transmission of wild polioviruses during infancy and childhood.

OPV had been included in regular childhood immunization programs throughout the world [38]. This has led to a gradual increase in vaccine coverage to 80% in the first 2 years of life and as associated gradual decrease in polio incidence. Maintaining a routine immunization schedule that virtually reaches all of the target population requires the year-round maintenance of a supply of viable vaccine constantly at the point of contact. This can be difficult in warm climates with limited cold-chain facilities, particularly if the vaccine has not been treated with a stabilizer equal to, or superior to, molar $MgCl_2$ stabilizer [39]. Such a stabilizer is heavy water [40], especially when used in conjunction with the $MgCl_2$ stabilizer.

Sabin had advocated that paralytic poliomyelitis in tropical countries might best be eliminated by mass administration of OPV repeated annually in all age groups in which cases are occurring. In addition to providing protection to the vaccinees, the program results in interrupting the circulation of wild virus [23]. The effectiveness of OPV has been particularly striking in the Western Hemisphere. No polio case has been reported there since 1993, but also there has been a precipitous decrease in circulation of wild polioviruses. Mopping-up operation consisting of intensive use of OPV in a few localities has been a factor in stopping the circulation of wild poliovirus in the Western Hemisphere [41].

In some countries, notably Denmark [42] and Israel [43, 44], it has been found advantageous to establish immunization schedules using both killed and live poliovirus vaccines. This strategy has been successfully carried out in some high-risk areas of the Middle East where there is very early and repeated exposure of infants to challenge by importations of wild virulent viruses. In a study conducted in one such high-risk area, IPV alone was inadequate [45]. This study involved an outbreak in 1988 among young adults in Israel; it taught us (1) that IPV alone does not interrupt the circulation of wild virus, which can single out susceptible contacts of all ages and (2) that OPV alone, when administered only in infancy, is not completely effective for life. Furthermore, it was already known that other enteroviruses may interfere with multiplication of OPV.

A combined schedule of both OPV and IPV offers substantial benefit, with the optimal times of vaccination yet to be determined for different localities. For ease of administration, giving parenteral IPV and oral OPV simultaneously or sequentially offers advantages. This procedure has been regularly used, starting in 1978 in the West Bank and Gaza where cases have been occurring, particularly in infants, despite extensive campaigns of immunization with OPV [43, 44]. The combined schedule included administration of both vaccines in the first year of life. The rationale for the combined schedule was that under conditions of regular and heavy importation of virus, resulting in frequent challenge from virulent wild polioviruses early in infancy, features of both types of vaccine are needed. OPV acts by inducing protective immunity both in the form of serum antibodies and in the form of intestinal immunity, thus breaking the chain of circulation of wild virus. Furthermore, the immunity is long-lasting, like that which follows the natural infection. IPV, on the other hand, provides an immediate immunogenic stimulus that is not subject to interfering factors. By administering both vaccines in a combined schedule, immediate protection can be provided in the critical first weeks of months of life and long-lasting protection – humoral and intestinal – also is provided. The result is that there have been no cases for years in the areas where the combined vaccines are used.

The schedule proposed for heavily contaminated areas is one which is compatible with the current WHO recommendations. This schedule would (1) introduce IPV into the current OPV immunization program, so that in the first year of life both IPV and OPV are given to all infants and (2) in subsequent years continue with the current EPI program which uses OPV alone for polio immunization.

Advances in our understanding of the molecular biology of poliovirus may lead to the preparation of live vaccine candidates of potentially greater genetic stability [46]. However, it will be difficult to field-test such new vaccine candidates, as it will be necessary to prove that the vaccines produce less than one vaccine-associated case per million susceptible recipients. The global application of the present OPV is fast achieving an interruption of the circulation of wild poliovirus, closing the window during which any newly developed vaccine strains can be properly field-tested.

Eradication of polio is part of the EPI goal of universal immunization of children. There have been significant increases in recent years with regard to immunization with OPV – from the situation at the beginning of 1977, when 5% of all children in the world received the required three doses of OPV in the first year of life, to that in 1995, when this percentage had increased to 80%. Based on an expected worldwide polio incidence of 5 per 1,000 infants (prior to the availability of vaccines), the global OPV program is currently preventing hundreds of thousand cases of paralytic disease per year. As mentioned earlier, the disease seems to have been eliminated from the Western Hemisphere, and eradication of poliomyelitis from the world is now an achievable goal. Thus, a new perspective is in view for polio. The goal is not simply the control of epidemics caused by the virus, but the very eradication of the disease itself.

This leads to a new question soon to be considered, should we simply *stop* vaccinating for a disease that no longer exists in the world? In the case of smallpox, the answer was clear. No harm was done to those vaccinated and to their healthy contacts. With OPV, one in every 500,000 children receiving their first dose will develop paralysis. The recovered virus from such cases is virulent and has produced devastating paralysis in some healthy contacts. Another factor is the long period during which vaccinees are healthy carriers and remain contagious. It is essential that the contacts of vaccinated children be immunized together with or prior to immunization of the vaccinated child.

How can this be done with no risk? I believe there is an important place here for the injectable vaccine, IPV. The virus is inactivated so healthy carriers cannot be the result. There will be no contact cases and no vaccine-associated cases. The virus being excreted by those who are OPV immunized will fade away. To lessen a potential risk involved in manufacture, the virulent strains in presently available IPV can be replaced by Sabin's attenuated strains of OPV. At least one manufacturer is already testing such an IPV. After a year or two of IPV use alone, there will not be any poliovirus about, and the world can say good-bye to an unwanted virus.

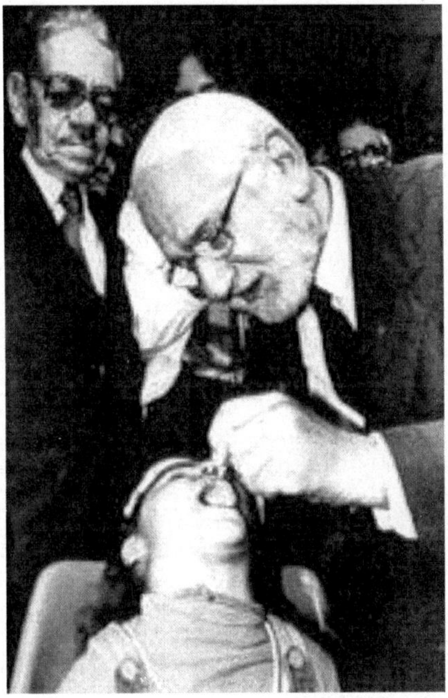

Albert Sabin giving OPV to a child

References

1. Paul JR. *A History of Poliomyelitis.* New Haven, CT: Yale University Press, 1971
2. Melnick JL. Poliomyelitis virus in urban sewage in epidemic and nonepidemic times. *Am J Hyg* 1947;45:240–53
3. Enders JF, Weller TH, Robbins FC. Cultivation of the Lansing strain of poliomyelitis virus in cultures of various human embryonic tissues. *Science* 1949;109:85–87
4. Melnick JL, Horstmann DM. Active immunity to poliomyelitis in chimpanzees following subclinical infection. *J Exp Med* 1947;85:287–303
5. von Magnus H, Melnick JL. Antibody response in monkeys following oral administration of poliomyelitis virus. *J Immunol* 1948;60:583–596
6. Bodian D, Horstman DM. Polioviruses. In: Horsfall FL, Tamm I eds. *Viral and Rickettsial Infection of Man,* 4th ed, Philadelphia: Lippincott, 1965;430–473
7. Melnick JL, Ledinko N. Development of neutralizing antibodies against the three types of poliomyelitis virus during an epidemic period. The ratio of inapparent infection to clinical poliomyelitis. *Am J Hyg* 1953;58:207–222
8. Brodie M, Park WH. Active immunization against poliomyelitis. *Am J Public Health* 1936; 26:119.
9. Kolmer JA. Vaccination against acute anterior poliomyelitis. *Am J Public Health* 1936; 26:126.

10. Robbins FC. The history of polio. Chapter 2 in Vaccines, 4th ed. Editors: Plotkin SA and Orenstein WA, Offit, PA. Saunders, Philadelphia, 2004:17–30
11. Koprowski HL, Jervis GA, Norton TW. Immune responses in human volunteers upon oral administration of a rodent-adapted strain of poliomyelitis. *Am J Hyg* 1952;55:108–126
12. Cabasso VJ, Jervis GA, Moyer AW, Roca-Garcia M, Orsi EV, Cox HR. Cumulative testing experience with consecutive lots of oral poliomyelitis vaccine. In: *Live Polio Virus Vaccine.* PAHO: Washinton DC, 1959:102–134
13. Dulbecco R, Vogt M. Biological properties of poliomyelitis viruses as studied by the plaque technique. *Ann NY Acad Sci* 1996;61:790–800
14. Sabin AB, Hennessen WA, Winsser J. Studies on variants of poliomyelitis virus: experimental segregation and properties of avirulent variants of three immunologic types. *J Exp Med* 1954;99:551–576
15. Sabin AB. Immunization of chimpanzees and human beings with avirulent strains of poliomyelitis virus. *Ann NY Acad Sci* 1955;61:1050–1056
16. Sabin AB. Behavior of chimpanzees avirulent poliomyelitis viruses in experimentally infected human volunteers. *Am J Med Sci* 1962;230:1–8
17. Sabin AB. Recent studies and field tests with a live attenuated poliovirus vaccine. *First Inst Conf on Live Poliovirus Vaccines,* Washington, DC, 1959:14–38
18. Chumakov MP, Voroshilova MK, Vasileva KA et al. Preliminary report on mass oral immunization of population against poliomyelitis with live virus vaccine from Sabin's attenuated strains. *First Int. Conf. on Live Poliovirus Vaccines,* Washington, DC, 1959:517–621
19. Sabin AB. Status of field trials with an orally administered, live attenuated poliovirus vaccine. *J Am Med Assoc.* 1959;17:863–868
20. Melnick JL, Brennan JC. Monkey neurovirulence of attenuated poliovirus vaccines being used in field trials. In: *Live Polio Vaccine.* PAHO: Washinton DC, 1959:65–101
21. Murray R, Kirschstein G, Van Hoosier G, Baron S. Comparative virulence for rhesus monkeys of poliovirus strains used for oral administration. In: *Live Polio Vaccine.* PAHO: Washinton DC, 1959:39–64
22. Sabin AB. Recent studies and field tests with live attenuated poliovirus vaccine. In: *Live Polio Vaccine.* PAHO: Washinton DC, 1959:33
23. Sabin AB. Paralytic poliomyelitis: Old dogmas and new perspectives *Rev Infect Dis* 1981;3:543–564
24. Melnick JL. Live attenuated and disadvantages of killed and live poliomyelitis vaccines. *Bull WHO* 1978;56:21–38
25. Salk J, Drucker J. Noninfectious poliovirus vaccine. In: Plotkin SA, Mortimer EA Jr., eds. *Vaccines,* 2nd ed, Philadelphia: WB Saunders, 1993;14–38
26. van Wezel AL, van Steenis G, van der Marel P, Osterhaus ADME. Inactivated poliovirus vaccine: current production methods and new developments. *Rev Infect Dis* 1984; 6 (suppl 2): S335–S340
27. Melnick JL. Live attenuated poliovaccines. In: Plotkin SA, Mortimer EA Jr., eds. *Vaccines,* 2nd ed, Philadelphia: WB Saunders, 1993;155–204
28. Melnick JL. Population genetics applied to live poliovirus vaccine. *Am J Public Health* 1962;52:472–483
29. Melnick JL. Variation in poliomyelitis virus on serial passage through tissue culture. *Cold Spring Harbor Symp Quant Biol* 1953;18:178–179
30. Nomoto A, Wimmer E. Genetic studies of the antigenicity and the attenuation phenotype of poliovirus. In: Russell WC, Almond JW, eds. *Molecular Basis of Virus Disease* 1987;107–134
31. Benyesh-Melnick M, Melnick JL, Rawls WE, Wimberly I, Barrera Oro J, Ben-Porath E, Rennick V. Studies of the immunogenicity, commincability and genetic stability of oral poliovirus administered during the winter. *Am J Epidemiol* 1967;86:112–136
32. Rico-Hesse R, Pallansch MA, Nottay BK, Kew OM. Geographic distribution of wild poliovirus type 1 genotypes. *Virology* 1992;160:311–322

33. Rezapkin GV, Chumakov KM, Zhengbin LU, Yuxin R, Dragunsky EM, Levenbrook IS. Microevolution of Sabin 1 in vitro and genetic stability of oral poliovirus vaccine. *Virology* 1994;202:370–378
34. McGoldrick A, Macadam AJ, Dunn G et al. Role of mutations G-480 and C-6203 in the Attenuation Phenotype of Sabin Type 1 Poliovirus. *J Virol* 1995;69:7601–7605
35. Ren R, Racaniello VR. Human poliovirus receptor gene expression and poliovirus tissue tropism in transgenic mice. *J Virol* 1992;66:296–304
36. Abe S, Yoshihiro O, Koike S, Kurata T, Horie H, Nomura T, Hashizume S, Nomoto A. Neurovirulence test for oral live poliovaccines using poliovirus-sensitive transgenic mice. *Virology* 1995;206:1075–1083
37. Assaad FA. The relation between acute persisting spinal paralysis and poliomyelitis vaccine (oral): results of a WHO inquiry. *Bull WHO* 1976;53:319–331
38. World Health Organization. *Poliomyelitis in 1985–1989. Weekly Epidemiological Record* 1987;62:273–280, 1989;64:273–279, 1991;66:49–53, 70–72
39. Melnick JL, Ashkenazisa A, Midulla VC, Berstein A. Immunogenic potency of $MgCl_2$ –stabilized oral poliovaccine. *JAMA* 1963;185:406–408
40. Crainic R. *Stabilization of Vaccine by Dueterion Ovide*. WHO Task Force on Vaccines, 1995
41. de Quadros CA, Henderson DA. Disease eradication and control in the Americas. *Biologicals* 1993;21:335
42. Von Magnus H, Peterson I. Vaccination with inactivated poliovirus vaccine and oral poliovirus vaccine in Denmark. *Rev Infect Dis* 1984;6:S471–S474
43. Lasch EE, Abed Y, Abdulla K et al. Successful results of a program combining live and inactivated poliovirus vaccines to control poliomyelitis in Gaza. *Rev Infect Dis* 1984;6: S467–S470
44. Goldblum N, Gerichter CB, Tulchinsky TH, Melnick JL. Poliomyelitis control in Israel, the West Bank and Gaza Strip: changing strategies with the goal of eradication in an endemic area. *Bull WHO* 1994;72:783–796
45. Slater PE, Orenstein WA, Morag A et al. Poliomyelitis outbreak in Israel in 1988: a report with two commentaries. *Lancet* 1990;335:1192–1198
46. Minor PD. The molecular biology of poliovaccines. *J Gen Virol* 1992;74:3065–3077

The Development of IPV

A. John Beale†

Jonas Salk died on 23 June 1995, and so is prevented from writing about the development of what he called "non-infectious polio vaccine". My credentials to undertake the task are not perfect. I never worked for Jonas directly and by the time I reached Toronto to take up a Research Fellowship in Virus Diseases under the directions of Dr. Andrew Rhodes at the Hospital for Sick Children in September 1954, the Francis Trial of injectable polio vaccine (IPV) was already in progress.

Toronto, at the time of working for Andrew Rhodes, was an excellent place to observe the development of both IPV and oral polio vaccine (OPV). Rhodes in 1947 had set up a polio research unit at the Connaught Laboratories associated with the School of Hygiene of the University of Toronto. This unit was to play a decisive role in the development of IPV.

The development of any vaccine against poliomyelitis looked a bleak prospect at the end of World War II. Vaccines made in the 1930s by Kolmer [1] and Brodie [2] from virus extracted from monkey central nervous system (CNS) tissue infected with poliovirus and either partially attenuated [1] or treated with formalin [2] proved to be unsafe. As a result of cell culture experiments carried out by Sabin and Olitsky [3], poliovirus was thought to be strictly neurotropic. This appeared to

† Deceased

S.A. Plotkin (ed.), *History of Vaccine Development*,
DOI 10.1007/978-1-4419-1339-5_20, © Springer Science+Business Media, LLC 2011

mean that production of large amounts of virus would be very difficult, if not impossible, as well as hazardous. Moreover, no one knew how many serotypes of poliovirus there were; Burnet and McNamara [4] showed that there were at least two types. Rather few strains had been studied because the neutralisation tests had to be performed in monkeys, so how many more serotypes there were remained an unanswered question. Moreover, the climate of scientific opinion then was that effective long-lasting immunity against virus infections depended on stimulating immunity in all its aspects as a result of infection with a live virus.

This opinion was based on the long, often life-long, immunity that follows recovery from the common virus diseases of childhood, and the success of immunisation with live vaccines against smallpox and yellow fever.

Salk saw the opportunity to intervene positively for a number of reasons. Like every virologist at the time, he realised Enders, Weller and Robbins [5] had made a major break through when they found that polioviruses would grow well on non-nervous system cells cultivated in vitro. At first, it was human embryo cells, but it soon established that other human and primate cells would serve as well. Salk had learned his virology with Tommy Francis, first as a medical student at New York University Medical School, and then as a graduate when he joined him at the School of Public Health, University of Michigan, working on the influenza virus.

He was fortunate because Francis was one of the few virologists at the time who thought it feasible to try and make a killed vaccine. Francis thought the prevailing view that a living virus vaccine was essential for long-lasting immunity was a prejudice, based on unwarranted extrapolation from two examples, smallpox and yellow fever. The possibility of testing this hypothesis was opened up for influenza by the ability to grow influenza viruses in the allantoic cavity of developing chick embryos, and Salk produced high titre virus for inactivation with formalin. He was soon able to make influenza vaccines which elicited high titres of specific antibodies to the virus. When he moved to Pittsburgh as a Research Professor, Salk continued his work on influenza viruses and vaccines, especially on the use of adjuvants, for example, calcium phosphate and mineral oil.

Following Ender's discovery, he started to play with polioviruses in his laboratory. He had first become interested in polio research when he joined the effort funded by the National Foundation for Infantile Paralysis (NFIP) research to type poliovirus in 1948. The Committee consisted of the following leading investigators in poliomyelitis research at that time: Charles Armstrong, David Bodian, Thomas Francis Jr, Louis Gebhardt, John Kessel, Charles F Pait, Albert Sabin, Jonas Salk and Herbert Wenner, and they confirmed the finding of Bodian et al. that there were just three types of poliovirus [6].

Salk now seized the opportunity that he has sensed. He was experienced at antibody typing and at producing antibodies to polioviruses in monkeys, and knew that killed virus vaccines were possible. He had helped to solve the first problem, namely to define the number of serotypes required for the vaccine, and in the process had assessed the immunogenicity of the different strains. He next tackled the problem of cultivating large quantities of virus for inactivation experiments and devising rapid and accurate methods of titrating the viruses and antibodies. In this work he was greatly helped by his colleagues, especially Julius Youngner. They

Fig. 1 Jonas Salk at the time IPV was first tested

devised roller-tube cultures of monkey testicular and kidney tissue to produce high titre virus. This was harvested and filtered to provide the virus for treatment with formalin. Salk experimented with various concentrations of formalin and temperatures for the inactivation and times of treatment before settling on his final formulation (Fig. 1). In this work, Salk was keenly aware of the disastrous experience with the Kolmer and Brodie vaccines, having been at the New York University Medical School when the debacle occurred. He persuaded himself and others that, when properly prepared and filtered to remove clumps or aggregates of virus and cell debris, the inactivation at 37°C using 1/4,000 formalin followed first-order kinetics [7], and therefore, there was a built-in safety factor in the preparation. This concept was not universally accepted and was severely tested when, shortly after the licensure of Salk's vaccine, the failure of inactivation of the virus in vaccine from one manufacturer (Cutter laboratories, hence the Cutter Incident) led to a serious disaster when a total of 260 cases of poliomyelitis and 10 deaths were attributed to residual live virus in the vaccine. Salk kept his nerve admirably during this episode: he was encouraged by the firm resolution of Defries in Canada who had complete confidence in the Connaught-prepared vaccine and successfully conveyed that to the Canadian Health Minister. It resulted in greater attention to the detail of filtration before the addition of formalin and the introduction of a second filtration step half-way through the treatment period to remove any aggregates that may have developed during treatment. It also led to an increase in the sample size required to be tested to ensure the safety of a lot of virus, and reinforced the concept of consistency of manufacture as a requirement for vaccine licensure. The use of monkeys as well as cell cultures for detecting residual virus was also reemphasised.

Salk tested his experimental formalin-treated vaccine in gradually increasing numbers of children, starting with his own, and was gratified to experience no adverse side-effects. Moreover, the vaccine induced the production of good levels of neutralising antibody. The stage was therefore set for a large-scale clinical trial under the auspices of the National Foundation for Infantile Paralysis. There were numerous problems. The first was the supply of vaccine, in particular, the supply of large quantities of all three types of poliovirus of high titre, not only sterile but also free from harmful contaminants, either living or non-living. Due to the foresight of Harry Weaver from the NFIP, the Connaught Medical Research Laboratories, Toronto, were given grants to develop large-scale methods of production of virus for vaccine development. This was first under the direction of Andrew Rhodes, and later under the direction of RD Defries. They made three major contributions to the IPV story as later recognised by Hart E van Riper, Medical Director of the NFIP [8]. First, Joseph Morgan and Helen Morton under the direction of Raymond Parker [9] developed a totally synthetic medium, "Medium 199" that would support cell growth. It was adapted by Franklin for producing poliovirus, so that serum and other ill-defined medium components were not present in the virus growth, giving rise to the "Toronto method", which used small pieces of monkey kidney tissue cut up with scissors and grown in Maitland style suspension culture [10] in large Povitsky flasks which were being continuously rocked. This rocking flask method was originally devised for the production of diphtheria toxin at Connaught. In the course of this work, they identified the danger from possible simian virus contamination, especially from Virus B, which they showed was readily inactivated by formalin. Finally, they prepared for their own use and for Parke Davis and Eli Lilly over 5,000 L of poliovirus for processing into Salk vaccine for the Francis Field Trial [11]. This Canadian contribution was kept low key because neither O'Connor at the NFIP nor RD Defries and the President of the University of Toronto were anxious for publicity that Canadian resources were being paid for by the March of Dimes for the ultimate benefit, at least in the first instance, of the US public. For this and other contribution to Public Health, especially his steadfastness in Canada at the time of the Cutter incident, RD Defries was given the Albert Lasker Award by President Truman in 1955.

The successful production of vaccine, due in large measure to the work of the Connaught team [12], enabled the Francis trial to be conducted in 1954. Fortunately, Francis accepted the offer from the NFIP to run the trial, when he was assured that the ultimate responsibility for the design of the trial was his, and that his team at the School of Public Health at the University of Michigan would have a free hand in the conduct of the trial as well as in the analysis and preparation of the report of the results. There was considerable controversy and anxiety about the placebo component of the trial, including pressure to abandon this feature. Francis and his team remained firm that a placebo (the placebo was medium 199) controlled trial was essential, and were thus able to produce emphatic evidence for the effectiveness of the vaccine in their final report [13].

Personal Experience in IPV Development

I was involved in the evaluation of IPV in Toronto. I also gave my then pregnant wife a dose of the early Connaught IPV about a week before the Cutter incident. As you can imagine this was a shattering experience and led to a transatlantic telephone call, in the day when they were very rare, from my mother-in-law enquiring whether I knew what I was doing. Fortunately, all was well and my good relations with both my wife and her mother were restored. This episode forcefully reminded me of the adage "behind every successful man is an astonished mother-in-law!"

When I returned to England in 1956, it was the time of the Suez crisis, and public service salaries were frozen, mine at the level it had been when I had left England 2 years earlier. I therefore looked at alternative employment and, after some soul searching, joined Glaxo Laboratories to make IPV. At that time, jobs in industry were looked upon with much suspicion, especially in medical circles. However, I was joining Bill Wood who has preceded me working with A Rhodes in Toronto. At Glaxo, I found a desperate situation: residual living virus in 7.5% of monovalent vaccine lots; after processing over 100 monovalent pools successfully this was a shattering blow. Indeed, so shattering that my colleague who had recruited me decided to retire and left me holding a very difficult baby. At that time, we took samples on day 6, 9 and 12, after addition of formalin, and to our surprise, the proportion of positive samples were the same in each set of sample. Moreover, the sample did not yield their virus until over 2 or 3 weeks of observation in culture. We eventually extended the observation period to 28 days, with subcultures at 7, 14, 21 and 28 days. These observations and the finding that our initial problems started when a key craftsman, who made the sintered glass filters, retired led us to conclude that the virus that escaped inactivation was embedded in aggregates or clumps of cell debris. As the clumps and aggregates slowly broke down, the infectious virus was released. We overcame the problem by improving the filtration procedure, using first Seitz type EKS2 filter pads and later 0.22-μm membrane filters. We also improved the quality assurance on the filter pads, weighing them and subjecting them to X-ray examination, for example. The filter presses were also redesigned, so that they were either single filters or, if employing a multi-filter system, they were redesigned so that the separating plates were either between two input channels or between two filtrate collection chambers. This was necessary because it was found that some steel separation plates had minute pinholes. Thereafter, Glaxo had no problems with residual poliovirus in killed vaccine. We did, however, experience a problem with potency. Karl Fantes [14] solved this by concentrating and purifying the vaccine. The related problem of in vitro measure of potency was then solved by adapting Le Bouvier's method [14, 15] of measuring the specific D-antigen content associated with full virus particles by immunoprecipitation in agar gels [16]. This test proved a rapid and reliable in vitro guide to potency, which enables us to monitor and avoid the loss of antigen during processing, and to blend a trivalent vaccine of balanced potency. This also enabled Glaxo to introduce a high-potent IPV vaccine with the diphtheria–tetanus–pertussis (DTP) vaccine in the 1960s. At Glaxo, we also replaced the Maitland-type Toronto method of culture by converting to monolayer cell cultures in large roller cultures, so that the ratio of cells to culture fluid could be optimised. The problem of simian viruses which was well analysed and classified by Robert Hull at Eli Lilly [17] plagued us as it did all manufacturers, and we experimented with HeLa cells, but the time was not right for their adoption for virus growth in vaccine production. The most important simian virus (SV40) was discovered by Sweet and Hilleman [18]. It was found to be resistant to formalin and later, to be an oncogenic virus in animals. It caused the termination of a Merck initiative to prepare a highly purified and killed polio vaccine by ultra-centrifugation of the virus. Unfortunately, the living SV40 was copurified, and as a result the project was abandoned, but not before it had been given to a small number of newborns, fortunately without ill effect in the short or long term.

Later Developments

Van Wezel's Improvements at the Rijkinstituut voor Volksgezonheit et Milieuhygien

Jonas Salk dedicated the chapter he wrote with Jacques Drucker for Plotkin's and Mortimer's books on vaccines to Anton van Wezel from the Netherlands' Rijkinstituut voor Volksgezonheit et Milieuhygien (RIVM), who introduced the technology to produce enhanced potency IPV (e-IPV) [19]. He first used trypsinised monkey kidney cells from home-bred monkeys in a largely successful effort to overcome the problem of simian viruses. Using these monkeys, it was possible to employ cells subcultured once or twice for vaccine production. The RIVM workers also made improvements in the trypsination process by perfusing the kidneys with trypsin solution in vivo under anaesthesia, immediately prior to surgical removal of the kidneys. These two measures increased the yield of cells, so that each pair of kidneys could provide the virus for a million doses of the vaccine. The cells were cultivated on microcarrier beads in suspensions culture in large stainless-steel tanks. This technique enables the virus production to be on a truly industrial scale. Later, workers at Mérieux led by Dr. Montagnon [20], who had adopted and scaled up van Wezel's process, used the continuous vervet monkey cell line Vero, from a pedigreed cell bank, to provide the cells, thus solving the problem of viral contamination. Serum is now the only biological component of the production system that is not fully controlled.

Strains of Poliovirus

Salk chose Type I Mahoney, Type 2 MEF1 and Type 3 Saukett on the basis of immunogenicity observed in the collaborative typing experiments of the late forties, and on the yield in cell cultures in the fifties. The choice has stood the test of time since five of the six remaining licensed manufacturers of IPV still use these strains. The choice of the very virulent Mahoney strain has always been controversial, and after the Cutter Incident even more so. In Sweden, Sven Gard was a powerful advocate for IPV, but he preferred the Brunenders strain for Type 1, as did the UK authorities. He was also critical of Salk's claim that inactivation by formalin was a first-order reaction. This criticism was partly due to Gard's use of 25°C as the temperature for formalin treatment. Later experience in Sweden, using inactivation at 37°C and well-filtered virus, confirmed the satisfactory experience elsewhere with the viral inactivation procedure. Joseph Melnick has advocated the use of Sabin strains for IPV production to lessen the impact of any future failure to completely inactivate the virus, but the suggestion has not found favour.

Purification of the Virus

Van Wezel decided to concentrate and purify the virus before treatment with formalin. He and his colleagues devised a simple ion-exchange column scheme that leads to essential pure virus preparations that can be consistently inactivated by formalin. Since this procedure has been introduced, there have been no failures of the inactivation process, thus justifying in most extensive practice Salk's confidence in the process he developed.

Measurement of D Antigen to Standardise Vaccine

Van Wezel and his team decided to use the D antigen test introduced by Beale and Mason [21] to control and monitor the production of the vaccine, from live virus to totally not infectious antigen. The D antigen test is also used to standardise the content of antigen for all three types in the trivalent e-IPV; that is Type I, 40; Type 2, 8; Type 3; 32-antigen units. The final potency test is performed in animals.

Performance in Controlling Poliomyelitis

IPV was shown to be conclusively effective in raising antibodies and preventing paralytic poliomyelitis in the placebo-controlled famous Francis field trial conducted in 1954–1955. Since then, in all countries where it has been used, it has had a pronounced effect on the incidence of poliomyelitis. Thus, in the USA, the incidence was reduced from over 20 (and up to 35 in some years) per 100,000 to around 2 before the introduction of OPV. In the countries where it has remained or become the mainstay of the immunisation, the result has been the elimination of polio. IPV has been particularly valuable in the control of the disease in developing countries where three doses of OPV often give disappointing results. In the Gaza strip, administered by Israel, the problem of controlling poliomyelitis was intractable for many years, until a joint programme of combined IPV and OPV was used to give good antibody levels from those given IPV and swamping of the

Fig. 2 Jonas Salk late in life

environment with OPV to oust any wild poliovirus from any remaining niche. Perhaps Salk (Fig. 2) would have received the most gratification and pleasure from the announcement at the end of 1995 that IPV was recommended again officially by the Centers for Disease Control for primary childhood immunisation against poliomyelitis in the USA. Three men were most responsible for keeping faith with IPV: Jonas Salk, Charles Mérieux and Hans Cohen.

Acknowledgments I am grateful to Professor Frances Doane for putting me in touch with Paul Bator, the author of the history of the Connaught Medical Research Laboratory, and, Christopher Rutty who made his thesis on "The history of poliomyelitis in Canada from 1927–1962" available to me.

Further Reading

Books

The following books have provided me with much information about the background of the development of IPV.

Bator PA, Rhodes AJ. *Within Reach of Everyone: A History of the University of Toronto School of Hygiene and the Connaught Laboratories, Volume 1, 1927–1955*. Ottawa, Canada: The Canadian Health Association, 1990

Benison S. Tom Rivers: *Reflections on a Life in Medicine and Science. An Oral History Memoir*. Cambridge Mass and London, England: the MIT Press, 1967

Defries RD. *First Forty Years 1914–1955*. University of Toronto: Connaught Medical Research Laboratories, 1968

Rutty CJ. *Do Something!... Do Anything! Poliomyelitis in Canada 1927–1962*. PhD Thesis, Department of History, University of Toronto, 1995

Salk J, Drucker J. Non-infectious Poliovirus Vaccine, Chapter 8 In: Plotkin SA, Mortimer EA, eds. *Vaccines*. Philadelphia: WB Saunders, 1988

References

1. Kolmer JA. Vaccination against acute anterior poliomyelitis. *Am J Public Health* 1936;26: 125–35
2. Brodie M, Park WH. Active immunisation against poliomyelitis. *Am J Public Health* 1936;26: 119–25
3. Sabin AB, Olitsky PK. Cultivation of poliomyelitis virus in human embryonic nervous tissue. *Proc Soc Exp Biol Med* 1936;31:357–9
4. Burnet FM, McNamara J. Immunological differences between strains of poliomyelitis virus. *Br J Ex Path* 1931;12:57–61
5. Enders JF, Weller TH, Robbins FC. Cultivation of the Lansing strain of poliomyelitis virus in cultures of various human embryonic tissues. *Science* 1949;109:234–45
6. Bodian D. Morgan IM, Howe HA. Differentiation of types of poliomyelitis viruses; the grouping of fourteen strains into three basic immunologic types. *Am J Hyg* 1949;49:234–45
7. Salk J. Poliomyelitis Vaccine in the fall of 1955. *Am J Public Health* 1956;46:1–14
8. Van Riper HE. Progress in the control of paralytic poliomyelitis through vaccination. *Can J Public Health* 1955;46:427–48
9. Morgan JF, Morton HJ, Parker RC. Nutrition of animal cells in tissue culture. 1. Initial studies on a synthetic medium. *Proc Soc Exp Biol and Med* 1950;73:1–8
10. Maitland HB, Maitland MC. Cultivation of vaccinia virus without tissue culture. *Lancet* 1928;2:596–7
11. Wood W, Shimada FT. Isolation of strains of Virus B from tissue cultures of cynomolgus and rhesus kidney. *Can J Publ Hlth* 1954;45:509–18
12. Farrell LN, Wood W, Franklin RE, Shimada F, Macmorine H, Rhodes AJ. Cultivation of poliomyelitis virus in tissue culture: VI methods for quantity production of poliomyelitis viruses in cultures of monkey kidney. *Can J Public Health* 1953;44:273–80
13. Francis TM Jr, Napier JA, Voight RB et al. Evaluation of the 1954 field trial of poliomyelitis vaccine. *Final Report Ann Arbor University of Michigan* 1957
14. Fantes KH. Concentrations and partial purification of poliomyelitis viruses. *J Hygiene* 1962;60:123–33
15. LeBouvier GL. The D to C change in poliovirus particles. *Br J Exper Path* 1959;40:605–20
16. Beale AJ. The D-antigen content in polio vaccine as a measure of potency. *Lancet* 1961;2: 1166–8
17. Hull R, Minner JR, Mascoli GC. New agents recovered from tissue culture of monkey kidney cells: III Recovery of additional agents both from cultures of monkey kidney tissues and directly from tissues and excreta. *Am J Hyg* 1959;68:31–44
18. Sweet DH, Hilleman MR. The vacoulating virus, SV40. *Proc Soc Exp Biol Med* 1960;105: 420–7
19. Van Wesel AL4, van Steenis G, van der Marel P, Osterhouse ADM. Inactivated poliovirus vaccine: current production methods and new developments. *Rev Infect Dis* 1984; 6 (suppl 2): 335–40
20. Montagnon B, Fanget B, Vincent-Falguet JC. Industrial scale production of inactivated poliovirus vaccine prepared by culture of Vero cells and microcarriers. *Rev Infect Dis* 1984; 6S:341–4
21. Beale AJ, Mason PJ, The measurement of D-antigen in poliovirus preparations. *J Hyg* 1962;60 :113–21

The Long Prehistory of Modern Measles Vaccination

Constant Huygelen[†]

During the first decades after its introduction into Western Europe, the practice of smallpox inoculation or variolation had only limited applications, mainly because of the high morbidity and mortality rates it caused. Gradually, however, the inoculation procedures were improved, resulting in milder reactions and, towards the middle of the eighteenth century, variolation was becoming more popular. The time seemed then ripe to extend the inoculation principle to other scourges of humans and domestic animals, like measles and rinderpest. Both diseases are caused by closely related morbilliviruses, but in the eighteenth century they were seen as belonging to the pox group of diseases; this view encouraged those who wanted to explore the feasibility of inoculations against them in analogy with smallpox.

At the end of 1754, the first report on rinderpest inoculation was published [1]; it inspired Stephanus Weszprémi, a Hungarian physician living in England in those days, to write his *Tentamen de inoculando peste* in which he advocated the application of the inoculation approach to measles and human plague [2]. His book was

†Deceased

S.A. Plotkin (ed.), *History of Vaccine Development*,
DOI 10.1007/978-1-4419-1339-5_21, © Springer Science+Business Media, LLC 2011

published in 1755. In the same year Charles Brown published his dissertation on measles in Edinburgh, in which he gave a detailed description of the disease and referred to the possibility of transmitting it by rubbing a piece of wool or wood on the skin lesions of a measles patient, or by putting it in the axillary region to have in impregnated with perspiration. The wool or wood should then be rubbed on the skin or put in the armpits of another person. He concluded: "I am in no doubt that measles could be safely propagated in this way, if somebody wants to do the experiment [3]." Two successive years later, Alexander Monro Secundus made the following suggestive statement: "How successful that inoculation of smallpox has turned out is known to all, but I regard it as altogether certain that inoculation of measles will be more useful and successful. For it is well-known how liable this disease is to infest the lungs, and how great destruction it causes there. This seems in the first place to be due in large part to the contagion which flies about in the air, and is drawn into the lung cells in breathing, and persistently clings to them, and causes a cough there, or in other words, excites an attempt by nature to drive off the noxious matter. If measles were really to be induced by inoculation artificially produced, it is very likely that the lungs would be more free from inflammation, and in general that the disease would attack the skin only. If this should turn out so, what a great profit and utility it would bring to mankind! The experiment can bring about no inconvenience or loss. It is probably that inoculation can be performed, if only the pustules and spots of matter can be rubbed on cotton, and if this (either fresh or put on glass carefully covered and preserved) be applied to a little wound, exactly in the same was a variolous matter [4, 5].

None of the three authors mentioned here carried out inoculation trials himself; Francis Home was the first to put the idea into practice [6–8]. In the winter of 1758 a measles epidemic struck Edinburgh and Home decided to try inoculation so as to mitigate the pulmonary form of the disease. He found great difficulty in finding subjects for inoculation; some he certainly had to pay [9]. In June 1758, the following announcement appeared in the *Gentlemen's Magazine*: "Tuesday 23. A discovery of the highest utility has lately been made at Edinburgh, and already sufficiently confirmed by a number of successful experiments: Dr Francis Home has inoculated for the measles, and has produced a disease free from all alarming symptoms." Home published his results in detail in 1759 [10]. He had first tried to obtain enough infectious material from the skin, but had failed. He then started to use blood, "the magazine of all epidemic diseases." In fact he did not use blood alone but a mixture of blood and material from skin lesions, as we can conclude from the following statement: "I thought that I should get the blood more fully saturated with what I wanted, if it was taken from the cutaneous veins amongst the measles, than I took it from a large vein, where there was a much greater proportion of blood from the more internal parts, that from the skin. I therefore ordered a very superficial incision to be made among the thickest of the measles, and the blood which came slowly away was received upon some cotton [11]." His publication contained an accurate and detailed description of his experiments; they have been analyzed in detail and put in table from by Hektoen [5] and Zeiss [6]. The inoculation was performed in 15 children, three of whom received the material intranasally without any effect. In the other ones, the cotton was placed on the scarified skin for 3 days. The age of the children varied between 7 months and 13 years, but the older children had the measles 2 years previously. In most children symptoms started on the sixth day: they were hot, feverish, sneezing, their eyes were watery, and there was some coughing. These symptoms were followed by some measles spots, but overall the disease produced by inoculation was very mild. One child "took measles again" a few weeks after inoculation.

The reactions from the medical opinion leaders of those days were scarce. Tissot referred to the experiments as follows in 1763: "Measles may have been inoculated in countries where it is very bad and this method may have great advantages in this disease, but as in smallpox, it can only be useful to people by means of a hospital" [12]. In 1766 Böhme devoted his "inaugural dissertation" to the potential legal problems related to measles inoculation and concluded that there would be no legal objections to it [13]. Apparently the only enthusiastic reaction appeared in a letter by Cook to the *Gentleman's Magazine* in 1767, that is, 9 years after the experiments: "The measles, though not so fatal as smallpox, is yet attended in the natural way with many dangerous symptoms, and often produces very troublesome effects. I would therefore beg leave to recommend to the public the practice of inoculation in this distemper as well as in the other, and am confident that by this method many may be preserved from that malignant sort which often proves mortal and is always dangerous. Dr. Francis Home was the first who attempted this practice in Edinburgh about 9 years ago, since which, many physicians in that country have followed his example, though I do not find it is much encouraged in England, though in smallpox it is now become universal. The method is easy, may be performed with safety by a careful nurse, and is not attended with the remotest of danger. Dip only a little bit of cotton, or lint, in a watery humour that stands in the eyes of persons ill of the measles, at the time of the crisis, make a slight scratch in the skin of the arm, above the elbow, of the person to be inoculated, put the wetted pledget upon the incision and cover it with a bit of sticking plaister to keep it on; and this, without further trouble will produce the measles in a gentle and favourable degree, which during the whole course of it, will want no other care by that of keep the patient moderately warm, not any attendance but that of watching the fever, and encouraging the crisis, which in a few days well carry off the infection, and complete the cure. The epidemic disease should be communicated to those young subjects who have not yet had it, when it first makes its appearance in any neighbourhood, by which the dangerous symptoms that often attend it will effectually prevented" [14]. It is noteworthy that Cook recommended the use of conjunctival secretions as inoculum instead of blood.

One year later Spry wrote: "The method of procedure in inoculating for measles does not differ from that which, as we have before described, is to be followed in the case of smallpox. I think there is only one thing which in this case deserves some notice as being peculiar, viz, that the lined threads were are used for introducing the contagion ought to be impregnated with blood, for matter is rarely found, drawn from pustules of the measles near the tip or a little away from it. In this method of treatments, not yet so common as the earlier one, all the symptoms are found to be less serious – a fact of which I am quite certain, not from observation but from the study of many cases" [15].

Home himself did not publish anything else on measles inoculation except for a brief reference to it in his "Principia Medicinae" in the 1770s, which read as follows: "Measles, as experimentation has confirmed to me, is transmitted by inoculation. On the sixth day usually as light fever is noticed, accompanied by a slight cough. The patient suffers neither from sleeplessness nor from inflammatory symptoms, and after his recovery he is not exposed to hectic fever, cough, or inflammation of the eyes" [16]. Most other contemporary authors, like Cullen, did not mention measles inoculation at all or referred to it only in passing [5, 6]. Cullen was living and teaching in Edinburgh like Home himself. The well-known Swedish author, Rosen von Rosenstein [17], mentioned Home's method in his book without further comment. In a footnote, Murray, the German translator of Rosen von Rosenstein's book, complained that in spite of Cook's letter about the success of the method in Scotland, nobody knew whether it had become popular in England.

The Dutch medical opinion leader, Thomassen à Thuessink, who had visited Edinburgh in 1784 and in 1785, had not been convinced by what he saw and thought that the children in Home's experiments had contracted measles not from inoculation, but from exposure to natural infection [6].

Around the turn of the century, a few more inoculation attempts were carried out. In 1789, Dr. Green is reported to have successfully inoculated three young persons with blood taken from the eruptive surface in a severe case measles. Others obtained negative results, like Willan, the pioneer of modern dermatology, and Chapman who used "the mucus of the nostrils and bronchia, the eruptive matter in the cuticle, properly moistened" [5].

In 1816, a severe measles epidemic struck the area of Groningen, and Themmen, a pupil of Thomassen à Thuessink's, wrote a thesis about it [18]. The third part of his work was devoted to inoculation. In the introduction he briefly mentioned some previous experiments in this field done by Panzani of Istria, and also the comments of contemporary authors. He wondered why contemporaries and colleagues of Home's like Cullen, Black, Duncan, and others did not mention Home's work. He also questioned why Home never extended his experiments later, since measles continued to be a very dangerous disease in Scotland. Themmen also referred to measles inoculation, and thought that the procedure often did more harm than good, because it caused severe pulmonary and other symptoms. Themmen then described his own experiments in five children using blood and other material from patients. None of the five contracted measles; however, his experiment was probably worthless since he admitted in his conclusion that the children he had used were apparently not very susceptible to measles: they had been living in an environment in which measles was prevalent and yet had remained free of the disease. The main argument used by Themmen and his teacher Thomassen à Thuessink in their attacks against Home's conclusion was that the incubation period of around 6 days was too short, and hence the symptoms have not been induced by the inoculation, but by previous exposure to natural infection. This argument was surprising since these authors must have been familiar with the well-known fact that in smallpox the incubation period after inoculation was much shorter than in the natural disease. To us today it may seem even more surprising that Hektoen [5], in critically reviewing Home's work in 1905 and writing almost a century after Themmen, used the same wrong argument in his evaluation of Home's work: "Examination of this tale cannot but lead to the conclusion that probably not a single one of the 15 cases inoculated by Home has measles 25 a result of the inoculation. In support of this view I may point out that in no single case is the period between inoculation and the appearance of the rash given as more than 10 days, but generally less, and even so short as 7 days, whereas we now know definitely that the period between exposure and eruption in measles is 13–14 days," and further: "On the whole there seems to be no escape from the conclusion that Home's claim to have produced measles by inoculation is without foundation [19]." It is well known that measles, when given parenterally, has a much shorter incubation period that after infection by the natural route [20].

In 1822, Speranza tried to revive the inoculation procedure during an epidemic in Mantua. Six children were inoculated with blood taken from the most prominent measles spots on the skin of measles patients and put into superficial incisions on the arm of the children. The six children and four other persons inoculated likewise developed a mild form of measles 5–6 days later, followed by some mild respiratory symptoms and a few spots on the skin; on the 9th–10th day, the patients had completely recovered. Speranza also mentioned three other authors (Negri, Grigori, Rasori) who carried out small experiments and produced a mild form of measles, but no details were available [5, 6]. In 1830, Albers tried in vain to inoculate four patients with blood. He concluded that blood did not contain the measles agent [21]. In Jörg's handbook on childhood diseases in 1836, we find the following negative judgment referring to Home's experiments: "Experience has thought that measles transmitted in this way, are not milder that when contracted naturally and therefore the inoculation has been rightly abandoned" [22].

By far the largest experiment was carried out by von Katona in Hungary in 1841–1842 [23]. He inoculated 1,122 persons with a 93% success rate during a severe epidemic. His report is only one-and-half pages long and few details are given. Seven percent had no symptoms and were probably not infected. The measles symptoms produced in the others were very mild and not a single fatal case was recorded. He inoculated with a needle dipped in vesicle fluid mixed with blood at the height of the rash. Fever mostly developed on the seventh day and typical measles symptoms on the ninth or tenth day, except for two cases in which they appeared only on the 13th day. In view of these excellent results it seems surprising that no further experiments appear to have been carried out as a result of this publication.

A few years later in 1850–1851, McGirr inoculated a group of children in an orphanage in Chicago, using blood from "a vivid exanthematous patch"; inoculated measles were mild and symptoms appeared on the 4th–9th day [24].

Around the middle of the century, von Mayr [25] and Bufalini [26] succeeded in transmitting measles to small groups of children, but neither of them became an advocate of general inoculation because, a von Mayr wrote, it was impossible to limit the infection to local skin eruption as in vaccinia.

A few decades later, Hugh Thompson, vaccinator appointed by the Faculty of Physicians and Surgeons of Glasgow, became interested in measles inoculation [27].

He did not use blood but liquid from small blisters he induced on the skin of a measles patient. Despite his initial failures, he apparently continued his work, but was "not universally believed," as reported in the Proceedings of the Royal Society [28]. After this abortive attempt to revive the interest in measles inoculation, the issue became dormant again for many years.

In the mean time, as a result of Behring's discovery of diphtheria antiserum, the attention shifted from measles virus inoculation to the administration of antiserum. Some authors continued, however, to explore the possibility of actively immunizing with virus alone. In 1915, Herman in the USA inoculated 40 infants intranasally with nasal secretions taken from infected children before the onset of the rash. The infants were between 2 and 5 months old, most of them between 4 and 5 months; it was well known that infants before 5 months were generally immune to measles. Most of his patients had no specific reactions: 15 showed a slight temperature rise and few had some spots between the 14th and 18th day. Four out of the 40 came into contact with measles when over 1 year of age and did not become ill. Two of the children were reinoculated by Herman when they were 21–23 months old and remained healthy. He concluded his paper with the wish that a method could be found to maintain the infectivity of nasal secretions for more than 24 h and to use this material for general immunization purposes [29]. In 1922, the author summarized the results he had obtained up to that date: 165 infants had been inoculated in their fifth month. Of these, 75 had been followed for 4–8 years, and only five of these 75 contracted measles; 70 had remained measles-free [30]. Unknowingly, Herman had applied a method which had been used with relative success against the closely related rinderpest virus infection in calves one-and-half century before by Reinders in the Netherlands [31]. This author had developed a procedure by which calves with maternally derived immunity were inoculated through incisions in the skin with nasal secretions from diseased animals. He inoculated the calves three times at difference stages, so that he increased his chances of inoculating them at a time when the maternally derived antibody tiers had declined to a level which was not too high to completely inhibit the virus replication and not too low to prevent full-blown symptoms with rinderpest [32].

In 1921, two Japanese workers, Hiraishi and Okamoto [33], attempted to obtain active immunity by inoculating small amounts of diluted citrated blood taken at the beginning of the disease, and determine the minimum immunizing dose. They came to a conclusion that a dose of higher than 0.002 mL produced the disease, whereas a dose below 0.001 mL did not. Based on an experiment in 44 children, they concluded that 0.0001 mL of blood induced at least a partial immunity in most recipients. They recommended injecting children under 5 years with 0.001 mL followed by a second dose injection of 0.01 mL.

In applying the principle of the Japanese workers, Debré et al. first inoculated on scarified skin blood diluted one-tenth, but soon abandoned this method because of its unreliability and switched to subcutaneous injection [34]. They considered the virus titer in the blood to be sufficiently constant to obtain consistent results. In 2- to 4-year-old children they inoculated 1/400 mL of blood diluted in saline and produced mild measles symptoms 7–10 days later. They claimed that with 1/800 mL of blood they did not produce overt clinical symptoms, but only a leukopenia which started 3–4 days post inoculation and peaked on the 8th–10th day. They recommended administering the 1/800 mL dose first, and 3 weeks later the 1/400 mL dose. In their second article the authors reported their experience on thousands of children [35].

Papp, who collaborated in the Paris trials, applied the method when back in Hungary in the 1930s but using infants with maternally derived immunity, as pioneered by Herman. She used diluted plasma with leucocytes from measles patients. A total of 1,802 infants were inoculated: 849 received one injection, 493 two, and 460 three. The reactions to the "vaccine" were generally mild, but as could be expected, the percentage of reactions increased with age, but remained low [36]. In the large majority of cases the reactions were subclinical. Based on her results, she recommended starting the inoculation at the age of 5 months using three injections at 1-month intervals. One of her main problems was the lack of a titration method for measles antibody, and hence the impossibility to know with certainty whether the mother was immune; moreover, the wide individual variations in maternally derived immunity made concrete recommendations hazardous. Based on a survey she did later among the mothers of the children, she concluded that 79% of the "vaccinated" children did not contract measles during a follow-up period until the age of 5 or 6 years.

In the first half of the twentieth century, measles antibody of different origins was widely used to prevent or attenuate symptoms after exposure. Some attempts were made at "servovaccanation". In 1919, Richardson and Connor published the results of a trial in three children who were given convalescent serum, followed by a swabbing of the nasal and pharyngeal mucosae with secretions from a measles patient. Two of the three children had no symptoms and one had an abortive case of the measles [37]. The immunization experiments had been suggested to Richardson and Connor by CV Chapin based on the use of virus and serum for the immunization against hog cholera (swine fever), which was being used successfully in those days.

Little further progress in active immunization against measles was made until the advent of modern tissue culture methods. The histories of these vaccines and the results obtained with them have been extensively reviewed at the International Conference on Measles Immunization in 1961 [38]. The results of several decades of use were reviewed by Preblud and Katz [20].

Conclusion

Two hundred years have elapsed between the initial trials preformed by Francis Home and the advent of modern vaccines of reliable safety and immunogenicity. As summarized in this paper, several attempts were made in a variety of countries to induce active immunity against measles. The attempts undertaken during those 200 years all suffered from the same basic problems:

- The authors were all using virulent virus with the inherent risk of severe reactions.
- Neither blood nor nasal/conjuctival secretions were reliable sources: titration or standardization of the inoculum was impossible.
- The virus was much more labile than smallpox virus, as Home had already noted in his early trials.
- In the case of "serovaccination," the authors had to work with two unknowns since neither the virus inoculum nor the serum antibodies could be adequately quantitated.
- Similar problems of individual variations beset the inoculation of infants with maternally derived immunity.
- The basic idea of inducing a primarily cutaneous disease with reduced pulmonary involvement had worked relatively well in the inoculation against smallpox, but, because of their different pathogenesis, morbillivirus diseases like measles and rinderpest did not lend themselves to this approach.

References

1. *Letters in Gentleman's Magazine*. 1754;24:493. 549
2. Weszpremi S. *Tentamen de inoculanda peste.* London: Tauch J, 1755, 3
3. Brown C. *Dissertatio medica inauguralis de morbillis.* Edinburgh: Hamilton and Balfour, 1775, 3
4. Monro A (secundus). *De venis lymphaticis valvulosis et de earum imprints origine*. Berolini, 1757, quoted by Hektoen, 239
5. Hektoen L. Experimental measles. *J Infect Dis* 1905;2:238–55
6. Zeiss H. Die experimentelle Masernübertragung. *Ergeb Inn Med Kinderheilkd* 1921;20:425–510
7. Enders JF. Francis Home and his experimental approach to medicine. *Bull Hist Med* 1964;38:101–12
8. Plotkin SA. Vaccination against measles in the 18th century. *Clin Pediatr (Phila)* 1967;6:312–5
9. Home WE. Francis Home (1719–1813). First professor of material medica in Edinburgh. *Proc R Soc Med* 1927;21:1013–5
10. Home F. *Medical facts and experiments*. London: Millar, 1759, 253–88
11. Hektoen. *Op cit*, 239–40
12. Tissot. *Avis au people sur sa santé.* Nouvelle ed, Liège, 1763, 160
13. Boehme JAB. Disseratio inauguralis medica de nonnullis ad morbillorum insitionem spectantibus. Halle-Magdeburg (Halae-Magd): Hendel, 1766
14. Cook J. *Letter in Gentleman's Magazine*. 1767;37:163
15. Spry E. *De variolis ac morbillis iisque inoculandis.* Dissertatio medico-practica inauguralis, de variolis ac morbillis iisque inoculandis Lugduni Batavorum (Leiden): Luchtmans, 1768, 54
16. Home F. *Principes de medicine traduits du latin en française par M Gastellier: auxquels on a joint un extrait d'un ouvrage du meme auteur, intitulé Experiences et Observations de Médecine;* [translated from English]. Paris: Vincent, 1772, 184
17. Rosen von Rosenstein N. *Anweisung zur Kenntniss und Kur der Kinderkrankheiten; aus dem Schwidischen übersetzt und mit Anmerkungen erläutert von J A Murray.* Vienna, 1787

18. Themmen CJ. Dissertatio medica inauguralis epidemiae morbillosae, Groningae anno 1816 observatae historiam sistens. Groningae, 1817
19. Hektoen. *Op cit*, 244
20. Preblud SR, Katz SL. Measles vaccine. *In:* Plotkin SA, Mortimer EA, eds. *Vaccines*. Philadelphia: Saunders, 1988
21. Albers. Die Ueberimpfung der Masern. *J Chirur Augenheilk* 1834;21:541
22. Jörg JCG. *Handbuch zum Erkennen und Heilen de Kinderkrankheiten.* 2nd ed. Leipzig: Cnobloch, 1836, 876
23. von Katona M. Nachricht von einer im Grossen erfolgreich vorgenommenen Impfung der Masern während einer epidemischen Verbreitung derselben. *Oesterr Med Wschr* 1842;29:697–8
24. McGirr JE. Inoculation in rebeola. *Northwest Med Surg J* 1850–51;7:434 (quoted by Hektoen)
25. von Mayr F. Beobachtungen über die Masern, ihre Complicationen, Nachkrankeiten und epidemische Verbreitung Z. *Gesellschaft Aerzte [Vienna]* 1852;i:6–24
26. Bufalini B. Sull' epidemia di morbillo, che ha dominato in Sienna dal Gennaio fin presso la metà di Guigno 1869. *Riv Sci R Acad Fisiocritici* 1869;I:111–26
27. Thompson H. Inoculation with suggestions for it further application in medicine, especially in mitigating the severity of measles. *Glasgow Med J* 1890;33:428
28. Copeman WSC. On some recently developed methods for measles prophylaxis. *J Hyg* 1925;24:427–41
29. Herrman C. Immunization against measles. *Arch Pediatr* 1915;32:503
30. Herrman C. Immunization against measles. *Arch Pediatr* 1922;39:607–10
31. Reinders G. *Waarneemingen en proeven meest door inëntinge op het rundvee gedaan, dienende ten bewyze, dat wij onze kalvers van gebeterde koejen geboren, door inëntinge tegen de veepest kunnen beveiligen.* Gronongen: L Huisingh, 1776
32. Huygelen C. Immunization of cattle against rinderpest in eighteenth century Europe. *Med His*, 1996, in press
33. Hiraishi S, Okamoto. On prophylactic inoculation against measles. *Jpm Med World* 1921;1:10
34. Debré R, Joannon P, Papp K. L'immunisation active contre la rougeole. 1re mémoire. *Ann Med (Paris)* 1926;20:343–61
35. Debré R, Papp K, Cross-Decam J. L'immunisation active contre la rougeole. 2e mémoire. *Ann Med (Paris)* 1928;23:119–35
36. Papp K. L'immunisation active des nourrissons contre la rougeole. *Arch Franç Pediatr* 1947;4:241–60
37. Richardson DL, Connor H. Immunization against measles. *JAMA* 1919;72:1046–8
38. International Conference on Measles Immunization. Bethesda, MD, 7–9 November 1961. *Am J Dis Child* 1962;103(3):335–517

The History of Measles Virus and the Development and Utilization of Measles Virus Vaccines

Samuel L. Katz

John Enders and Samuel Katz

Introduction

Although Rhazes, a Persian physician, is credited with the first written description of the measles [1], and perhaps with distinguishing between it and smallpox, earlier Hebrew physicians (such as Al Yehudi) had recognized the illness, but without distinction between it and other rash disorders. As urbanization occurred in subsequent centuries, the proximity or larger populations nurtured epidemics with continued circulation of virus in cities. By the seventeenth century, measles was more clearly recognized as a distinct entity as described in 1670 by the London physician,

S.L. Katz (✉)
Duke University School of Medicine, Box 2925, Durham, NC 27710, USA
e-mail: katz0004@mc.duke.edu

S.A. Plotkin (ed.), *History of Vaccine Development*,
DOI 10.1007/978-1-4419-1339-5_22, © Springer Science+Business Media, LLC 2011

Thomas Sydenham [2]. In 1758, nearly 40 years before Jenner's description of smallpox vaccine, a Scottish physician, Francis Home, attempted to produce mild measles by mimicking the variolation process used to mitigate smallpox [3]. Because, in contrast to smallpox, there were no vesicles or pustules, he chose to inoculate blood from an infected patient and was able to successfully pass infection with rash to ten of his 12 childhood subjects. Thus, he demonstrated the presence of viremia more than a century before the concept of virus had even been set forth.

In 1846, a Danish medical graduate, Peter Panum [4], observed and accurately described an outbreak of measles in the Faroe Islands and was able to define the incubation period, the infectiousness of the illness and the life-long duration of immunity among individuals who had contracted measles more than 50 years previously. The infectivity for susceptible monkeys was demonstrated in 1911 by Goldberger and Anderson who transmitted the infection to monkeys with blood and pharyngeal washings from measles patients [5]. One unsustained series of experiments by Rake, Shaffer and Stokes in the early 1940s suggests that they were able to cultivate measles virus serially in chick embryos and to transmit a modified infection to monkeys and to a small proportion of children; however, no protection was offered in late exposure to measles [6].

Attenuated Virus Vaccine

It remained for John Enders [7] and his laboratory colleagues to propagate the virus successfully in roller tube cultures of human renal epithelial cells in the mid-1950s. The cytopathic effects observed were those of cell fusion with large syncytia containing multiple nuclei which revealed easinophilic nuclear and cytoplasmic inclusions when fixed and stained. These were identical to those observed in lungs, gastrointestinal tract and other organs of patients who had died of measles. This isolate, named for David Edmonston, the 13-year-old youngster from whom Thomas Peebles had obtained blood and pharyngeal washings, was passed serially in human kidney cells, human amnion and subsequently in fertilized hen's eggs [8], and eventually in chick embryo cell cultures [9]. This became the progenitor for measles vaccines used subsequently throughout the world (Table 1). Its attenuation was first demonstrated in susceptible monkeys which

Table 1 Summary of the development of Edmonston measles virus vaccine

1954	Initial laboratory isolation of measles virus
1955	"Measles" in monkeys infected with early laboratory isolates
1956–1958	Virus adaptation to human amnion cells, chick embryos, chick embryo cell cultures
1958–1959	Attenuated, immunizing infection of monkeys inoculated with chick cell virus; resistant to challenge with "virulent virus"
1959	First susceptible children immunized
1963	Licensure of attenuated vaccine in USA

developed antibodies after inoculation with the chick cell virus, but no detectable viremia or illness, in marked contrast to other monkeys inoculated with the early kidney cell-passaged material. After intracerebral inoculation of susceptible monkeys with chick cell virus revealed no histopathology, this attenuated virus was administered to immune adults [10].

Early Vaccine Studies

With its innocuity having been demonstrated in adults, studies were then undertaken in a small group of institutionalized youngsters in whom measles occurred annually with high morbidity and mortality. In parallel with the earlier cell culture and animal studies, serologic assays had been developed to test complement-fixing, virus-neutralizing and hemagglutination-inhibiting antibodies to measles virus. Susceptible children were chosen on the basis of absent antibodies and, after parental permission, they were the first children to receive the attenuated virus subcutaneously [11]. The success of these studies led to further trials among home-dwelling children in five US cities [12]. Although significant numbers of the recipients developed fever, and some a mild rash, this occurred with apparent well-being and without discomfort. However, in attempts to modify this further, the vaccine was later accompanied by a small dose of immunoglobulin further attenuating its clinical reactivity. Subsequent laboratory passages of Edmonston virus at lower temperatures produced a "further attenuated" level of virus which was administered successfully without globulin. This vaccine or similar materials have now been used throughout the world for more than 30 years in an attempt to control and eventually eliminate measles [13].

Host Factors in Measles and Its Pathogenesis

The clinical manifestations of measles virus infection are greatly influenced by host factors. Healthy, well-nourished children usually have a self-limited illness which may be accompanied by complications (otitis media, pneumonia, gastroenteritis), but carries a low mortality rate (1 per 500). In contrast, malnourished children, particularly those with protein deficiencies, are subject to severe illness with mortality rate that may reach 20% or higher [14].

In contrast to the self-limited disease of healthy children, they may have prolonged gastroenteritis, desquamation, negative nitrogen balance for weeks or months thereafter, cutaneous bacterial abscesses, cancrum oris and other debilitating complications. In general, patients at either extreme of age (infants in the first 18 months of life and adults) are apt to be more ill. Recent studies have demonstrated the efficacy of oral Vitamin A administration in preventing the severe keratoconjunctivitis as well as other lower respiratory tract infections accompanying measles [15]. Rare central nervous system (CNS) complications have been some of the most dreaded aspects of measles [16]. Postinfection encephalitis occurs in approximately 1 per 1,000 cases and a delayed subacute sclerosing panencephalitis (SSPE) in 1 per 100,000. The latter is of particular interest because it may not become apparent until 7–10 years after the acute measles [17] and is the result of defective measles virions within the neuronal and glial cells. An intermediate form has been observed in children with severe immune compromise. Severe progressive giant cell pneumonia has occurred in some immune deficient patients [18].

Early observations of the loss of delayed cutaneous hypersensitivity to tuberculoproteins among infected patients who developed measles correlated with an exacerbation of the underlying mycobacterial disease [19]. This loss of cell-mediated immunity continues to be one of the more provocative aspects of measles pathogenesis [20, 21].

Inactivated Virus Vaccine

Simultaneously with the licensure of live attenuated virus vaccines in the early 1960s, several pharmaceutical firms marketed a "killed" measles vaccine which was prepared by formalin inactivation of Edmonston virus. Two or three injections of this vaccine produced a detectable antibody response, but when such patients subsequently were exposed to wild measles virus, they developed a severe atypical illness with CNS obtundation, marked pneumonia and a centrifugal rash that was quite unlike that of natural measles [22]. In addition to the acute nodular pneumonia with effusion that many suffered, they were later found to have persistent pulmonary nodules that remained for years after this syndrome [23]. After approximately 4 years (1963–1967), the inactivated vaccine was removed from the US market. Studies suggest that formalin inactivation that denatured the fusion protein (F) of the virion, so that patients with atypical measles lacked antibodies to this protein.

Measles Control in the Vaccine Era

Various countries have approached measles control in differing fashions, but the most common has been to administer vaccine shortly before or after the first birthday, by which time most maternal transplacental measles antibody has been catabolized, so that the attenuated virus replicated successfully. With its high immunogenicity and prophylactic efficacy, a single dose of measles vaccine produces seroconversion in 95–97% of susceptible individuals and field efficacy of at least 90%. Nonetheless, with annual birth cohorts of millions of children, even if all receive measles vaccine, there would still be an annual increment of significant number (3–5%) who would fail to respond to that initial dose. As a result, many countries have now instituted two dose schedules with the second dose given at varying times or ages after initial inoculation. With such a strategy, it has been possible to eliminate measles from some countries (Finland, Sweden) and to reduce the annual number of reported cases from more than two million to less than 100 in the USA. Because the virus is so highly communicable, importations from countries where measles vaccination is not widely practiced continue to provide introduction of virus among those few who remain susceptible.

Genetic Stability of Measles Virus

One reassuring aspect of measles virus has been its genetic stability. Viruses recovered from patients throughout the world and at various times from the 1950s to the present are all neutralized by vaccine-induced antibodies. Genomic analysis had revealed changes in only a few nucleotides of the hemagglutinin (H) and the nucleoprotein (N) genes, but this minor genetic drift had not altered its overall antigenicity. The entire genome of the Edmonston strain has been sequenced and demonstrated to contain six genes from which the six major structural proteins of the virus are coded. Using genomic analysis, it is possible to identify distinct strains from various parts of the world [24]. By such molecular epidemiology, Bellini et al. [25] and others have described in detail the replicative biology and biochemistry of the virus. They have demonstrated that the sequence among the vaccine strains used throughout the world differ by no more than 0.5–0.6% at the nucleotide level.

Global Aspects of Measles Prevention

The World Health Organization (WHO) estimates that in the early 1980s as many as 2.5 million children died annually from measles [26]. By 1994, with 78% global coverage by measles vaccine before 1 year of age, reported cases had dropped significantly and deaths were calculated to be less than 750,000. At least three different approaches have been utilized. Most commonly, an initial age of administration has been added to the immunization schedule for children in the Program on Immunization (EPI) of the WHO to 12–15 months of age (the USA). A number of nations in attempts to completely eliminate the indigenous circulation of measles virus have added a second dose of vaccine [27, 28] to cover those children who were primary vaccine failures or unusual secondary failures. A third, completely different approach had been that of mass immunization utilizing several days annually during which all children in varying age cohorts (usually up to 5 years) have been immunized nationwide. Successful examples of this last approach have been demonstrated in Cuba and Brazil [29].

An unsolved problem had been the challenge of infants who acquire wild measles before 9 months of age in a number of countries where the virus continues to circulate widely, particularly some of the sub-Saharan African nations. The "window of susceptibility" between the loss of protection afforded by maternal transplacental antibody and the age at which attenuated vaccine will successfully replicate leaves a hiatus in which wild measles continues to occur. In attempts to overcome this resistance to earlier immunization, various measles vaccines were administered at unusually high titers to children at 4–6 months of age. The successful seroconversion and protection by this approach led the WHO in 1989 to recommend administration of vaccine at high titer to children as young as 6 months of age. Although it was technologically difficult to produce large amounts of such high-titered vaccine, these programs were initiated. However, in 1990–1991 it became apparent that there was an increased mortality rate later in infancy among recipients of these high-titer materials, not from measles itself but from other infections (respiratory, gastrointestinal, malaria). Of even greater perplexity was the observation that this increased mortality occurred most often in girls. The boys, in contrast, remained well protected by the vaccine, without enhanced morbidity of mortality [30]. The biological basis of these epidemiological observations remains incompletely explained, but the immune suppression that transiently follows natural measles and measles virus vaccine has been implicated, although its pathogenesis is not fully elucidated. As a result, the early use of high-titered vaccine has been abandoned. In many countries, it appears that vaccine coverage of 85–90% on 9–12-month-old children has been successful in markedly reducing measles cases and the subsequent spread of the virus within communities [31]. The presence of successfully immunized older infants and children reduces the likelihood of younger infants' exposure.

Current Diagnostic Dilemmas

As measles has become less frequent in many areas, physicians and other health care workers have lost familiarity with the illness, so that diagnosis of rash disease is less reliable. In attempts to assist with field diagnosis, Bellini and others have devised rapid measles-specific IgM antibody tests that can be done on heel stick blood as a slide agglutination test [personal communication, 1995]. Further studies explore the possibility of similar tests on salivary and/or urine samples [32]. Field availability of such rapid, sensitive and specific tests will greatly facilitate the epidemiological surveillance of measles in varied locales.

Conclusion

It remains to be determined whether the circulation of measles virus can be totally controlled by the currently available vaccine or whether newer products will be required that permit immunization of younger infants under the cover of maternal antibody. A number of laboratories have returned to the original monkey models to investigate the possibility of such approaches using experimental vaccine prepared in ISCOMS, vectored recombinants or other more current preparations. The EPI of WHO has announced a global target of reduction of measles incidence by 90% and measles mortality by 95% from pre-EPI (1976) levels. Any sustained approach to these goals will require the establishment of a continuing effort as new susceptible cohorts is emerging each year. Until the virus has been totally eliminated, its high degree of communicability will threaten any programs that are not sustained. Although there is no nonhuman reservoir of measles virus, its ready transmission from one to another susceptible poses this major obstacle.

A number of unanswered questions remain as challenges to current and future investigators. These include the development of strategies to successfully overcome transplacental immunity, a better understanding of measles-induced immunosuppression [33], the identification of the molecular basis for virus attenuation and the explanation of gender differences in the pathogenesis of high-titered vaccine administrations in the youngest infants.

References

1. Abu Becr M. A discourse on the smallpox and measles [trans R Mead]. London: J Brindley, 1748
2. Sydenham T. *The works of Thomas Sydenham*. London: Sydenham Society, 1922;2:250–1
3. Enders JF. Francis Home and his experimental approach to medicine. *Bull Hist Med* 1964;38:101–12
4. Panum PL. Observation made during the epidemic of measles on the Faroe Islands in the year 1846. *Med Classics* 1939;3:839–86
5. Goldberger J, Anderson JF. An experimental demonstration of the presence of the virus of measles in the mixed buccal and nasal secretions. *JAMA* 1911;57:476–8
6. O'Neil GC, Stokes J Jr., Shaffer MF, Rake G. Virus of measles grown in developing chick embryo. *Am J Dis Child* 1940;60:757
7. Enders JF, Peebles TC. Propagation in tissue cultures of cytopathogenic agents from patients with measles. *Proc Soc Exp Biol Med* 1954;86:277–86
8. Milovanovic MV, Enders JA, Mitus A. Cultivation of measles virus in human amnion cells and developing chick embryo. *Proc Soc Exp Biol Med* 1957;95:120–7
9. Katz SL, Milovanvic MV, Enders JF. Propagation of measles virus in cultures of chick embryo cells. *Proc Soc Exp Biol Med* 1958;97:23–9
10. Enders JF, Katz SL, Milovanvic MV, Holloway A. Studies on an attenuated measles-virus vaccine. I. Development and preparation of the vaccine: technics for assay of effects of vaccination. *N Engl J Med* 1960;263:153–9
11. Katz SL, Enders JF, Holloway A. Studies on an attenuated measles-virus vaccine. II. Clinical, virologic and immunologic effects of vaccine in institutionalized children. *N Engl J Med* 1960;263:159–61
12. Katz SL, Kempe HC, Black FL, Lepow ML, Krugman S, Haggerty RJ, Enders JF. Studies on an attenuated measles-virus vaccine. VIII. General summary and evaluation of the results of vaccination. *N Engl J Med* 1960; 263:180–4
13. Markowitz LE, Katz SL. Measles vaccine. In: Plotkin SA, Mortimer EA Jr., eds. *Vaccines*, 2nd ed. Philadelphia: WB Saunders Co, 1994;229–76
14. Morley DC. Measles in the developing world. *Proc R Soc Med* 1974;67:1112–5
15. Hussey GD, Klein M. A randomized, controlled trial of Vitamin A in children with severe measles. *N Engl J Med* 1990;323:160–4

16. Johnsson RT, Griffin DE, Hirsch RL, Wolinsky JS, Roedenbeck S, Lindo de Soriano I, Vaisberg A. Measles encephalomyelitis – clinical and immunologic studies. *N Engl J Med* 1984;310:137–41
17. Modlin JF, Jabbour FT, Witte JJ, Halsey NA. Epidemiologic studies of measles, measles vaccine, and sub-acute sclerosing panencephalitis. *Pediatrics* 1977;59:505–12
18. Mitus A, Enders JF, Craig JM, Holloway A. Persistence of measles virus and depression of antibody formation in patients with giant cell pneumonia after measles. *N Engl J Med* 1959;261:882–9
19. von Pirquet CE. Das Verhalten der kutanen Tuberkulin-reaktion wahrend de Masern. *Dtsch Med Wochenschr* 1908;34:1297–300
20. McChesney MB, Oldstone BA. Virus-induced immunosuppression: infections with measles virus and human immunodeficiency virus. *Adv Immunol* 1989;45:335–80
21. Griffin DE, Ward BJ. Differential CD4 T cell activation in measles. *J Infect Dis* 1989;168:275–81
22. Fulginiti VA, Eller JJ, Downie AW, Kempe CH. Altered reactivity to measles virus. Atypical measles in children previously immunized with inactivated measles virus vaccines. *JAMA* 1967;202:1075–80
23. Annunziato D, Kaplan MH, Hull WW, Ichinose H, Lin JH, Balsam D, Paladino VS. Atypical measles syndrome: pathologic and serologic findings. *Pediatrics* 1982;70:203–9
24. Rota JS, Hummel KB, Rota PA, Bellini WJ. Genetic variability of the glycoprotein genes of current wild-type measles isolates. *Virology* 1992;188:135–42
25. Bellini WJ, Rota JA, Rota P. Virology of measles virus. *J Infect Dis* 1994;170 (Suppl 1): S15–23
26. Assaad F. Measles: summary of worldwide impact. *Rev Infect Dis* 1983;5:452–9
27. Bottiger M, Christenson B, Romanus V, Taranger J, Strandell A. Swedish experience of two dose vaccination programme aiming at eliminating measles, mumps and rubella. *Br Med J* 1987;295:1264–7
28. Peltola H, Heinonen OP, Valle M et al. The elimination of indigenous measles, mumps, and rubella from Finland by a 12 year, two-dose vaccination program. *N Engl J Med* 1994;331:1397–402
29. Anonymous. Measles elimination in the Americas. *Bull Pan Am Health Organ* 1992;26:271–5
30. Gellin BG, Katz SL. Measles: state of the art and future direction. *J Infect Dis* 1994;170 (Suppl 1):S3–14
31. Cutts FT, Markowitz LE. Successes and failures in measles control. *J Infect Dis* 1994;170 (Suppl 1):S32–41
32. Gresser I, Katz SL. Isolation of measles virus from urine. *N Engl J Med* 1960;263:452–4
33. Griffin DE, Ward BJ, Esolen LM. Pathogenesis of measles virus infection: a hypothesis for altered immune responses. *J Infect Dis* 1994;170 (Suppl 1):S24–31

The Development of Live Attenuated Mumps Virus Vaccine in Historic Perspective and Its Role in the Evolution of Combined Measles–Mumps–Rubella

Maurice R. Hilleman[†]

Early History

Diseases with distinctive clinical features can often be identified with occurrence in early history. Hippocrates has been credited with clinical descriptions of mumps, parotitis and orchitis in thea fifth century BC [1]. Hamilton [2] is credited with the first recognition, in 1790, of central nervous system involvement in mumps and with the first description of neuropathology in fatal cases. Johnson and Goodpasture [3, 4] provided the first definitive evidence of viral etiology in mumps through monkey and human experimentation. Cultivation of mumps virus in embryonated hens' eggs was accomplished by Habel in 1945 [5].

[†] Decessed

S.A. Plotkin (ed.), *History of Vaccine Development*,
DOI 10.1007/978-1-4419-1339-5_23, © Springer Science+Business Media, LLC 2011

Virus

Mumps virus [6] is placed in the genus *Paramyxovirus* and within the family Paramyoxviridae. The virus is "pleomorphic" and consists of a single negative strand RNA genome surrounded by a nucleocapsid and an envelope from which spikes protrude that bear the hemagglutinin/neuraminidase glycoproteins and fusion proteins. The particle size is 100–300 nm. Although the virus is highly pantropic with respect to the organs and tissues it can infect [7, 8], the principal clinical effects are on the parotids, the gonads and the central nervous system.

Genome

There are seven genes separated by intergenic sequences. As shown in Table 1 (see [6]), they are tentatively believed to encode six structural proteins (N, P, M, L, HN and F proteins), two nonstructural proteins (NS1 and NS2) and the SH protein. Important among the structural genes are those coding for the F protein that is essential to viral penetration into the host cell, the HN proteins essential to viral attachment and release and the N proteins of the nucleocapsid. The highly variable SH protein is hydrophobic and its structural or nonstructural status is unknown. The specific functions of the SH, NS1 and NS2 proteins are still undefined.

Mumps Vaccine Developments

Background

The earlier history of attempted development of both killed and live mumps virus vaccines has been marked by abortive and inconsequential efforts to evolve vaccines having the needed attributes of high-level efficacy, lasting duration and

Table 1 Proteins encoded by the mumps virus genome

Structural	
N	Nucleocapsid
P	Phosphoprotein
M	Matrix
F	Fusion protein
HN	Hemagglutinin-neuraminidase
L	Large protein
Nonstructural	
NS1	Protein 1
NS2	Protein 2
Structural or nonstructural	
SH	Small hydrophobic

apathogenicity [1, 6, 7]. The Jeryl Lynn strain live attenuated mumps virus vaccine [9, 10], which was developed in our laboratories and brought to market in 1967, has been a paradigm for successful vaccine development because of its desirable properties of clinical nonreactivity, high-level protective immunity and durable immunity following a single dose of vaccine that may prove to be lifelong [11, 12]. Cochi et al. [1] listed ten live attenuated mumps vaccines being marketed worldwide as of 1988. Some, however, have been withdrawn from problems of reactogenicity (see later). The Leningrad strain of mumps vaccine developed by Smorodintsev [13] was used extensively in the Soviet Union, but causes mumps and meningitis is some recipients [14]. Wolinsky et al. [7] pointed out the difficulty of developing killed mumps vaccines that achieve an immune response directed against the proteins in their native conformation. A killed vaccine used by the Finnish army [15] gave protection against mumps, but immunity was not of long duration.

Development of the Jeryl Lynn Strain Mumps Vaccine

Basic Concepts

Work toward the development of a live attenuated mumps virus vaccine [10, 16–19] was started in our laboratories in 1959; at the time when the development of live measles virus vaccine was already well along. It was conceived at the time that a new era for pediatric vaccinology might be created through the development of combinations and permutations of possible single-dose live virus vaccines (including measles, mumps, rubella, varicella and hepatitis A), even though all but the measles were still only theoretical possibilities. Mumps was chosen in what was conceived to be a possible initial combined vaccine containing measles and mumps components.

Problems in Selection of Parent Virus Strains for Vaccine

Clinical studies of individual measles, mumps, rubella and varicella vaccines shared the lack of suitable animal models that would prove a reliable preclinical appraisal of safety and efficacy when given to human beings. Hence, the development process was guided more by judgment than actual data. Determination of the proper passage level to assure safety in human subjects was made possible by the creation of numerous lots of vaccine, at sequential embryonated egg or egg cell culture passage levels that were presumed to correlate with increased attenuation. This was immensely difficult and costly since each test vaccine was akin to a vaccine production lot that needed to comply with the rigorous standards of the Federal Regulatory Authority (now FDA) applied to commercially distributed products. The largest problem was the need to make the judgment call as to which distant point in the attenuation cycle would serve to start first clinical tests. Once safety at a particular level was established, then sequential progressive work backing for determining the optimal clinical reaction/efficacy relationship was more routine.

A second hurdle in vaccine strain development lay in the fact that various clinical isolates of a particular virus may genetically be different, with many failing to provide the selectivity needed to achieve a satisfactory balance between immunogenicity and attenuation. This was an especially difficult problem for mumps vaccine [17] since the virus attenuated rapidly in chick embryos/cells and acceptability for a clinically useful product is confined to a very narrowly restricted passage history. A further contribution to the research problem was that only the laborious serum neutralization test provided the sensitivity and reliability [20] needed to assess antibody responses to vaccination.

Attenuated Jeryl Lynn Mumps Virus

Throat washings were taken on March 23, 1963, from a 5-year-old child (Jeryl Lynn) with mumps and were used to develop the mumps vaccine. After a series of clinical tests, passage level A (collectively 12 passages in chick embryos and in chick embryo cell culture) and level B (collectively 17 or 18 passages, 12 in chick embryos and 5 or 6 in cell culture) were selected for further clinical study. Table 2 presents a synopsis of the findings [17]. Clearly both vaccines A and B were immunogenic, but only level B vaccine caused no overt clinical reactions and was chosen for further investigations. Vaccine at collective passage level 27 was not adequately immunogenic and was not tested further. All vaccines were dried.

Proof of Safety and Efficacy

Studies to determine the safety and protective efficacy of the vaccine were carried out among 867 initially seronegative children, in families or in schools, who resided in the Havertown-Springfield area of Philadelphia, PA, USA [18, 19]. Ninety-eight percent of the vaccinated children developed neutralizing antibodies following vaccination. A synopsis presented in Table 3 shows that the level or protective efficacy was about 98% and this correlated with the positive neutralizing antibody responses.

Table 2 Clinical and immunological response to Jeryl Lynn mumps vaccines A and B

Clinical[a]				
Passage level	Parotitis	Virus isolation	Serum amylase elevation	Neutralizing antibody response[a]
A (12)	4/16	3/16	0/16	16/16
B (17)	0/14	0/14	0/14	14/14

[a] No children/total

Table 3 Occurrence of proved mumps cases in controlled studies[a] for protective efficacy of mumps vaccine in children exposed to mumps virus infection

Cases of proved mumps/total exposed			
Venue for exposure	Vaccinated	Nonvaccinated controls	Protective efficacy (%)
Classroom	0/14	22/24	100
Families	2/86	39/76	96

[a]The total initially seronegative study population was 362 vaccinated and 505 controls

The clinical reactions to vaccination were inconsequential and were mainly limited to local inflammation at the injection site.

Subsequent Studies

In additional studies, the vaccine proved equally immunogenic and as nonreactogenic in adults as in children [21]. It was further determined [22, 23] that the minimal amount of virus per dose required to immunize 97% or more of child or adult recipients was about 317 fifty percent tissue culture infectious units ($TCID_{50}$). Five thousand $TCID_{50}$ per dose are routinely used in the vaccine. The dried vaccine is stable on storage in the refrigerator [9, 24], and there is no spread of virus by contact between vaccinated and susceptible children [9]. Immunity has proven durable and long lasting to the present [11, 12].

Little is known about subclinical reinfection on exposure to mumps and about the basis of immunity. Possibly, clinically discernable reinfection following natural infection has been reported, although with low frequency [25]. It may be conjectured, however, that subclinical reinfection may occur at respiratory mucosal surfaces as for measles and rubella. Virus neutralization by antibody may be presumed to the principal mechanism for immunity. However, in more viral infections, reinfection is abortive, and cytotoxic T cell responses clear virus-infected cells. This is probably true also for mumps. It may be conjectured also that long-term immunity is based on immunologic memory resident in memory B, T-helper and cytotoxic T cells.

Special Problems in Developing Suitable Mumps Virus Vaccines

Indigenous Viruses of Primary Chick Cell Cultures

A major problem in developing the Edmonston B measles virus vaccine in our laboratories in 1961 [26] and its more attenuated successor [27] was the ubiquitous occurrence in chick cells, in culture, or viruses of the avian leukosis complex, including leukemia. Such viruses did not provide evidence for lack of safety for humans, based on many years of use of chick embryo-grown yellow fever virus vaccine. We were, however, restive and determined not to put that human population at possible, or even, remote risk to side effects due to leukemia virus.

An answer to the leukosis problem was provided by the development of the resistance-inducing factor (RIF) test [28] that permitted in vitro detection of viruses of the avian leukosis complex. Acting on the finding that eggs laid by hens that had antibodies against leukosis, Hughes et al. [29] develop experimental chicken flocks that were free of leukosis virus. Use of chick embryos from these starter flocks to prepare cell cultures permitted our laboratories to develop and produce a measles vaccine that was free of leukosis viruses. Once leukosis-free chick cell cultures were available, it was also possible to develop and produce mumps vaccine free of the contaminating avian leukosis virus(es).

Central Nervous System Involvement in the Clinical Application of Certain Live Attenuated Mumps Vaccines

It was observed around the turn of the present decade that certain mumps virus vaccines used singly or in combination with measles and rubella viruses were insufficiently attenuated and cause aseptic meningitis or meningoencephalitis. The Leningrad-3 strain was reported as early 1986 [14] to be underattenuated, causing about one case of meningitis per 1,000 vaccinated persons. Prime attention in Western Europe [30–33], Canada [34] and Japan [35, 36] was focused on the neuropathogenicity of the Urabe mumps strain-containing vaccine made by European and Japanese vaccine manufacturers. The World Health Organization (WHO) [30, 32] estimates the occurrence of about one case of meningitis per 4,000 vaccinees. Identity of the Urabe vaccine strain in viruses isolated from vaccinated patients was established by genetic sequencing of the viral nucleic acid. In Japan [35, 36], the attack rate for Urabe vaccine meningitis was about 1:400–1:1,200. In addition, there was one report of possible virus transmission from a vaccinated child to a susceptible sibling [37]. The high frequency of Urabe virus meningitis may have been associated with changes in the manufacturing process [36]. Use of the Urabe strain vaccine was discontinued in many countries [30, 33, 34] worldwide, even though the risk-to-benefit ratio from use of Urabe vaccine was still highly favorable, considering the severity of natural infection.

There is no definitive evidence to indicate that the Jeryl Lynn virus is ever causally associated with meningitis, certainly no more than one case per million vaccinated persons. For this reason, the Jeryl Lynn vaccine had been frequently substituted, when available, for the Urabe product [32, 34].

Mumps Virus Quasi-Species and Plurality Between and Within Clusters

Viruses, especially RNA viruses belonging to any particular species, may greatly differ in their genetic composition. Individual strains of virus may vary substantively even within patients, and in vitro culture. Individual viral isolates may be assembled into distinctive clusters based on the degree of genetic relatedness. Collectively, these differences are the basis for the designation of multiple quasi-species rather than strict species composition.

Based primarily on genetic analyses of the hypervariable SH region, the Jeryl Lynn mumps virus was shown [38] to consist of two very similar but distinguishable quasi-species entities designated JL2 and JL5. It is not known whether their variants were present in the original patient or whether they arose on serial passage in chick embryos and in cell cultures during attenuation. It is also unknown how many such SH variants may be present in the Jeryl Lynn vaccine strain or in other mumps virus vaccines as well. The important matter is that variation in the SH region is without significance to safety since it has not demonstrated effect on either virulence or protective efficacy. With respect to vaccine safety, Afzal et al. [38] of the British Regulatory Authority noted that the master seed-controlled passage system for production "ensures a consistently reproducible immunogenicity, safety and tolerability profile" and "the product will remain consistent in quality and its excellence record of clinical safety and efficacy will be maintained."

Importantly, a number of attenuated and wild mumps viruses have been shown to fall into three [38] or four [39] distinctive clusters based on genetic composition. It is of importance that the Jeryl Lynn, Leningrad-3 and Urabe vaccine strains each reside in a different cluster. It is possible that the distinctive genetic profiles of these viruses may account for their differences in neurovirulence.

Individual and Combined Live Measles, Mumps and Rubella Vaccines

The combined live measles–mumps–rubella (MMR) vaccine was the realization of a long-term plan to create single-dose bi- and trivalent combinations of these three vaccines [40]. Development of the combined vaccines depended on prior development of the individual vaccines. Measles virus vaccine (Rubeovax) [26], which was licensed in 1963 using the original Enders' Edmonston B strain [41], was so virulent that reactogenicty needed to be reduced by giving virus to children simultaneously with measles immune globulin. The Edmonston strain was further attenuated biologically in our laboratories to create the more attenuated Enders' line, Moraten [27], which was licensed (Attenuvax) in 1968. The development of the Jeryl Lynn mumps virus was described earlier. Licensed rubella virus vaccines were the sequel to our pioneering studies on the Merck Benoît strain [42–47] of duck cell culture attenuated virus isolated from a child names Benoît. The HPV-77 strain that was grown in monkey kidney cultures by Parkman and Meyer [48] was substituted by us for Benoît in order to respond to Mary Lasker's (Lasker Foundation) request to concentrate on a single vaccine that could be developed more quickly than two separate vaccines, in order to avert an anticipated rubella epidemic in the United States. The original HPV-77 monkey virus [48] was excessively virulent and was never licensed. However, HPV-77 monkey cell strain was further attenuated in or laboratories to develop the HPV-77 duck cell line which was licensed (Meruvax) in 1989 [46, 47, 49, 50]. Later, HPV-77 duck cell vaccine was substituted by the RA27/3 human diploid cell grown virus [51] and was licensed under the name Meruvax II in 1978 [52, 53].

The only prior precedent for combined MMR vaccines was developed in our laboratories starting with combined MMR (Biavax, 1970) [54], followed by measles–rubella (M-R-Vax, 1971) [55], MMR (M-M-R, 1971) [56–58] and measles–mumps vaccine (M-M-Vax), which was licensed in 1973 [59]. All the proprietary names for the products were registered trademarks and an II was added to the proprietary name of each vaccine in which the RA27/3 rubella strain was substituted for HPV-77 duck virus. M-M-R® [56–58] and M-M-R® II [52, 53] were the final and successful culmination of long-term effort and became both the driving force for immunization against measles, mumps and rubella vaccines in the United States, Canada and Europe and are now being introduced into some developing countries.

The development of combined vaccines was not without difficulty and the major investigative hurdles included efforts to obtain (1) as good immune responses as with the individual vaccines give alone, (2) lack of increased reactogenicity, (3) compatibility of formulation to assure stability of potency on drying and storage, and (4) capability in quality control to quantify the infectivity (potency) of each viral component.

The Legacy of Measles, Mumps and Rubella Vaccines

The benefits from use of mumps vaccine, singly or in combination, are best illustrated in Fig. 1, which show the reduction in occurrence in the United States of mumps, following introduction of the vaccine against them. All measles, mumps and rubella vaccines used in the United States have been exclusively of Merck & Co source, except for short-term use of the Schwartz strain measles vaccine and the Cendehill strain rubella vaccine. It is seen that each of the three diseases was dramatically reduced to inconsequential or near inconsequential levels within the decade following their introduction. It is also significant that measles, mumps and rubella each are caused by a single virus of constant immunologic specificity. Therefore, all the three diseases should be able to be eradicated worldwide, especially through the use of the combined vaccine, and this maybe a worthy target for the future. Protective immunity following proper initial response is of long, if not permanent, duration [11, 12].

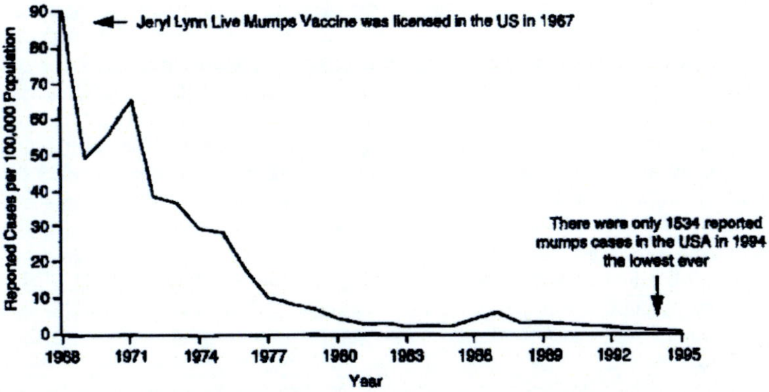

Fig. 1 Mumps cases, by year, in the United States, 1968-1994. Adapted from Summary of notifiable diseases, United States, 1994, *MMWR Supplement* 6 October 1995; 43(53):42

Editor's Note

Dr. Hilleman's description of the origin of the Jeryl Lynn strain fails to mention the poignant fact that the Jeryl Lynn isolate was from his own daughter. On the negative side, his optimism that mumps vaccination would lead to permanent immunity has been vitiated by subsequent observations of partial effectiveness in the field and epidemics of mumps in young adults despite prior vaccination [60, 61]. Nevertheless, even in those epidemics the effectiveness of vaccination was shown to be considerable [62, 63].

References

1. Cochi SL, Wharton M. Plotkin SA. Mumps vaccine. In: Plotkin SA, Mortimer ES Jr, eds. *Vaccines* Philadelphia: WB Saunders, 1994:277–301
2. Hamilton R. An account of a distemper by the common people of England vulgarly called the mumps. *London med J* 1790; (1) 190–211
3. Johnson CD, Goodpasture EW. An investigation of the etiology of mumps. *J Exp Med* 1934;59:1–19
4. Johnson CD, Goodpasture E. The etiology of mumps. *Am J Hyg* 1935;21:46–57
5. Habel K. Cultivation of mumps virus in the developing chick embryo and its application to studies of immunity to mumps in man. *Public Health Rep* 1945;60:201–12
6. Rima BK. Mumps Virus. In: Webster RG, Granoff A, eds. *Encyclopedia of Virology*. London: Academic Press, 1994;2:876–83
7. Wolinsky JS, Waxham MN. Mumps virus. In: Fields BN, Knipe DM, eds. *Virology*. New York: Raven Press, 1990;2nd ed:989–1011
8. Nussinovitch M, Volvovitz B, Varsana I. Complications of mumps requiring hospitalization in children. *Eur J Pediatr* 1995;154:732–4
9. Hilleman MR, Buynak EB, Weibel RE, Stokes J Jr. Live, attenuated mumps-virus vaccine. *N Engl Med* 1968;278:227–32

10. Hilleman MR. Past, present, and future of measles, mumps and rubella virus vaccines. *Pediatrics* 1992;90:149–53
11. Weibel RE, Buynak EB, McLean AA, Hilleman MR. Follow-up surveillance for antibody in human subjects following live attenuated measles, mumps and rubella virus vaccines. *Proc Soc Exp Biol Med* 1979;162:328–32
12. Weibel RE, Buynak EB, McLean AA, Roehm RR, Hilleman MR. Persistance of antibody in human subjects for seven to ten years following administration of combined live attenuated measles, mumps and rubella virus vaccines. *Proc Soc Exp Biol Med* 1980;165:260–3
13. Smorodintsev AA, Klyachko NS. Mumps live vaccine in the USSR: a summary of recent developments. *Prog Med Virol* 1961;3:273–86
14. Cizman M, Mozetic M, Furman-Jakopic M, Pleterski-Rigler D, Radescek-Rakar R, Susec-Michieli M. Aseptic meningitis following combined vaccination with the Leningrad-3 strain of mumps virus and the Edmonston-Zagreb strain of measles virus. *Zdrav Vestn* 1986;55:587–91
15. Penttinen K, Cantell K, Somer P, Poikolianen A. Mumps vaccination in the Finnish defense forces. *Am J Epidemiol* 1968;39:363–71
16. Buynak EB, Hilleman MR. Live attenuated mumps virus vaccine. I. Vaccine development. *Proc Soc Exp Biol Med* 1966;123:768–75
17. Stokes H Jr., Weibel RE, Buynak EB, Hilleman MR. Live attenuated mumps-virus vaccine. II. Early clinical studies. *Pediatrics* 1967;39:363–71
18. Weibel RE, Stokes J Jr., Buynak EB, Whitman JE Jr., Hilleman MR. Live attenuated mumps-virus vaccine 3. Clinical and serological aspects in a field evaluation. *N Engl J Med* 1967; 276:245–51
19. Hilleman MR, Weibel RE, Buynak EB, Stoke J Jr., Whitman JE Jr. Live attenuated mumps-virus vaccine. 4. Protective efficacy as measured in a field evaluation. *N Engl J Med* 1967; 276:252–8
20. Buynak EB, Whitman JE Jr., Roehm RR, Morton DH, Lampson GP, Hilleman MR. Comparisons of neutralization and hemagglutination-inhibition techniques for measuring mumps antibody. *Proc Soc Exp Biol Med* 1967;125:1068–71
21. Davidson WL, Buynak EB, Leagus MB, Whitman JE Jr., Hilleman MR. Vaccination of adults with live attenutatied mumps virus vaccine. *JAMA* 1967;201:995–8
22. Hilleman MR. Mumps vaccination. In: Heath RB, Waterson AP, eds. *Modern Trends in Medical Virology*. London: Butterworths, 1970;241–61
23. Buynak EB, Hilleman MR, Leagus MB, Whitman JE Jr., Weibel RE, Stokes J Jr. Jeryl Lynn strain live attenuated mumps virus vaccine: Influence of age, virus dose, lot, and γ-globulin administration on response. *JAMA* 1968;203:9–13
24. McAleer WJ, Markus HZ, McLean AA, Buynak EB, Hilleman MR. Stability on storage at various temperatures of live measles, mumps, and rubella virus vaccines in new stabilizer. *J Biol Stand* 1980;8:281–7
25. Gut JP, Lablache C, Behr S, Kim A. Symptomatic mumps virus reinfections. *J Med Virol* 1995;45:17–23
26. Hilleman MR, Stokes J Jr., Buynak EB, Weibel R, Halenda R, Goldner H. Enders' live measles virus vaccine with human immune globulin. 2. Evaluation of efficacy. *Am J Dis Child* 1962;103:373–9
27. Hilleman MR, Buynak EB, Weibel RE, Stokes J Jr., Whitman JE Jr., Leagus MB. Development and evaluation of the Moraten measles virus vaccine. *JAMA* 1968;206:587–90
28. Rubin H. A virus in chick embryos which induces resistance in vitro to infection with rous sarcoma virus. *Proc Natl Acad Sci USA* 1960;46:1105–19
29. Hughes WF, Watanabe DH, Rubin H. The development of a chicken flock apparently free of leukosis virus. *Avian Dis* 1963;7:154–65
30. World Health Organization: Meningitis associated with measles-mumps-rubella vaccines. *Weekly Epidemiological Record* 1992;67:301–2
31. Forsey T, Bentley ML, Minor PD, Begg N. Mumps vaccines and meningitis. *Lancet* 1992;340:980

32. Miller E, Goldacri M, Pugh S et al. Risk of aseptic meningitis after measles, mumps, and rubella vaccine in UK children. *Lancet* 1993;341:979–82
33. Schmitt HJ, Just M, Neiss A. Withdrawal of a mumps vaccine: reasons and impacts. *Eur J Pediatr* 1993;152:387–8
34. Fures J. Vaccine-related mumps meningitis-Canada. *Can Dis Weekly Rep* 1990;16–50:253–4
35. Motegi Y, Fijinaga T, Tamura H, Kuroume T. A survey of children with meningitis following measles, mumps, and rubella (MMR) immunization. *Pediatr Res* 1990;27:95A (Abstr 558)
36. Yawata M. Japan's troubles with measles-mumps-rubella vaccine. *Lancet* 1994;343:105–6
37. Sawada H, Yano S, Oka Y, Togashi T. Transmission of Urabe mumps vaccine between siblings. *Lancet* 1993;342:371
38. Afzal MA, Pickford AR, Forsey T, Heath AB, Minor PD. The Jeryl Lynn vaccine strain of mumps virus is a mixture of two distinct isolates, *J Gen Virol* 1993;74:917–20
39. Künkel U, Driesel G, Henning U, Gerike E, Willers H, Schreier E. Differentiation of vaccine and wild mumps viruses by polymerase chain reaction and nucleotide sequencing of the SH gene: brief report. *J Med Virol* 1995;45:121–6
40. Hilleman MR, Weibel RE, Buynak EB, Villarejos VM. Practical aspects concerning combined live viral vaccines. *Schweiz Rundschau Med (PRAXIS)* 1975;64:98–108
41. Katz SL, Enders JF. Immunization of children with a live attenuated measles virus. *Am J Dis Child* 1959;98:605–7
42. Stokes J Jr., Weibel RE, Buynak EB, Hilleman MR. Clinical and laboratory tests of Merck strain live attenuated rubella virus vaccine. First International Conference on Vaccines against Viral and Rickettsial Diseases of Man, Washington, DC, 7–11 November 1966. *Pan America Health Organization Sci Publ No 147.* 1967;402–5
43. Buynak EB, Hilleman MR, Weibel RE, Stokes J Jr. Live attenuated rubella virus vaccines prepared in duck embryo cell culture. I. Development and clinical testing. *JAMA* 1968; 204:195–200
44. Weibel RE, Stokes J Jr., Buynak EB, Whitman JE Jr., Leagus MB, Hilleman MR. Live attenuated rubella virus vaccines prepared in duck embryo cell culture. II. Clinical tests in families and in an institution. *JAMA* 1968;205:554–8
45. Hilleman MR, Buynak EB, Weibel RE, Stokes J Jr. Current concepts. Live, attenuated rubella-virus vaccines. *New Engl J Med* 1968;279:300–3
46. Hilleman MR, Buynak EB, Whitman JE Jr., Weibel RE, Stokes J Jr. Summary report on rubella virus vaccines prepared in duck embryo cell culture. *Symp Series Immunobiol Stand* 1969;11:349–56
47. Weibel RE, Stokes J Jr., Buynak EB, Hillman MR. Live rubella vaccines in adults and children. HPV-77 and Merck-Benoit strains. *Am J Dis Child* 1969;118:226–9
48. Meyer HM JR., Parkman PD, Panis TC. Attenuated rubella virus. II. Production of experimental live-virus vaccine and clinical trial. *N Engl J Med* 1966;275:575–80
49. Hilleman MR, Buynak EB, Whitman JE Jr., Weibel RE, Stokes J Jr. Live attenuated rubella virus vaccines. Experiences with duck embryo cell preparations. *Am J Dis Child* 1969;118:166–71
50. Buynak EB, Larson VM, McAleer WJ, Mascoli CC, Hilleman MR. Preparation and testing of duck embryo cell culture rubella vaccine. *Am J Dis Child* 1969;118:347–54
51. Plotkin SA, Farquhar JD, Katz M, Buser F. Attenuation of RA27/3 rubella virus in WI-38 human diploid cells. *Am J Dis Child* 1969;118:178–85
52. Weibel RE, Villarejos VM, Klein EB, Buynak EB, McLean AA, Hilleman MR. Clinical and laboratory studies of live attenuated RA27/3 and HPV-77-DE rubella virus vaccines. *Proc Soc Exp Biol Med* 1980;165:44–9
53. Weibel RE, Carlson AJ Jr., Villarejos VM, Buynak EB, McLean AA, Hilleman MR. Clinical and laboratory studies of combined live measles, mumps, and rubella vaccines using the RA27/3 rubella virus. *Proc Soc Exp Biol Med* 1980;165:323–6
54. Weibel RE, Stokes J Jr., Villarejos VM, Arguedas GJA, Buynak EB, Hilleman MR. Combined live rubella-mumps virus vaccine. Findings in clinical-laboratory studies. *JAMA* 1971;216:983–6

55. Villarejos VM, Arguedas GJA, Buynak EB, Weibel RE, Stokes J Jr., Hilleman MR. Combined live measles-rubella virus vaccine. *J Pediatr* 1971;79:599–604
56. Buynak EB, Weibel RE, Whitman JE Jr., Stoke J Jr., Hilleman MR. Combined live measles, mumps, and rubella virus vaccines. *JAMA* 1969;207:2259–62
57. Hilleman MR, Weibel RE, Villarejos VM, et al. Combined live virus vaccines. Proceedings of the International Conference on the Application of Vaccines against Viral, Rickettsial, and Bacterial Diseases of Man, Washington DC, 14–18 December 1970. *Pan American Health Organization Sci Publ No 226*. 1971;397–400
58. Stokes J Jr., Weibel RE, Villarejos VM, Arguedas GJA, Buynak EB, Hilleman MR. Trivalent combined measles-mumps-rubella vaccine. Findings in clinical laboratory studies. *JAMA* 1971;218:57–61
59. Weibel RE, Villarejos VM, Hernández CG, Stokes J Jr., Buynak EB, Hilleman MR. Combined live measles-mumps virus vaccine. Findings in clinical-laboratory studies. *Arch Dis Child* 1973;48:532–6
60. Plotkin SA, Rubin SA. Mumps vaccines. Vaccines 2008, 5th ed. pp: 435–466. W.B. Saunders, Philadelphia
61. Barskey AE, Glasser JW, LeBaron CW. Mumps resurgences in the United States: A historical perspective on unexpected elements. *Vaccine* 2009;27:6186–95
62. Dayan GH, Rubin S. Mumps outbreaks in vaccinated populations: are available mumps vaccines effective enough to prevent outbreaks? *Clin Infect Dis* 2008;47:1458–67
63. Dominquez A, Tormer N, Castilla J, et al. Mumps vaccine effectiveness in highly immunized populations. *Vaccine* 2010;28:3567–70

History of Rubella Vaccines and the Recent History of Cell Culture

Stanley A. Plotkin

The name of a disease is always a matter of some importance. It should be short for the sake of convenience in writing, and euphonious for ease in pronunciation... Rotheln is harsh and foreign to our ears... I therefore venture to propose Rubella as a substitute for Rotheln, or, as a name for the disease which it has been my object in this paper to describe.

Henry Veale

Rubella is not one of those diseases whose origins are lost in antiquity. Unknown until the end of the eighteenth century, it remained an unimportant rash disease for almost 200 years, when it was discovered to be a fetal teratogen. Twenty years later, the viral agent of rubella was isolated, and vaccines were developed and

S.A. Plotkin (✉)
University of Pennsylvania and Vaxconsult,
4650 Wismer Rd., Doylestown, PA 18902, USA
e-mail: stanley.plotkin@vaxconsult.com

S.A. Plotkin (ed.), *History of Vaccine Development*,
DOI 10.1007/978-1-4419-1339-5_24,

commercialized within 10 years. By 5 years postlicensure, an impact on rubella incidence was evident, but another 5 years were required, together with changes in public health policy towards more universal vaccination, before congenital rubella syndrome (CRS) became rare.

Rubella was first discussed in the medical literature under the name "rotheln," which reflects its original description by German physicians at the end of the eighteenth century [1]. Maton [2], in 1815, is credited with the first English language description of the disease, and it was Veale [3], a military physician writing from India, who conferred on it the eponym "rubella" – "little red" in Latin. Subsequent writers were mostly concerned with rubella as a problem in differential diagnosis of rash disease, particularly in relation to measles and scarlet fever, until a consensus was reached in 1881 that rubella was indeed a specific illness [1]. This clinical inference was confirmed in 1938, when two Japanese scientists, Hiro and Tasaka [4], transmitted the disease from human to human using throat washings.

Today, rubella is understood as a viral upper respiratory infection in which replication takes place initially in the nasopharynx and then in adjacent lymph nodes, from which a viremia is generated. After an incubation period of 14–21 days, there is a short febrile prodromal illness, followed by a fine maculopapular rash beginning on the face and extending over the entire body surface. Resolution of rash occurs rapidly, but there are three principal complications of acquired disease: arthritis, encephalitis, and thrombocytopenia. Arthritis occurs in 70% of adults, encephalitis in 1 of 6,000 infections (but occasionally at a much higher rate), and thrombocytopenia in 1 of 3,000 patients [5, 6].

However, the fourth and most important clinical complication was discovered by a remarkable Australian ophthalmologist, Norman McAlister Gregg (1891–1966). In 1939, Australia became involved in the Second World War, with resultant recruitment and mixing of large numbers of young men, creating the right conditions for rubella epidemics. These epidemics inevitably spread to the soldiers' young female consorts. In 1940, Gregg began to see an unusual number of infants with congenital cataracts. By taking accurate histories from the new mothers, Gregg found that many had had rubella early in pregnancy. From this, Gregg drew the correct inference that rubella had affected the ocular development of the fetus [7].

Space does not permit a full description of the CRS and its interesting pathogenesis. Table 1 summarizes the important clinical aspects of the disease. CRS is a disseminated viral infection of the fetus secondary to maternal viremia, which results in a myriad of anatomic and functional abnormalities.

Gregg's identification of rubella as the cause of congenital disease went through the usual period of criticism and doubt, but by the end of the 1940s, ample confirmation had been obtained throughout the world. Volunteer experiments confirmed that rubella is caused by a virus, but no success was reported in cultivating the agent by animal inoculation or by growth in cell culture.

Table 1 Prominent clinical findings in congenital rubella syndrome

Encephalitis	
Microcephaly	Intrauterine growth retardation
Mental retardation	
Autism	Metaphyseal rarefactions
Patent ductus arteriosus	
Peripheral pulmonic artery stenosis	Hepatosplenomegaly
Cochlear deafness	Thrombocytopenic purpura
Central auditory imperception	Interstitial pneumonitis
	Diabetes
	Hypothyroidism
Retinitis	
Cataracts	
Microphthalmia	
Glaucoma	

Modified from Alford CA, Griffiths PD. Rubella. In: Remington JS, Klein JO, eds. *Infectious Diseases of the Fetus and Newborn Infant.* Philadelphia: WB Saunders, 1983.

Breakthrough

In 1961, two laboratories succeeded where others had failed. One lab, at the Harvard School of Public Health, was headed by Tom Weller, who, in his usual painstaking way, noted a subtle cytopathogenic effect of rubella in human amniotic cells incubated for long periods of time [8]. Although this method uncovered the virus, it was never used much, owing to the difficulty that few were willing to take the same pains that Weller had.

The second laboratory was at the Walter Reed Army Institute [9]. It was headed by Edward Buescher and staffed by two young physicians, Paul Parkman and Malcolm Artenstein. They were investigating an outbreak of adenovirus disease at the Fort Dix New Jersey army base in the spring of 1961, but also came across many recruits who were hospitalized for rubella. The team set out to isolate rubella virus by inoculating throat washes into the usual cell cultures, with the important addition of African green monkey kidney (AGMK), which had just become available. In order to detect the presence of noncytopathogenic agents that might induce interferon, Parkman decided to challenge the cultures with an unrelated cytopathogenic agent. ECHO 11 was chosen for this purpose. This combination of techniques worked superbly: specimens from soldiers with rubella induced a state of refractoriness to the ECHO 11 and that interference effect could be passaged serially to other cultures, demonstrating that it was produced by a live virus.

Rubella virus has been photographed and sequenced [10]. Space does not permit a description of its molecular biology, which includes an RNA genome coding for two envelope proteins, a core protein and lipid envelope.

Table 2 Estimated morbidity associated with the 1964–65 rubella epidemic in the United States

Clinical events		
Rubella cases		12,500,000
Arthritis-arthralgia		159,375
Encephalitis		2,084
Deaths:		
Excess neonatal deaths	2,100	
Other deaths	60	
Total deaths		2,160
Excess foetal wastage		6,250
Congenital rubella syndrome:		
Deaf children	8,055	
Deaf-blind children	3,580	
Mentally retarded children	1,790	
Other congenital rubella syndrome	6,575	
Total congenital rubella syndrome		20,000
Therapeutic abortions		5,000

The isolation of rubella virus in 1961 could not have come at a more fortunate time, for in the spring of 1963 an epidemic of rubella occurred in Europe, which I witnessed because at the time I was a Registrar at the Hospital for Sick Children in London. The epidemic moved to the United States, where it spread from east to west during 1964. Cases of CRS were seen during 1964 and 1965, as expected. However, the size and the scope of the outbreak, together with the breadth of the clinical manifestations in the fetuses, were unexpected. Table 2 gives the numbers of pregnancies marred during the rubella outbreak in the US alone.

The Race Is On

Early in vaccine development, the difficulties of making an effective inactivated vaccine were appreciated [11] and several groups began to think rather of attenuated virus vaccines (Table 3). Paul Parkman had moved to the laboratory of Harry Meyer at the former Bureau of Biologics of the FDA, where he grew rubella virus in weekly serial passages in AGMK cells. Seventy-seven passages were required before in vitro markers were sufficiently changed to warrant tests in man [12]. When clinical trials were done, the results were quite satisfactory, in that nearly all previously seronegative subjects seroconverted to rubella, with a minimal degree of symptoms [13, 14]. Vaccine virus was excreted from the nasopharynx, but did not spread to contact subjects. The High Passage Virus (HPV-77) strain was made available to two manufacturers: Merck and Philips-Roxane. At Merck laboratories, the strain was adapted by 5 passages in duck embryo cell culture, whereas

Table 3 Rubella vaccines: attenuated rubella virus vaccine strains

Vaccine	Strain derivation	Attenuation
HPV77	Army recruits with rubella (1961)	AGMK (77)*
HPV77.DE5	As above	AGMK (77); duck embryo (5)
HPV77DK12	As above	AGMK (77); dog kidney (12)
Cendehill	Urine from a case of post-natally acquired rubella (1963)	AGMK (3); primary rabbit kidney (51)
RA27/3	Kidney of rubella-infected fetus (1964)	Human embryonic kidney (4); WI-38 fibroblast (17–25)

*Figures in parentheses indicate number of passages.

at Philips-Roxane dog kidney cell culture was chosen, in which HPV-77 was passaged 12 times.

Before receiving the HPV-77 strain, Merck, under the leadership of Maurice Hilleman, had been active on its own. They had obtained an isolate called Benoit in AGMK cell culture, and after 11–19 serial passages in the same substrate, they passaged the strain 20 times in duck embryo cell culture [15–17]. Although the immunogenicity of Benoit was superior to that of HPV-77 [18], Hilleman was subjected to political pressure by Mary Lasker to take the strain developed by the FDA on the grounds that it would be licensed more easily (M. Hilleman, personal communication). Rather than pursuing further research on Benoit, Merck elected to take the HPV-77 strain and adapted it to duck embryo cell culture (HPV-77-DEV5) (personal communication).

The dog kidney cell culture strain of HPV-77 (HPV-77-DK12) retained its immunogenicity in humans and was actively developed by Philips-Roxane [13].

Meanwhile, Smith-Kline had selected primary rabbit kidney cell culture as their substrate and had isolated a strain called Cendehill [19, 20]. Their marker studies suggested a change in the virus by the 51st passage, and satisfactory results were obtained in human studies at that level [21].

The Great Substrate Battle

Whenever you find yourself on the side of the majority, it is time to pause and reflect

Mark Twain

At this point, a digression is necessary concerning the choice of cell cultures for the manufacture of viral vaccines for man. The first viral vaccines were made in animal tissues. After development of cell culture technology in the late 1940s, the utility of cell culture passage for attenuation became evident. The question was which cell cultures?

Although human fibroblasts were the first cells used for viral culture in Enders laboratory [22], they were obtained from human embryos and were not considered sufficiently convenient to be taken into account for industrial use. Thus, the initial choices were primarily based on the susceptibility of certain cells to the replication of a particular agent and the availability of those cells. In the case of the inactivated polio vaccine developed by Salk, rhesus monkey kidney rapidly became the cell culture of choice, as monkeys were cheap and easily available and polio viruses could be grown in them to high titer.

At about the same time, Koprowski and Sabin were developing live oral attenuated polio vaccines. The choice of primary monkey kidney cells as the substrate was self-evident, particularly after it became possible to isolate single clones of viruses by plaquing those cells.

The choice made for measles vaccine was also evident. In the attempt to attenuate the virus, the Edmonston strain was adapted to grow in chick embryo cell culture, which seemed all the more logical since yellow fever vaccine and influenza vaccine had been prepared for many years in the embryonated hen's egg.

In retrospect, it is amazing that primary cells were thought to be the safest choices. Considering all the microbial agents to which animals are exposed, the subsequent events were predictable. Contaminants began to be found in monkey kidney cultures and were given the name of simian virus. The 40th such agent isolated, SV40, transformed cells in culture [23]. Later, many other troublesome viral contaminants were found in monkey kidney cell cultures, including Marburg virus and B virus [24].

Meanwhile, at the Wistar Institute where I was working, Leonard Hayflick and Paul Moorhead were studying the properties of human fetal fibroblasts [25]. They demonstrated that these cells could be grown and passaged in vitro, but only for a finite number of population doublings. Nevertheless, enough cells could be grown, frozen, and reconstituted at will, so that human fetal cells could be considered as cell populations with defined properties useful for isolation and cultivation of viruses. However, after the original publication in 1961, the use of human diploid cell strains (HDCS) immediately became controversial, with many arguing that unknown but dangerous agents might be present in those cells and that HDCS would spontaneously transform to aneuploid cells, as many other cell populations do. The Director of the Bureau of Biologics referring to primary cells and HDCS actually said, "Better the devil we know, than the devil we don't."

When HDCS became available, we set about growing various viruses in them and developed new rabies, polio, and other vaccines [26–29] (Table 4). Most importantly, it was demonstrated that when strains were obtained from the lungs of human fetuses aborted by maternal choice, no contaminating agents could be found. For the first time, cells became available which would support the growth

Table 4 Viral vaccines that have been commercially produced in human diploid fibroblast cell culture

Rubella
Oral Polio
Rabies
Hepatitis A
Varicella

of most human viruses, but which were free of contaminants [30, 31]. Ultimately, all vaccines produced commercially have been grown in cells derived from two features: WI-38 and MRC-5.

RA27/3

I now return to the development of rubella vaccine. In 1962–1963, I was at the Hospital for Sick Children in London, working with Alastair Dudgeon, a British virologist who was interested in congenital infections. When the news broke concerning the methods of cultivating rubella virus, we started studies on the natural history or congenital infection and its diagnosis [32]. I returned to Philadelphia in the fall of 1963, where the first effects of the rubella epidemic began to be felt in 1964. Gynecologists were besieged with requests to do abortions on women who developed rubella in pregnancy, and we gained access to numerous specimens of infected fetal tissue. For diagnostic purposes, we used AGMK to isolate the virus from the fetal tissues, but it was apparent that the strain grown only in human cells might be more useful as a vaccine strain. Accordingly, explant culture of the fetal organs was made and eventually passed in the HDCS called WI-38. A rubella virus that grew particularly well was isolated from a kidney explant obtained from the 27th rubella abortus submitted to our laboratory and was baptized RA27/3 [33].

We then set about to attenuate the virus, with two guiding ideas: one, to pass only in normal human cells, and second, to attenuate by cold adaptation, a technique learned in attenuating polio viruses. Table 5 gives the passage history which adapted the strain to grow at 30°C.

Table 6 shows that between the 17th and the 25th passages in HDCS, in vitro markers changed and a satisfactory level of attenuation in humans was reached [34–36]. It is remarkable that only 25 passages were required for attenuation, considering that other strains required 50 or more. I attribute that to the use of cold adaptation which was advantageous, as the low passage number permitted the virus to retain a good immunogenicity. Moreover, the RA27/3 was still immunogenic when given by the intranasal route, although the minimal infectious dose for 90% immunization was relatively large: 10,000PFU [37].

Table 5 Passage history of RA27/3

Passage no	1	2	3	4	5	6	7	8	9	10	11	12	
Dilution passed (c)	35	35	35	35	35	35	35	35	35	35	35	35	
Passage no	–	–	–	–	–	–	–	–	9	10	11	12	
Dilution passed (c)	–	–	–	–	–	–	–	–	33	33	33	33	
Type	C	C	C	C	S	S	S	S	S	S	S	S	
Passage no	13	14	15	16	17	18	19	20	21	22	23	24	25
Dilution passed (c)	33	30	30	30	30	35	30	30	30	30	30	30	35
Type	S	S	S	S	S	S	TD	S	TD	TD	TD	TD	S

C cell to cell; *S* supernatant fluid; *TD* terminal dilution of supernatant fluid

Table 6 Changes in markers with passage of RA27/3

Passage level	rct 30°C	BHK plaques	Markers Nt Ab induction	Rash induction	Pharyn exc
<8	–	–	+++	?	?
8–14	–	–	++	++	++
15–20	±	+	+	+	+
>20	+	+	+	-	±

rct reproductive at 30°C compared to 37°C; *BHK* baby hamster kidney; *Nt Ab* neutralizing antibodies

David and Goliath

Attenuation of RA27/3 was accomplished between 1964 and 1967, and between 1967 and 1969, clinical trials conducted in the US, Great Britain, Ireland, Switzerland, Iran, Taiwan, and Japan gave encouraging results [38–42]. In February of 1969, the NIAID held a three-day international conference on rubella immunization on the NIH campus in Bethesda, which was attended by a packed house of hundreds of interested parties, including Albert Sabin. Although Albert had not himself worked on rubella, he was there as a guru.

As the meeting progressed, I heard that Sabin had made a number of statements in private deprecating the use of a diploid cell vaccine, and on the last morning of the meeting, there was even an interview in the Washington Post, in which the opinion was stated. Finally, towards the end of the meeting, Sabin rose to inveigh against HDCV in his rabbinical style, darkly alluding to some unknown agents that might be lurking in WI-38. It sounds theatrical, but I remember that these words from the Bible came into my mind: "The Lord has delivered him into my hands." After he sat down, I took the microphone and criticized his statements one by one and at length, pointing out that they were strictly ex cathedra and without a factual basis [43].

Much to my surprise, I received a thunderous ovation, surely not for myself but rather because the audience was convinced that HDCS should be used. That was a famous victory, but do not mistake my purpose in recounting it; Albert Sabin was a great man, and afterwards, he came up to me and joked about the argument. Later in his life, he came around to accepting HDCS.

And Then There was One…

Nevertheless, official American acceptance of vaccine growth in HDCS was nil, whereas in Europe, and specifically in the United Kingdom, the attitude was much more open. In particular, the late Frank Perkins, the Head of the British control authority for vaccines, was a staunch proponent. So, although no American manufacturer was willing to take RA27/3, John Beale started developing it for the

Wellcome Laboratories in the UK, and Robert Lang at the Institut Merieux did the same in France. The first licensure of RA27/3 occurred on 29 December 1970 in the UK, and the use of this strain gradually widened in Europe and elsewhere.

Back in the US, three rubella vaccines had been licensed: HPV77-DE5, HPV77-DK12, and Cendehill. During the subsequent years, data were developed concerning the relative properties of the rubella vaccines. It gradually became evident that RA27/3 gave more satisfactory systematic and local immune responses to the vaccines licensed in the US [39, 42, 44]. In addition, the issue of reinfection arose, particularly through work of Dorothy Horstmann [45]. She confirmed that when vaccinees were exposed to wild rubella virus, those who received HPV-77 or Cendehill were likely to become reinfected, whereas RA27/3 vaccine showed a resistance similar to that after natural infection. Similar data had been previously generated, using artificial challenges with wild rubella virus [42, 44].

In addition, HPV77-DK12, once used in the general population, caused an unacceptable rate of reactions, particularly carpal tunnel syndrome and other neurasthesias. Accordingly, it was withdrawn from the market [46]. HPV77-DE5, although quite well tolerated in children, was less well tolerated in adult women, in whom it often caused arthralgia and arthritis, at a frequency greater than that of RA27/3 [47]. Cendehill was well tolerated, but its relatively lower immunogenicity created a problem, and ultimately it too was withdrawn.

Sometime in 1978, I believe I was sitting in my office when the phone rang and it was Maurice Hilleman; Maurice said that Dorothy Horstmann had convinced him that it would be a good idea to replace HPV77-DE5 with RA27/3. After recovering my faculty of speech, I readily agreed, and with Maurice's usual energy and efficiency, large scale clinical tests of material produced by Merck were rapidly done, and the RA27/3 vaccine received an American license in 1979.

Today, aside from the well-protected markets of Japan and China, RA27/3 is the only rubella vaccine used throughout the world [48].

The Effect on Public Health

L'avenir, tu n'as pas á le prèvoir, mais à le permettre.
Antoine de Saint-Exupéry, *Pilote de Guerre.*
(You don't have to forsee the future but only to permit it.)

The public health impact of rubella vaccination can be gauged by its impact in four countries. Figure 1 shows graphically the impact of rubella vaccination on the incidence of reported rubella and CRS in the United States. As not all cases of CRS are reported, Center for Disease Control (CDC) authors calculated the probable number of cases occurring in the United States since the licensure of rubella vaccines, using data from three sources. As shown in Table 7 and Fig. 1, 1,064 cases of CRS were estimated to occur between 1970 and 1979 or 106 per year. From 1980 to 1985, there was an average of 20 cases per year. Since then, rubella has been eliminated from the United States [49, 50].

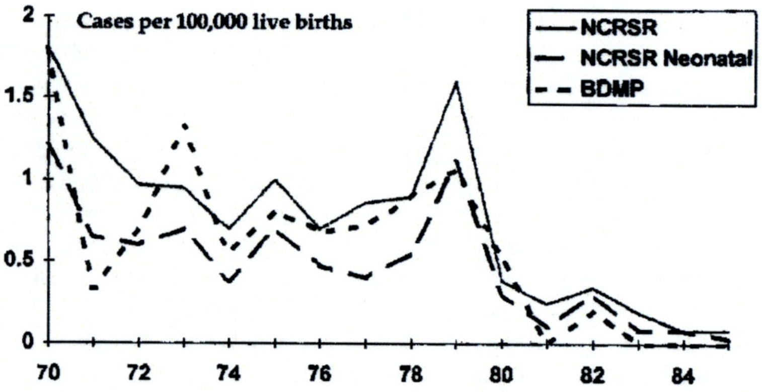

Fig. 1 Reported congenital rubella syndrome, United States, 1970–1985, NCRSR: National Congenital Rubella Syndrome Registry. BDMP: Birth Defects Monitoring Program

Table 7 Data for calculation of estimated actual congenital rubella syndrome births (N) in the United States, 1970–1985

Year	N*	95% confidence interval	Estimated incidence*
1970–1979	1,064	668–1,460	3.19
1980–1985	122	8–236	0.55
Aggregate	1,240	787–1,693	2.24

*Cases/100,000 live births/year

Finnish public health authorities have undertaken eradication of measles, mumps, and rubella from their country by using a two-dose schedule of immunization. This policy was started in 1985, and the result has been the disappearance of rubella from Finland [51, 52]. Similarly, progress in the elimination of CRS from the United Kingdom is shown in Fig. 2 [53]. Elimination of rubella from the western hemisphere is on the verge of being declared [54].

Any country that wishes to apply the vaccine thoroughly can eradicate CRS, and perhaps rubella itself [53], since immunity postvaccination is persistent [55]. Indeed, global elimination is within reach, using RA27/3 vaccine strain [56].

In retrospect, why did the RA27/3 strain that I developed become the accepted rubella vaccine strain? I would like to think that the first part of the answer is that it has the best clinical and immunological qualities, and that it works. The second part of the answer revolves around personalities and prespicacities. John Beale, Dorothy Horstmann, and Maurice Hilleman each exerted efforts to bring the vaccine to use and all can share the credit for its success.

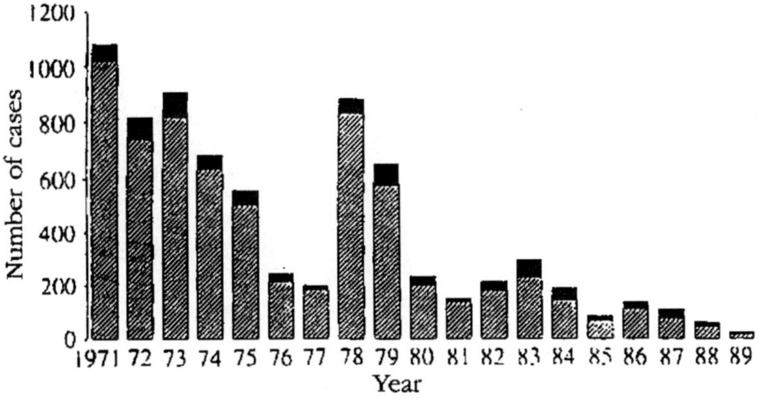

Fig. 2 Congenital rubella cases ( ) and termination of pregnancy for rubella ( ) disease contact. England and Wales 1971–1989

References

1. Forbes JA. Rubella: historical aspects. *Amer J Dis Child* 1969;118:5–11
2. Maton WG. Some account of a rash liable to be mistaken for scarlatina. *Med Transact Coll* 1815;5:149–65
3. Veale H. History of an epidemic of rötheln, with observations on its pathology. *Edinburgh Med J* 1866;12: 404–14
4. Hiro Y, Tasaka S. Die Rötheln sing eine Viruskrakheit. *Mschr Kinderheilk* 1938;76:328–32
5. Anonymous. Recommendation of the Immunization Practices Advisory Committee (ACIP). Rubella prevention. *MMWR* 1984;33:301–10
6. Moriuchi H, Yamasaki S, Mori K, Sakai M, Tsuji Y. A rubella epidemic in Sasebo, Japan in 1987, with various complications. *Acta Paedictr Jpn* 1990;32:67–75
7. Gregg NM. Congenital cataract following German measles in the mother. *Trans Ophtalmol Soc Aust* 1941;3:35–46
8. Weller TH, Neva FA. Propagation in tissue culture of cytopathic agents from patients with rubella-like illness. *Proc Soc Exp Biol Med* 1962;111:215–25
9. Parkman PD, Buescher EL, Artenstein MS. Recovery of rubella virus from army recruits. *Proc Soc Exp Biol Med* 1962;111:225–30
10. Nakhasi HL, Thomas D, Zheng D, Liu TY. Nucleotide sequence of capsid, E2 and E1 protein genes of rubella virus vaccine strain RA27/3. *Nucl Acids Res* 1989;17:4393–4
11. Sever JL, Schiff GM, Huebner RJ. Inactivated rubella virus vaccine. *J Lab Clin Med* 1963;62:1015
12. Parkman PD, Meyer HM, Jr., Kirschstein RL, Hopps HE. Attenuated rubella virus: I. Development and laboratory characterization. *N Engl J Med* 1966;275:569–74
13. Meyer HM, Jr., Parkman PD, Hobbins TE, Larson E, Davis WJ, Hopps HE. Attenuated rubella viruses. *Amer J Dis Child* 1969;118:155–65
14. Meyer HM, Jr., Parkman PD, Panos TC. Attenuated rubella virus: II. Production of an experimental live-virus vaccine and clinical trial. *N Engl J Med* 1966;275:575–80
15. Hilleman MR. Advances in control of viral infections by nonspecific measures and by vaccines with special reference to live mumps and rubella virus vaccines. *Clin Pharmacol Ther* 1966;7:752–62

16. Buynak EB, Hilleman MR, Weibel RE, Stokes J, Jr. Live attenuated rubella virus vaccines prepared in duck embryo cell culture: I. Development and clinical testing. *JAMA* 1968;204:195–200
17. Weibel RE, Stokes J, Jr., Buynak EB, Whitman JE, Jr., Leagus MB, Hilleman MR. Live attenuated rubella virus vaccines prepared in duck embryo cell culture: II. Clinical tests in families and in an institution. *JAMA* 1968;205:166–71
18. Hilleman MR, Buynack EB, Whitman JE, Jr. et al. Live attenuated rubella virus vaccines. *Amer J Dis Child* 1969; 118:166–71
19. Peetermans J, Huygelen C. Attenuation of rubella virus by serial passage in primary rabbit kidney cell cultures: I. Growth characteristics in virtro and production of experimental vaccines at different passage. *Arch Ges Virusforsch* 1967;21:133–43
20. Huygelen C, Peetermans J. Attenuation of rubella virus by serial passage in primary rabbit kidney cell cultures: II. Experiments in animals. *Arch Ges Virusforsch* 1967;21:357–65
21. Prinzie A, Huygelen C, Gold J, Farquhar J, McKee J. Experimental live attenuated rubella virus vaccine. Clinical evaluation of Cendehill strain. *Amer J Dis Child* 1969;118:172–7
22. Enders JF, Weller TH, Robbins FC. Cultivation of the Lansing strain of poliomyelitis virus in cultures of various human embryonic tissues. *Science* 1949;109:85–7
23. Giradi AJ, Sweet BH, Slotnick VB, Hilleman MR. Development of tumors in hamsters inoculated in the neonatal period with vacuolating virus, SV_{40}. *Proc Soc Exp Biol* 1962;109:649–60
24. Hilleman MR. Line cell saga – an argument in favor of production of biologicals in cancer cells. *Adv Exp Biol Med* 1979;118:47–58
25. Hayflick L, Moorhead PS. The serial cultivation of human diploid cell strains. *Exp Cell Res* 1961;26:585–621
26. Hayflick L. Cell substrates for human vaccine preparation. *Nat Cancer Inst Monogr* 1968;83–106
27. Plotkin SA. Vaccine production in human diploid cell strain. *Amer J Epidemiol* 1971;94:303–6
28. Plotkin SA, Eagle H, Hayflick L, Ikic D, Koproski H, Perkins F. Serially cultured animal cells for preparatiom of viral vaccines. *Science* 1969;165:1278–82
29. Hayflick L, Moorhead PS, Pomerat CM, Hsu TC. Choice of cell system for vaccine production. *Science* 1963;140:766–8
30. Hayflick L. Human virus vaccines: Why monkey cells? *Science* 1972;176:813–4
31. Hayflick L. History of cell substrates used for human biologicals. *Dev Biol Stand* 1988;70:11–26
32. Plotkin SA, Dudgeon JA, Ramsay AM. Laboratory studies on rubella and the rubella syndrome. *Brit Med J* 1963;2:1296–9
33. Plotkin SA, Cornfeld D, Ingalls TH. Studies of immunization with living rubella virus: trials in children with a strain culture from an aborted fetus. *Amer J Dis Child* 1965;110:381–9
34. Plotkin SA, Farquhar JD, Katz M, Ingalls TH. A new attenuated rubella virus grown in human fibroblasts: evidence for reduced nasophyrangeal excretion. *Amer J Epidemiol* 1967;86:486–77
35. Plotkin SA, Farquhar JD, Katz M, Buser F. Attenuation of RA27/3 rubella virus in WI-38 human diploid cells. *Am J Dis Child* 1969;118:178–85
36. Plotkin SA. Development of RA27/3 attenuated rubella virus grown in WI-38 cells. *Symp Series Immunobiol Standard* 1968;11:178–85
37. Plotkin SA, Ingalls TH, Farquhar JD, Katz M. Intranasally administered rubella vaccine. *Lancet* 1968;2: 934–6
38. Beasley RP, Detels R, Kim KSW, Gale JL, Lin TL, Grayston JT. Prevention of rubella during an epidemic in Taiwan. HPV-77 and RA27/3 rubella vaccines administered subcutaneously and intranasally HPV-77 vaccine mixed with mumps and/or measles vaccine. *Am J Dis Child* 1969;118:301–6
39. Dudgeon JA, Marshall WC, Peckham CS. Rubella vaccine trials in adults and children. Comparison of three attenuated vaccines. *Am J Dis Child* 1969;118:237–42

40. Furukawa T, Miyata T, Kondo K, Kuno K, Isomura S, Takekoshi T. Clinical trials on RA27/3 (Wistar) rubella vaccine in Japan. *Am J Dis Child* 1969;118:262–5
41. Saidi S, Naficy K. Subcutaneous and intranasal administration of RA27/3 rubella vaccine. Alone and in conjuction with live attenuated measles vaccine. *Am J Dis Child* 1969;118:209–12
42. Plotkin SA, Farquhar JD, Ogra PL. Immunologic properties of RA27/3 rubella vaccine. *JAMA* 1973;225: 585–90
43. Sabin AB, Plotkin SA. Discussion. *Am J Dis Child* 1969;118:378–80
44. Fogel A, Gerichter CH, Barnea B, Handser R, Heeger E. Response to experimental challenge in persons immunized with different rubella vaccines. *J Pediatr* 1978;92:26–9
45. Horstmann DM. Viral vaccines and their ways. *Rev Infect Dis* 1979;1:502–16
46. Schaffer W, Fleet WF, Kilroy AW et al. Polyneuropathy following rubella immunization: A follow-up study and review of the problem. *Am J Dis Child* 1974;127:684–8
47. Polk BF, White JA, DeGirolani PC. A controlled comparison of joint reactions among women receiving one of two rubella vaccines. *Am J Epidemiol* 1982;115:19–25
48. Best JM. Rubella vaccines: past, present and future. *Epidemiol Infect* 1991;107:17–30
49. Cochi SL, Edmonds LE, Dyer K et al. Congenital rubella syndrome in the United States, 1970–1985 On the verge of elimination. *Am J Epidemiol* 1989;129:349–61
50. Reef SE, Cochi SL. The evidence for the elimination of rubella and congenital rubella syndrome in the United States: a public health achievement. *Clin Infect Dis* 2006;43 Suppl 3:S123–5
51. Paunio M, Virtanen M, Peltola H et al. Increase of vaccination coverage by mass media and individual approach: intensified measles, mumps, and rubella prevention program in Finland. *Am J Epidemiol* 1991;133:1152–60
52. Ukkonen P, von Bonsdorff CH. Rubella immunity and morbidity: effects of vaccination in Finland. *Scand J Infect* 1988;20:255–9
53. Miller E. Rubella in the United Kingdom. *Epidmiol Infect* 1991;107:31–42
54. Dayan GH, Castillo-Solórzano C, Nava M, Hersh Andrus J, Rodriguez R, Reef SE. Efforts at rubella elimination in the United States: the impact of hemispheric rubella control. *Clin Infect Dis* 2006;43 Suppl 3:S158–63
55. O'Shea S, Woodward S, Best JM, Banatvala JE, Holzel H, Dudgeon JA. Rubella vaccination: persistence of antibodies for 10–21 years. *Lancet* 1988;2:909
56. Meissner HC, Reef SE, Cochi S. Elimination of rubella from the United States: a milestone on the road to global elimination. *Pediatrics* 2006;117:933–5

Three Decades of Hepatitis Vaccinology in Historic Perspective. A Paradigm of Successful Pursuits

Maurice R. Hilleman†

History of Hepatitis

Hepatitis is an apt subject for discussions of the history of vaccination, since it is clearly an example of recognizable diseases of ancient recorded history, as well as of clinical diseases of diverse and multiple etiologies. Hepatitis, with its distinguishing yellow jaundice, must have been a recognizable feature for an illness of human beings ever since the species discarded its hairy overcoat in favor of the buff.

† Deceased

S.A. Plotkin (ed.), *History of Vaccine Development*,
DOI 10.1007/978-1-4419-1339-5_25, © Springer Science+Business Media, LLC 2011

Table 1 Brief history of hepatitis

The disease recognized	
Early	Pre-history
East Mediterranean	400 BC
China	200s AD
Recognition of contagiousness	
Pope Zaccharius	700s AD
Epidemic jaundice	1700s AD
Obstructive vs catarrhal jaundice	
Nonlearned debate	1700–1900s
Campaign jaundice	1600–1944
Viral etiology	
Argumentation – some learned	1800s
Yellow fever vaccine and dirty synage cause serum hepatitis	1900–1950
Human experimentation	
Definitive experimentation	1942–1970
Hepatitis A and B named	

History records the occurrence of icteric disease as early as 400 BC (Table 1) [1, 2]. The concept of contagiousness was evolved during the millennium from 700 to 1700 AD. This was followed by two centuries of endless and illiterate argument about the causation of the disease, in spite of compelling examples of epidemics of campaign jaundice in the military that indicated its infectious nature. Infectious etiology was well established in the early 1900s, by observed jaundice in the recipients of yellow fever vaccines stabilized with human serum and in the reuse of nonsterilized syringes that had been employed in administering drugs against syphilis. The final fact came from experimentation in human volunteers during World war II, in which infectious and serum hepatitis were defined and were named hepatitis A and B.

Etiologic Discovery

The era of definitive etiologic discovery in hepatitis was initiated in the mid-1960s, with the discovery by Blumberg and coworkers [3, 4] and Prince [5] of the hepatitis B surface antigen that circulates in the bloods of carriers of the infection. This was followed by the discovery of hepatitis A virus by Deinhardt et al. [6], who reported reliable propagation of hepatitis A virus in marmosets and demonstrated the biochemical alterations and liver histopathologic changes that result from infection. This was confirmed by our group [7] in extensive laboratory and epidemiological investigations of hepatitis in Costa Rica.

At present, the hepatitis viruses that are designated A, B, C, D, and E [2, 4, 5, 8–10] fall into five established families (Table 2). In addition, there are at least four additional groups of viruses that are ill-defined and are less well established [11–16] (Jungsuh Kim, 1995, personal communication). The hepatitis viruses fall into two groups: those

Table 2 Viral hepatitis (contemporary appraisal)

Kind	Family	Transmission	Persistence	Vaccine	Occurrence**
Established					
A	Picomaviridac (RNA)	Feeal, oral	No	Yes	Epidemic
B	Hepadnaviridae (DNA)	Blood, secretions	Yes	Yes	Endemic
C	Flaviviridae (RNA)	Blood, secretions	Yes	Doubtful	Endemic
D	Unclassified (satellite) (RNA)	Blood, secretions	Yes	Not needed	Endemic
E	Calciviridae (RNA)	Fecal, oral	No	Experimental	Epidemic
Not Definitive					
F	Unknown (HFV agent)	(Sool isolates, non A	Unknown	Unknown	(Sporadic)*
(not confirmed)	(DNA)	Non B-non Chepatitis)			
G	Flaviviridae* (RNA)	Blood, secretions	Yes*	Possible	Endemic*
GB a/c	Flaviviridae* (RNA)	Blood, secretions	Yes*	Doubtful*	Endemic*
GB b	Flaviviridae* (RNA)	Blood, secretions	Yes*	Doubtful*	Endemic*

*Probable; **epidemic: recognizable outbreaks: endemic: contiuning smoldering prevalence.

transmitted by fecal/oral route that can be epidemic or endemic, and those transmitted by blood and body secretions that are of continuous endemic occurrence. Hepatitis A and B can be prevented by vaccination, using commercially licensed products. Feasibility for a vaccine against hepatitis E has been reported [17]. Hepatitis C viruses, caused by flaviviruses, are of multiple genotypes and are antigenically hypervariable, casting doubts as to whether an effective vaccine will be developed [18].

Hepatitis A

The history of hepatitis A research and the development of the vaccine is identifiable principally in the pioneering work carried out in our laboratories since 1969 [2]. An effective vaccine has also been developed by others, at the SmithKline Beecham Laboratories in Belgium [19].

Research studies in our laboratories on hepatitis A are separable into two time periods (Table 3): the marmoset era and the vaccine era [2]. The initial success was the isolation of the Costa Rican CR326 strain of hepatitis A in marmosets and its establishment as the cause of the human disease [7, 20].

The Virus

The demonstration by our group [20–25] of the presence of abundant virus in the livers of marmosets infected with CR326 virus led to the development of serologic assays for virus and antigen, seroepidemiologic investigations of the disease, and to characterization of the virus itself. The findings were presented in late 1974 [21, 22].

Table 3 Historic evolution of hepatitis A vaccine

Marmoset era	1965–1970
Virus discovery (Deinhardt)	1967
Virus purified from infected marmoset liver	1975
Serologic and diagnostic assays	
Seroepidemiology	
Standardization of human immune globulin	
Virus characterization – 'enterovirus-like'	
Vaccine era	
Prototype killed virus vaccine (marmoset liver)	1978
Cell culture breakthrough	1979
Live attenuated virus vaccine	
Marmoset	1982–1983
Human volunteer studies	1983–1991
Killed (cell culture) virus vaccines	
Animal studies	1986
Human studies	1991
Proof of protective efficacy	1991–1992

More specifically, serum neutralization, complement fixation, and immune adherence tests were developed for assay of hepatitis A antigen and antibody [21–25]. Appearance and persistence of specific antibody followed hepatitis A infection, and there was no serologic relationship to hepatitis B. Seroepidemiologic studies revealed early viral infection in most children in Costa Rica, as opposed to the lack of experience among most suburban inhabitants of the United States. The means of development and assay of precisely standardized hepatitis A human immune globulin was established by the use of the serologic assays. The virus itself was shown to be a 27 nm cytoplasmic particle containing RNA and having a density of 1.34 g/cm^3. It was determined to be "enterovirus-like" [22]. The 27 nm particle size of CR326 virus corresponded to that of similar particles described by Feinstone et al. [26], who found "fecal virus" particles of several different sizes in patients' stools which were identified by immune microscopy and were considered by them to be "parvovirus-like." Our findings, together with the demonstration of ether, ether, acid, and heat stability, showed the CR326 agent to be an enterovirus-like entity [22, 24, 25] that was uniquely stable to heat and low pH, differentiating it from true enteroviruses. The agent was subsequently named *hepatovirus* [27].

Prototype Vaccine

The vaccine era [2] of hepatitis A research was launched with the preparation in our laboratory of a prototype formalin-killed vaccine (Table 4) [28], using virus purified from infected marmoset liver. The vaccine was highly potent in stimulating antibody in marmosets and its protective efficacy in controlled challenge experiments in this species was 100%.

Live Virus Vaccine

The road to feasibility for practical development of a vaccine for use in human beings came in 1979, with the breakthrough propagation, in our laboratories, of hepatitis A virus in cell culture [29, 30]. Following initial isolation in fetal rhesus monkey kidney cells, the virus was quickly adapted to growth in WI-38 and MRC-5 diploid human fibroblast cell cultures that are acceptable for use in vaccines in humans.

Table 4 Protective efficacy of killed hepatitis A vaccine is controlled experiments in marmosets

	Time (no of animals/total)			
	Before challenge		After challenge	
Group	Antibody elevation	Enzyme elevation	Antibody elevation	Enzyme elevation
Normal liver (control)	0/8	0/8	8/8	8/8
Hepatitis A	8/8	0/8	0/8	0/8

Table 5 Trials of F and F live hepatitis A vaccines in human volunteers

		No positive/total			
			Antibody response		
Vaccine	Dose of Virus $TCID_{50}$	Increased ALT	HAVAB	HAVAB-M	Neutralization
F	$10^{5.6}$	5*/20	15/20	17/20	
F	$10^{6.3}$	0/11	6/11	10/11	
F	$10^{7.3}$	0/10		10/10	8/8

*ALT values ranged from 29 to 155 depending on laboratory.

Initial studies showed that appropriate serial passage of hepatitis A virus in cell cultures [31–36] resulted in the selection of variants of attenuated virulence for marmosets and chimpanzees that induced antibody and conferred protection against challenge with virulent virus. Attenuated hepatitis A variants designated F and F′ [32–35] have been tested in human volunteers (Table 5) who were given the virus by parenteral injection. Serial transmission studies of the F′ virus in animals showed no evidence for reversion to virulence [36]. Nearly all subjects given a $10^{6.3}$ or $10^{7.3}$ 50% tissue culture dose ($TCID_{50}$) of the live vaccine consistently developed antibodies as measured by neutralization and antibody binding assays [33, 35]. There was no important systemic illness among the recipients and no clinical evidence of hepatitis or significant liver dysfunction in any of the volunteers. Other workers have conducted studies of live attenuated virus vaccines, but space limitations preclude description here (see [2] for references).

Killed Virus Vaccine

Although live vaccine is highly promising, killed vaccine presented a quicker path to production and regulatory approval (Table 3). In 1986, Provost et al. [37] successfully prepared a killed hepatitis A vaccine, using virus grown in cell culture that was safe and protective in marmosets. Lewis et al. [38] subsequently reported development and early clinical testing of vaccine made from killed attenuated CR326 virus grown and purified from cell cultures of MRC-5 strain human diploid fibroblasts. The vaccine was more than 95% pure and was inactivated by formaldehyde and formulated in alum adjuvant.

Clinical Studies

Extensive clinical studies showed the vaccine to be safe and highly immunogenic, using 100 or 200 ng of antigen per dose [39]. The now-classic double-blind, placebo-controlled trial (Table 6) conducted by Werzberger et al. [40] showed the vaccine to be 100% effective. Protection was established within 3 weeks or less after the first vaccine dose and the vaccine has been widely tested in clinical studies [41]. Other workers have also tested experimental hepatitis A vaccines [42–44]. The SmithKline Beecham vaccine has also proved to be safe and effective and has been licensed for distribution in a number of countries [19, 42].

Table 6 Protective efficacy of hepatitis A vaccine in study at Monroe, New York

	Cases of hepaticis A for the period				
	Vaccinated		Placebo controlled		
Days following vaccination	Cases/total	Rate (%)	Cases/total	Rate (%)	Protective efficacy (%)
5–20	7/505*	1.4	3/508*	0.6	0
21–49	0/498	0	9/505	1.8	100
50–137	0/498	0	25/496	5.0	100
21–137	0/498	0	34/505	6.7	100

* Initially seronegative at the time vaccine was given or seronegative in the initial screening test and not retested.

Hepatitis B

Hepatitis B virus infection is endemic worldwide. It rarely causes severe disease on primary infection, and the infection is rapidly resolved in most persons. Difficulty arises in persons who develop a chronic persistent infection and who may show remissions and exacerbations of severe disease for the remainder of their lives.

Cirrhotic destruction of the liver may be slow and progressive due to host immune response to the virus, and the integration of segments of the virus into the liver cell genome is associated with the development of hepatocarcinoma [45]. It was deemed important to develop an effective vaccine against hepatitis B virus infection in order to prevent the serious and fatal illnesses that occur in the human population.

The Virus

Hepatitis B virus cannot be reliably propagated in cell culture. The door to a hepatitis B vaccine was opened by the discoveries by Blumberg and coworkers [3, 4] and Prince [5] of the 22-nm surface antigen particles of hepatitis B virus in the plasma of human carriers. Pioneering studies on hepatitis B vaccines were carried out in our laboratories at the Merck Institute [45, 46] in Pennsylvania, USA, and by Maupas et al. [47] at the Institute of Virology in Tours, France.

Plasma-Derived Vaccine

Work toward a plasma-derived hepatitis B vaccine (Table 7) was started in our laboratories in 1968 ([45], see [46] for extensive review) with two initial efforts, the first one being to purify the antigen from human plasma and the second one to measure the amount of antigen in carrier plasma to determine whether a vaccine would be practically feasible. There were also the problems of not only achieving high-level purity, but also assuring removal from or killing of hepatitis B virus and all the large

Table 7 Hepatitis B vaccine development at the Merck Institute

Plasma-derived vaccine	
1968	Initial efforts
	Purification of antigen
	Feasibility, quantification and yield
	Process development
	Purification
	Inactivation of all possible agents in human blood
1974	Viral inactivation tests
	Preclinical safety
	Mouse potency assay
	Chimpanzee
	Safety
	Immunogenicity
	Protective efficacy (controlled challenge)
1975	Safety and immunogenicity tests in human volunteers
	Protective efficacy in controlled clinical studies
1981	Szmuness et al
1982	Francis et al
1981	Vaccine licensed for general use
Recombinant yeast vaccine	
1975	Initiation of collaborative efforts to develop a recombinant antigen-derived vaccine
	Processes developed for purification of antigen and for preparation of the alum adjuvant product
1984	Clinical testing
1986	Vaccine licensed for general use

Table 8 Purification and/or inactivation in preparing antigen for plasma-derived hepatitis B vaccine

Centrifugation
Isopycnic banding
Rate zonal sedimentation
*Pepsin digestion
*Urea denaturation and renaturation
*Treatment with formaldehyde

*Critical inactivation steps

menagerie of possible known and unknown viral agents that circulate in human blood and that could be present in the preparations of purified antigen.

A purification and inactivation process (Table 8) consisting of physical separation followed by pepsid digestion, denaturation with urea, and renaturation and treatment with formaldehyde was developed. Each of the chemical steps alone was capable of inactivating viruses of a diversity of viral families [46], and all the steps collectively gave a failsafe assurance of viral inactivation. Hepatitis B virus does not grow in cell culture; the chimpanzee that became infected with hepatitis B served as the means of testing for final safety. Protective efficacy was established in controlled challenge experiments in chimpanzees [46].

In addition to safety and efficacy, our targeted program was aimed at developing a killed purified hepatitis B vaccine that would be affordable for the patient and conducted without knowledge of the amount of surface antigen that would be needed per human dose. We learned in 1971 of the Krugman et al. [48] experiments carried out in human subjects at the Willowbrook State School. These studies showed that boiled plasma from hepatitis carriers induced antibody and gave viral protection in some human beings against challenge with hepatitis B virus present in untreated carrier plasma. Our retrospective analysis of a sample of the boiled plasma that Krugman administered showed the presence of about 1 μg of antigen per mL. Although we could not be sure that live virus was not also present in the material and contributed to the immune response, these early probes did suggest that our vaccine could be developed in the cost-practical sense. The ultimate dose of the purified plasma vaccine for commercial distribution was established at 10 or 20 mg for adults and 5 mg for children.

Extensive preclinical tests (Table 7) for immunogenicity and safety of hepatitis B vaccine formulated in alum adjuvant were carried out [46]. Clinical trials began during 1975 [46]. These trials established the immunogenicity and safety of the vaccine for human beings. Protective efficacy was established in the now-classic placebo-controlled studies carried out by Szumuness et al. [49] and Francis et al. [50] in 1980–1982. The plasma-derived vaccine was licensed for general use in 1981.

It was evident from the start that the supply of suitable hepatitis B carrier plasma would not be sufficient to meet the needs for vaccine manufacture, and another source of antigen needed to be sought. Molecular biology came into being about 1975 and collaborative arrangements made between Rutter, Valenzuela and Hall [51], and our group [52] led to the development of in vitro expression systems for hepatitis B antigens. Recombinant yeast vaccine that contained the 226 amino acid S component of the surface antigen was developed in our laboratories [52]. Following extensive studies, the recombinant vaccine that contained the 226 amino acid component of the surface antigen was licensed in 1986. Michel et al. (see ref. [53]) developed an expression system for hepatitis B in mammalian Chinese hamster ovary (CHO) cells at the Pasteur Institute, and a vaccine was developed by Adamowicz et al. [53] at Pasteur Vaccines, now part of the Sanofi Pasteur Laboratories. A second yeast recombinant hepatitis B vaccine was developed at the SmithKline Beecham Laboratories and has been widely distributed.

Plasma-derived and recombinant yeast hepatitis B vaccines (Table 9) perform roughly the same in human subjects [46, 49, 54]. Both vaccines induce antibody responses in 95% or more of seronegative individuals. Antibody equates with immunity, and protective immunity lasts for 15 years or longer, even when the antibody is no longer detectable. Long-term protection is based on immunologic memory and anamnestic recall.

Table 9 Hepatitis B vaccine performance

95% or greater response after primary and secondary dosing
Antibody response means immunity
Protective immunity lasts for 15 years or longer, even if antibody is no longer detectable
Long-term immunity is based on immunologic memory and anamnestic recall
75% or more of infants born to e antigen positive carrier mothers are protected by immediate postnatal vaccination alone
Protective efficacy may be increased to 90% or more if hepatitis B immune globulin is also given to the babies at the time of delivery
Early protection before antibody appears is explained by rapid development of a cell-mediated response
Hepatitis B vaccine was the first licensed: i) subunit, ii) recombinant vaccine, iii) vaccine against virus-caused human cancer

Table 10 "Immune carriers" in hepatitis B

Persistent hepatitis B virus infection may occur in spite of antibody against the native virus
"Immune carriers" circulate an escape mutant that is antigenically different from the native virus used to prepare vaccine
Mutation most commonly represents substitution of adenosine for guanine and change to arginine in place of glycine

Vaccine Performance

Seventy-five percent of babies who are born to *e* antigen-positive mothers are protected from perinatal infection and the hepatitis B carrier state when vaccination is initiated within hours after birth [55, 56]. Effectiveness may be increased to 90% or more by coadministration of hepatitis B-immune globulin along with the vaccine. The mechanisms for protection of newborn through early postnatal vaccination cannot result from a humoral response since antibody does not appear for weeks or months after vaccination is started. A reasonable explanation [57] lies in the recent demonstration [58, 59] in animals that the liposomal hepatitis B surface antigen may be processed by the type I, as well as the type II pathway and may induce a cell-mediated immune response within days following vaccination. The hepatitis B vaccines collectively represent the first licensed subunit vaccine, the first vaccine against virus-induced human cancer, and the first licensed recombinant vaccine.

It has been found in recent years (Table 10) that vaccinated individuals, especially infants, may be persistently infected with hepatitis B virus in spite of successful immunization [60]. Such "immune carriers" typically circulate escape mutant virus that is different antigenically from the native virus.

The principal variant (Fig. 1) is a mutant in which arginine substitutes for glycine in the *a* antigen loop [60, 61] and poses the possibility that a second infectious serotype may evolve that will necessitate development of a bivalent vaccine. It is not known whether persons who are carriers of the mutant virus are contagious to others.

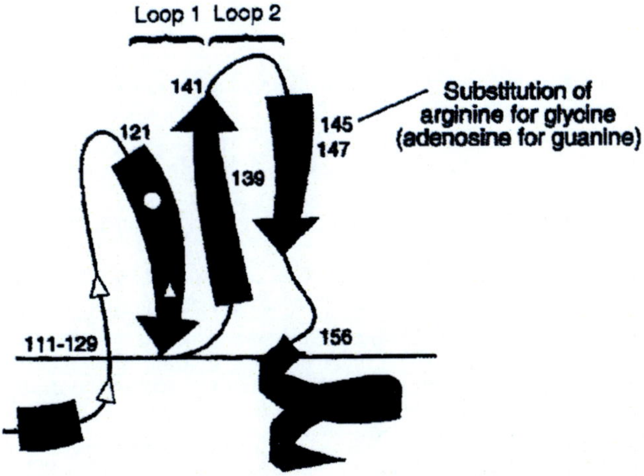

Fig. 1 Variation on the *a* antigen loop of hepatitis B S antigen (after Howard et al. [61])

Conclusion

The more than two millennia since the first recognition of clinical hepatitis has changed the disease from a poorly recognized clinical entity to one of defined etiology causes by more than five different viral agents. Both hepatitis A and B infections are now preventable by highly effective vaccines that induce long-term memory [46]. Since the viruses are antigenically stable and without an animal reservoir, both are eradicable if the vaccine is sufficiently and widely applied for a long time.

Hepatitis A vaccine is now in widespread use to prevent the disease in high-risk individuals and travelers. Hepatitis B vaccine is being given routinely to newborns in many countries. Widespread use of the vaccine to immunize individuals at high risk of infection and to immunize universally all newborn infants could reduce hepatitis B to insignificance within two or three generations. The process of virus elimination could be greatly accelerated by added immunization of all currently susceptible persons.

Vaccines against new and presently unknown viruses may be anticipated for the future. The most likely present candidate is hepatitis E. Hepatitis C and other agents belonging to the Flaviviridae may find substantive difficulty in vaccine development because of the multiplicity of serotypes and because of hypervariability in surface antigen which resembles that for the retroviruses that cause acquired immune deficiency syndrome (AIDS). RNA viruses, which are highly prone to errors in genetic transcription and give rise to variants in which the dominant neutralizing epitope is not conserved, may present insuperable problems for vaccine development.

References

1. Zuckerman, A J The history of viral hepatitis from antiquity to the present. In: Deinhardt F, Deinhardt J (eds) Viral Hepatitis: Laboratory and Clinical Science 1983: 2–32. New York: Marcel Dekker
2. Hilleman M R Hepatitis and hepatitis A vaccine: a glimpse of history. *J Hepatol* 1993;18 (suppl 2): S5–S10
3. Blumberg BS, Alter HJ, Visnich S A "new" antigen in leukemia sera. *JAMA* 1965;91: 541–46
4. Blumberg BS Australia antigen, hepatitis and leukemia. *Tokyo J Med Sci* 1968;76:1
5. Prince AM An antigen detected in the blood during the incubation period of serum hepatitis. *Proc Natl Acad Sci USA* 1968;60:814–21
6. Deinhardt F, Holmes AW, Capps RB et al Studies on the transmission of human viral hepatitis to marmoset monkeys. I. Transmission of disease, serial passages and description of liver lesions. *J Exp Med* 1967;125:673–88
7. Mascoli CC, Ittensohn OL, Villarejos VM et al Recovery of hepatitis agents in the marmoset from human cases occurring in Costa Rica. *Proc Soc Exp Biol Med* 1973;142:276–82
8. Houghton M, Weiner A, Han J et al Molecular biology of the hepatitis C viruses: implications for diagnosis, development and control of viral disease Hepatology 1991;14:381–88
9. Lazinski DW, Taylor JM. Recent developments in hepatitis delta virus research. In: Maramorosch K, Murphy FA, Shatkin AJ, eds. *Advances in Virus Research.* Elsevier, New York 1994;43:187–231
10. Bradley DW. Hepatitis E virus: a brief review of the biology, molecular virology, and immunology of a novel virus. *J Hepatol* 1995;22 (suppl 1):140–5
11. Deka N, Sharma MD, Mukerjee R. Isolation of the novel agent from human stool samples that is associated with sporadic non-A, non-B hepatitis. *J Virol* 1994;68:7810–5
12. Simons JN, Pilot-Matias TJ, Leary TP et al. Identification of two flavivirus-like genomes in the GB hepatitis agent. *Proc Natl Acad Sci USA* 1995;92:3401–5
13. Zuckerman AJ. The new GB hepatitis viruses. *Lancet* 1995;345:1453–4
14. Denis F, Nicot T. Découverte de nouveaux virus des hépatites, les "GBV": quelle est leur place et quel est leur pouvoir pathogène? *Médicine/Science* 1995;92:3401–5
15. Craske J. Hepatitis C and non-A and non-B hepatitis revisited: Hepatitis E, F, and G. *J Infection* 1992;25:243–50
16. Fields HA, Favorov MO, Margolis HS. The hepatitis E virus: a review. *J Clin Immunoassay* 1993;16:215–23
17. Tsarev SA, Tsarev TS, Emerson SU, Govindarajan S, Shapiro M, Gerin JL, Purcell RH. Successful passive and active immunization of cynomolgus monkey against hepatitis E. *Proc Natl Acad Sci USA* 1994;91:10198–202
18. Zuckerman AJ, Zuckerman JN. Prospects for hepatitis C vaccine. *J Hepatol* 1995;22 (suppl 1): 97–100
19. André FE, D'Hondt E, Delem AD, Safary A. Clinical assessment of the safety and efficacy of an inactivated hepatitis-A vaccine – rationale and summary of findings. *Vaccine* 1992;10:S106–68
20. Provost PJ, Ittensohn OL, Villarejos VM, Arguedas JA, Hilleman MR. Etiologic relationship of marmoset-propagated CR326 hepatitis A virus to man. *Proc Soc Exp Biol Med* 1973; 142:1257–67
21. Hilleman MR, Provost PJ, Miller WJ, Villarejos VM, Ittensohn OL, McAleer WJ. Immune adherence and complement fixation tests for human hepatitis A. Diagnostic and epidemiologic investigations. International Symposium on Viral Hepatitis, Milan, Italy, 4 December, 1974. *Dev Biol Stand* 1975;30:383–9 (S Karger, Basel)
22. Hilleman MR, Provost PJ, Wolanski BS, Miller WJ, Ittensohn O, McAleer WJ. Characterization of CR326 human hepatitis A virus, a probable enterovirus, International Symposium on Viral Hepatitis, Milan, Italy, 4 December, 1974. *Dev Biol Stand* 1975;30:418–24 (S Karger, Basel)

23. Provost PJ, Ittensohn OL, Villarejos VM, Hillerman MR. A specific complement-fixation test for human hepatitis A employing CR326 virus antigen. Diagnosis and epidemiology. *Proc Soc Exp Biol Med* 1975;148:962–9
24. Miller WJ, Provost PJ, McAleer WJ, Ittensohn OL, Villarejos, Hilleman MR. Specific immune adherence assay for human hepatitis A antibody. Application to diagnostic and epidemiologic investigations. *Proc Soc Exp Biol Med* 1975;49:254–61
25. Provost PJ, Wolanski BS, Miller WJ, Ittensohn OL, McAleer WJ, Hilleman MR. Physical, chemical and morphologic dimensions of human hepatitis A virus strain CR326. *Proc Soc Exp Biol Med* 1975;148:532–9
26. Feinstone SM, Kapikian AZ, Purcell RH. Hepatitis A: Detection by immune electron microscopy of a virus-like antigen associated with acute illness. *Science* 1973;182:1026–8
27. Francki RB, Fauquet CM, Knudson DL, Brown F, eds. Classification and nomenclature of viruses. Fifth Report of the International Committee on Taxonomy of Viruses, Vienna, Austria. *Arch Virol* 1991; (suppl 2)
28. Provost PJ, Hilleman MR. An inactivated hepatitis A virus vaccine prepared from infect marmoset liver. *Proc Soc Exp Biol Med* 1978;159:201–3
29. Provost PJ, Hilleman MR. Propagation of human hepatitis A virus in cell culture in vitro. *Proc Soc Exp Biol Med* 1979;160:213–1
30. Provost PJ, McAleer WJ, Hilleman MR. In vitro cultivation of hepatitis A virus. In: Szumness W, Alter H, Maynard J, eds. *Viral Hepatitis. International Symposium 1981*. Philadelphia: The Franklin Institute Press, 1982;21–30
31. Provost PJ, Banker FS, Geisa PA, McAleer WJ, Buynak EB, Hilleman MR. Progress toward a live, attenuated human hepatitis A vaccine. *Proc Soc Exp Biol Med* 1982;170:8–14
32. Provost PJ, Conti PA, Geisa PA, Banker FS, Buynak EB, Hilleman MR. Studies in chimpanzees of live, attenuated hepatitis A vaccine candidates. *Proc Soc Exp Biol Med* 1983;172:357–63
33. Provost PJ, Bishop RP, Gerety RJ, Hilleman MR, McAleer WJ, Scolnick EM, Stevens CE. New findings in live, attenuated hepatitis A vaccine development. *J Med Virol* 1986;20:165–75
34. Provost PJ, Emini EA, Lewis JA, Gerety RJ. Progress toward the development of a hepatitis A vaccine. In: Zuckerman AJ, ed. *Viral Hepatitis and Liver Disease*. New York: Alan R Liss, 1988;83–6
35. Midthun K, Ellerbeck E, Gershman K et al. Safety and immunogenicity of a live attenuated hepatitis A virus vaccie in seronegative volunteers. *J Infect Dis* 1991;163:735–9
36. Provost PJ, Banker PS, Wadsworth CW, Krah DL. Further evaluation of a live hepatitis A vaccine in marmosets. *J Med Virol* 1991;34:227–31
37. Provost PJ, Hughes JV, Miller WJ, Giesa PA, Banker PS, Emini EA. An inactivated hepatitis A viral vaccine of cell culture origin. *J Med Virol* 1986; 19:23–31
38. Lewis JA, Armstrong ME, Larson VM et al. Use of a live, attenuated hepatitis A vaccine to prepare a highly purified, formalin-inactivated hepatitis A vaccine. In: Hollinger FB, Lemon SM, Margolis H, eds. *Viral Hepatitis and Liver Disease*. Baltimore: Williams & Wilkins, 1991; 91–3
39. Ellerbeck E, Lewis J, Midthun K, Gershman K, Nalin D, McCaughtry M, Provost P, Calandra G. Safety and immunogenicity of an inactivated hepatitis A vaccine. In: Hollinger FB, Lemon SM, Margolis H, eds. *Viral Hepatitis and Liver Disease*. Baltimore: Williams & Wilkins, 1991;91–3
40. Werzberger A, Mensch B, Kuter B et al. A controlled trial of formalin-inactivated hepatitis A vaccine in healthy children. *New Engl J Med* 1992;327:453–7
41. Nalin DR, Kuter BJ, Brown L et al. Worldwide experience with the CR326F-derived inactivated hepatitis A virus vaccine in pediatric and adult populations: an overview. *J Hepatol* 1993;18:S51–5
42. Peetermans J. Production, quality control and characterization of an inactivated hepatitis-A vaccine. *Vaccine* 1992;10:S99–S101
43. Flehmig B, Heinricy U, Pfisterer M. Immunogenicity of a killed hepatitis A vaccine in seronegative volunteers. *Lancet* 1989;1:1039–41
44. Loutan L, Bovier P, Althaus B, Glück R. Inactivated virosome hepatitis A vaccine. *Lancet* 1994;343;322–4

45. Hilleman MR, Bertland AU, Buynak EB et al. Clinical and laboratory studies of HB_sAg vaccine. *In:* Vyas GN, Cohen SN, Schmid R, eds. *Viral Hepatitis.* Philadelphia: Franklin Institute Press, 1978;525–37
46. Hilleman MR. Plasma derived hepatitis B vaccine: a breakthrough in preventive medicine. *In:* Ellis R, ed. *Hepatitis B Vaccines in Clinical Practice*. New York: Marcel Dekker, 1993;17–39
47. Maupas Ph, Goudeau A, Coursaget P, Drucker J, Brian F, André M. Immunization against hepatitis B in man: a pilot study of two years duration. *In* Vyas GN, Cohen SN, Schmid R, eds. *Viral Hepatitis.* Philadelphia: Franklin Institute Press, 1978;539–56
48. Krugman S, Giles JP, Hammond J. Viral hepatitis, type B (MS-2 strain): studies on active immunization. *JAMA* 1971;217:41
49. Szumuness W, Stevens CE, Zang EA, Harley EJ, Kellner A. A controlled clinical trial of the efficacy of the hepatitis B vaccine (Heptavax B): a final report. *Hepatology* 1981;1:377–85
50. Francis DP, Hadler SC, Thompson SE et al. The prevention of hepatitis B with vaccine: report of the Centers for Disease Control multi-center efficacy trial among homosexual men. *Ann Intern Med* 1982;97:362–6
51. Valenzuela P, Medina A, Rutter WJ, Ammerer G, Hall BD. Synthesis and assembly of hepatitis B virus surface antigen particles in yeast. *Nature* 298:1982;347–50
52. McAleer WJ, Buynak EB, Maigetter RZ, Wampler DE, Milier WJ, Hilleman MR. Human hepatitis B vaccine from recombinant yeast. *Nature* 1984; 307:178–80
53. Adamowicz Ph, Tron F, Vinas R et al. Hepatitis B vaccine containing the S and the PreS-2 antigens produced in Chinese hamster ovary cells. *In:* Zuckerman AJ, ed. *Viral Hepatitis and Liver Disease*. New York: Alan R Liss, 1988;1087–90
54. Hilleman MR. Yeast recombinant hepatitis B vaccine. *Infection* 1987;15:2–6
55. Stevens CE, Taylor PE, Tong MJ, Toy PT, Vyas GN, Zang EA, Krugman S. Prevention of perinatal hepatitis B virus infection with hepatitis B immune globulin and hepatitis B vaccine. *In:* Zuckerman AJ, ed. *Viral Hepatitis and Liver Disease*. New York: Alan R Liss, 1988;982–8
56. Hilleman MR, Calandra GB, Hesley TM, Nalin DR, Ryan JL, West DJ. Vaccines against hepatitis A and B. *J Gastroenterol Hepatol* 1993;8:S21–6
57. Hilleman MR. Comparative biology and pathogenesis of AIDS and hepatitis B viruses: related but different. *AIDES Res Hum Retroviruses* 1994;10:1409–19
58. Schirmbeck R, Melber K, Kuhröber A, Janowicz ZA, Reimann J. Immunization with soluble hepatitis B virus surface protein elicits murine H-2 Class I-restricted $CD8^+$ cytotoxic T lymphocyte responses in vivo. *J Immunol* 1994;152:1110–9
59. Schirmbeck R, Melber K, Mertens T, Reimann J. Selective stimulation of murine cytotoxic T cell and antibody responses by particulate or monomeric hepatitis B virus surface (S) antigen. *Eur J Immunol* 1994;24:1088–96
60. Carman W, Thomas H, Domingo E. Viral genetic variation: hepatitis B virus as a clinical example. *Lancet* 1993; 341:349–53
61. Howard CR, Stirk HJ, Brown SE, Steward MW. Towards the development of synthetic hepatitis B vaccines. *In:* Zuckerman AJ, ed. *Viral Hepatitis and Liver Disease*. New York: Alan R Liss, 1988;1094–1101

Vaccination Against Varicella and Zoster: Its Development and Progress

Anne Gershon

Anne Gershon

One ought never to turn one's back on a threatened danger and try to run away from it. If you do that, you will double the danger. But if you meet it promptly and without flinching, you will reduce the danger by half.

Churchill

At one time in history, it seems, rash diseases tended to be lumped together under the term "pox" and these illnesses must have been very common. Syphilis was known as the "great pox" and smallpox was considered yet another "pox" illness. The well known quotation from Shakespeare's play, Romeo and Juliet, is usually given as "A pox on both your houses." In fact however, "A plague on both your houses" is what Shakespeare actually wrote; plague killed 25% of the European population between 1347 and 1348. "Pox" diseases included plague perhaps, as well as smallpox, syphilis, measles, rubella, chickenpox, and more. Eventually, scientific advances would considerably delineate the diversity of the myriad of pathogens that cause human infections manifested by fever and rash.

A. Gershon (✉)
Department of Pediatrics, Columbia University, New York, NY, USA
e-mail: aag1@columbia.edu

S.A. Plotkin (ed.), *History of Vaccine Development*,
DOI 10.1007/978-1-4419-1339-5_26, © Springer Science+Business Media, LLC 2011

Varicella, the primary infection with varicella-zoster virus (VZV), was for many centuries confused with smallpox, and was not recognized to be a separate illness until the mid-eighteenth century. The origin of the lay name, chickenpox, has been attributed to chickenpox being a milder form of smallpox (therefore the moniker "chicken"). Another possibility regarding nomenclature is that the typical vesicular skin lesions have been said to resemble chick peas. Nevertheless, these are only speculations and one must conclude that no one knows for certain how the name "chickenpox" originated [1, 2].

Long before the identification of filterable agents or viruses as pathogens, it was recognized that a medical connection must exist between varicella and zoster (shingles). Cases of varicella in children were often noted to have followed close exposure to a person with zoster. In the early twentieth century, therefore, investigators attempting to develop a vaccine as had been tried for smallpox (variolation), injected vesicular fluid from zoster patients into children, who usually developed a mild form of varicella or no illness at all [1, 2]. The attempt did not really succeed in that some children developed full-blown varicella, although most had no illness whatsoever. Later, Takahashi, who developed varicella vaccine would make use of this observation (see below). Obviously, this risky approach to human experimentation and vaccine development would not be carried out or even considered today.

It was suspected in the mid-twentieth century, by the pediatrician Joseph Garland, that zoster might be due to reactivation of latent VZV, based on what was known about primary herpes simplex virus (HSV) infection and reactivation of

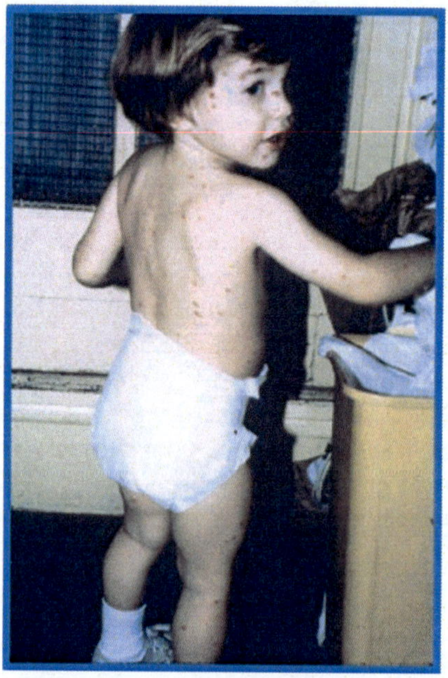

A child with typical Varicella

that virus from apparent latent infection [3]. This hypothesized relationship between varicella and zoster could not really be proven, however, until the era of varicella vaccine and development of molecular biological techniques [4–6]. Because the vaccine strain (Oka) does not circulate, the only way that zoster could be caused by VZV Oka is if it became latent following immunization, and then subsequently reactivated. Thus when a few vaccinated children developed zoster caused by Oka, the only explanation was that the vaccine virus had become latent and then subsequently reactivated.

VZV was not identified as a member of the herpesvirus family until the middle of the twentieth century; prior to that it was presumed to be a poxvirus, probably because it caused chickenpox. A major milestone in virology was the propagation of poliovirus in non-neuronal cell cultures, an achievement that brought John Enders, Fredrick Robbins, and Thomas Weller, at Harvard, the Nobel Prize in 1954. Weller then went on to be first to grow VZV in monolayers of human fibroblast cell cultures [7]. Subsequently, Albert Coons and Weller, by then both Harvard Professors, employing what they called their newly developed assay, the "fluorescent antibody" technique, were able to show that the viruses that caused both varicella and zoster, when propagated in cell cultures were antigenically identical [7]. Later, electron micrographs of VZV by Cook and Stevens [8] clearly indicated the agent to be a typical herpesvirus. More recently, these electron microscopic findings have been confirmed and extended in the laboratories of Drs. Anne and Michael Gershon, both Professors at Columbia University [9].

Varicella is a systemic infection with viremic phases, which follows exposure of an individual not previously infected with VZV. Spread occurs by the airborne route from patients with VZV infections with whom they have close contact. VZV is shed mainly from the skin of patients with VZV infections; skin vesicles are full of highly infectious cell-free virions that can be aerosolized [9]. Possibly in some instances spread also occurs from the respiratory tract [10], although in contrast to the skin, VZV is notoriously difficult to isolate from respiratory secretions. Infection is probably initiated in the tonsils, with spread to lymphocytes (viremia) and then to the skin [11]. Following an incubation period of about 2 weeks, a mild, brief prodrome of malaise and low grade fever may occur, especially in adults. Subsequent symptoms are mainly a generalized, pruritic vesicular rash with fever, which lasts about a week. Complications are unusual but may be severe, such as encephalitis (1/10,000) and bacterial skin infections due to *Staphylococci* or *Streptococci* [12]. Particularly serious are those caused by Group A *beta-hemolytic Streptococci*, which may lead to life-threatening necrotizing fasciitis. As shown by Dr. Ann Arvin and her colleagues at Stanford University, during the incubation period, innate immunity at first may play a role in controlling skin lesions, but eventually the virus wins this struggle and vesicles are formed. Fortunately, adaptive immunity is recruited which controls viral multiplication [13, 14]. The main host defense against VZV is not antibodies but sensitized CD4 and CD8 lymphocytes. Severe varicella is likely to occur in patients with inadequate cellular immunity, such as those undergoing treatment for cancer, who have had transplantation, and who have infection with human immunodeficiency virus (HIV) or congenital immunodeficiency diseases.

Cellular immunity may be of primary importance for the host because in the body VZV spreads only from one cell to another; and infectious virus is not released from cells where it would be neutralized by specific antibodies [2]. Cell-free VZV is produced possibly exclusively in skin vesicles; otherwise VZV is an intracellular pathogen.

Herpesviruses have the unique ability to establish a latent infection following the primary infection. For VZV, it is suspected that most or all individuals develop latent infection during chickenpox, but that for 75% of these people, the virus remains latent for that person's lifetime. Zoster results when latent VZV is reactivated and spreads from the site of latency (sensory ganglia) down the nerve to the skin. An early and insightful epidemiological investigation of zoster was carried out in Britain by a general practitioner, Edward Hope-Simpson, in the 1940s. He determined that zoster occurred only in individuals with a history of varicella and that the incidence of zoster increased steeply with increasing age. Reactivation of VZV is also associated with being immunocompromised. In both the aged and the immunocompromised, zoster occurs when the cell-mediated immune response to the virus is suboptimal for its control; humoral immunity does not seem to play a role in control of latent VZV infection [1, 2]. These observations eventually led to the development of a vaccine against zoster, which is discussed below.

By the middle of the twentieth century, it was clear that VZV is highly contagious, with attack rates of susceptibles approaching 80–90% after exposure to the virus in a household, although it was not yet possible to determine serologically who was immune or susceptible to varicella [15]. Nevertheless, the Pediatrician Avron Ross, practicing in Long Island, New York, was able to examine transmission of VZV in adults and children within families, and in that way showed that varicella is a disease of children, that adults are protected from it if they have a history of chickenpox, and that chickenpox in households is almost as contagious as measles and smallpox. Although Philip Brunell and Helen Casey, working at the Centers for Disease Control and Prevention (CDC), were able to detect humoral responses to VZV as a result of varicella using complement fixation, these antibodies seemed to be present only transiently and therefore were of no use to indicate immunity [16]. Enormous gains were made in diagnostic approaches to VZV infections in the 1970s and 1980s. Although Weller and colleagues had isolated VZV in the 1950s, the virus remained very difficult to propagate in cell culture, which impeded laboratory diagnosis and preparation of antigens for measurement of immunity to the virus. In the late twentieth century, various techniques for measuring VZV antibodies such as enzyme-linked immunosorbent assay (ELISA) [17] and the fluorescent antibody to membrane antigen (FAMA) [18] assay were developed. For the first time it became possible to identify individuals who had experienced natural varicella in the past and thus were immune to this disease (The FAMA assay, which was validated during a 30-year period of clinical experience, remains the best assay for identifying immune individuals after varicella or varicella vaccination.) [19]. The FAMA assay was developed in the laboratory of

Michiahi Tskahashi

Philip Brunell, then Professor of Pediatrics at New York University (NYU) Medical Center, and Anne Gershon, M.D. his postdoctoral fellow. It was critical to the evaluation of the varicella vaccine because it could accurately distinguish between individuals who were immune or susceptible to varicella [19]. Anne Gershon trained in infectious disease at NYU in the Brunell laboratory and also with Saul Krugman, and eventually came to direct the Division of Pediatric Infectious Diseases at Columbia University.

During this period, it also became possible to measure cell-mediated immunity (CMI) to VZV by the technique of lymphocyte stimulation [20]. Most individuals who have had varicella and are less than 50 years of age have positive lymphocyte stimulation responses (>5 times the control) to VZV as well as positive FAMA antibody titers. It thus became clear that individuals over age 50 years are the ones who are likely to develop zoster and that they have detectable antibodies to VZV but have lost demonstrable CMI to the virus [21, 22].

Success is the ability to go from one failure to another with no loss of enthusiasm.

Churchill

In this scientific setting, in the early 1970s, continuing the beginnings of major advances in VZV research, Takahashi and his colleagues began experiments to attenuate VZV in the hopes of developing a live vaccine to prevent varicella. Michiaki Takahashi M.D. was for many years the Director of the Microbiology Department at Osaka University. Prior to this, while spending a postdoctoral year in the laboratory of Dr. Joseph Melnick in Houston, TX, Takahashi, whose son had developed severe chickenpox, was motivated to try to develop a way to prevent the illness by attenuating the virus and using it as a live vaccine. He modeled his approach on that used by Sabin to attenuate polioviruses. Takahashi first isolated VZV from an otherwise healthy child who had typical varicella; the family name of the child was Oka. He passed the wild type parental Oka strain 11 times at 34°C in human embryonic lung fibroblasts, followed by 12 intermediate passages in guinea-pig embryo cells at 37°C, and roughly 10 final passages in human cells

(WI-38 and MRC-5) at 37°C. He used passages in guinea-pig cells because he reasoned that passages in nonhuman cells might be critical for attenuation. After approximately 35 serial passages of the virus, he reasoned that it was likely to have been attenuated. He tested the virus for safety in various animals. Because there was (and still is) no animal model of clinical varicella, however, he had to give the candidate vaccine to humans in order to show attenuation. He knew that passage of viruses in cell cultures was highly likely to result in an attenuated virus rather than a virulent one, and he also knew that when children were inoculated with vesicular fluid from zoster patients, they were unlikely to develop clinical varicella [23, 24]. He and his colleagues administered the candidate vaccine to healthy individuals without adverse consequences. These investigators were able to terminate a nosocomial outbreak of varicella by immunization [23]. Because there was no animal model of varicella in which to first test the vaccine prior to administering it to humans, one would have to conclude that Takahashi was both extremely clever in his approach as well as lucky, since his vaccine eventually proved to be both safe and effective. Maurice Hilleman and his colleagues working at Merck and Co. in Pennsylvania had tried a similar approach, but were unable to produce a virus that was both suitably attenuated and immunogenic [25, 26]. Early passages of the candidate vaccine were highly reactogenic and later ones were very well tolerated but not sufficiently immunogenic. The story is reminiscent of the childhood tale of the Three Bears; the Baby Bear's food and furniture were "just right" for Goldilocks. So was Takahashi's vaccine "just right." Eventually Merck was able to obtain the Oka strain, as was Smith/Kline (now Glaxo Smith/Kline). Today, all live attenuated varicella vaccine available worldwide is the Oka strain.

It is interesting to relive those years in the early- and mid-1970s when varicella vaccine was newly developed. In those days no one considered that the vaccine might also be used to prevent zoster, which now in the twenty-first century is a major use of VZV vaccine (see below). The vaccine was clearly very efficient in preventing varicella. Early Japanese studies hinted it would be close to 100% effective in preventing this disease. Most exciting was the prospect of preventing varicella in immunocompromised children [27–30]. In the late 1970s, some 80% of young children with leukemia were being cured of that disease by extended periods of chemotherapy, but some 5–10% were doomed to severe infection or death from varicella during this treatment. Despite the promise of improving the lives of leukemic children, however, varicella vaccine was highly controversial when it was first introduced into the United States in clinical trials. There was fear that VZV was potentially oncogenic (HSV was hypothesized to be a cause of cervical cancer at the time), and that vaccination might well increase the incidence and/or severity of zoster in comparison to natural disease. There was malaise about immunization of children with a virus that seemed to be likely to cause a latent infection that would last a lifetime and might have undesirable consequences. Others argued that varicella was too mild an illness to merit a vaccine against it, and that after vaccination immunity might only be transient, leading to a large group of varicella-susceptible adults, who were at high risk to develop severe varicella. At that time, few people were cognizant of the congenital varicella syndrome, and the risk of severe group A beta hemolytic streptococcal infection

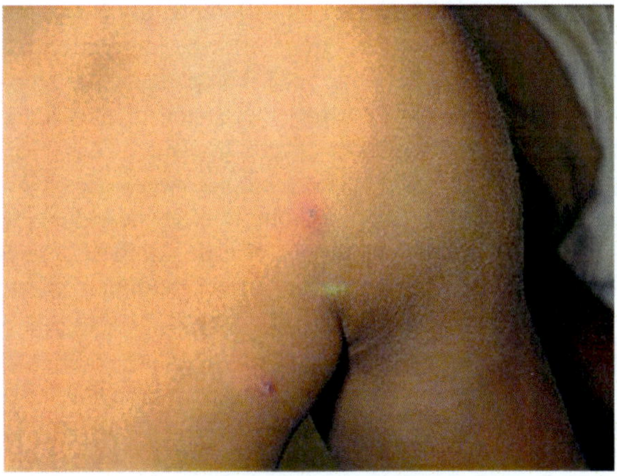

"Break through" Varicella

after varicella had not yet been appreciated. To look back at these years, it is amazing that early clinical trials of varicella vaccine were able to be carried out in the United States because there was so much controversy about its potential use [31]. The Japanese investigators, however, continued their studies with great apparent success, and finally it was deemed in the United States that the vaccine simply had to be tried here, because the potential advantages were so great and the risks appeared to be small or inconsequential, especially for immunocompromised children

...Give us the tools and we will finish the job.

Churchill

Fortunately for medicine and VZV research, molecular techniques for identification of VZV without having to propagate the virus were developed in the 1980s, so that it was possible to rapidly and accurately distinguish between wild type VZV and the Oka strain, by isolating the virus in cell culture and then applying the new technique of polymerase chain reaction (PCR) [32, 33]. Development of some form of rash was not uncommon after vaccination, particularly when the wild type virus was circulating at high levels, and it was therefore critical to have a rapid means of diagnosing VZV and determining whether a VZV rash is wild type or Oka. While at first it was necessary to propagate the virus and then analyze the DNA, it soon became possible to diagnose VZV and type the virus rapidly by employing PCR and analysis of the amplification products with restriction enzymes, avoiding the need for viral culture. These studies were led by Drs. Philip LaRussa in the Department Pediatrics and Saul Silverstein in the Department of Microbiology at Columbia University. The presence of a Pst splicing site in gene 54 of wild type VZV but not in Oka was exploited by these investigators and has been critical in interpreting clinical events such as possible zoster and breakthrough chickenpox in vaccinees in the vaccine era [34]. More recently, in the laboratory of Scott Schmid at the CDC, restriction sites in gene 62 that are specific to vaccine type have been identified

[35–37]. It is recognized, however, that differentiating between vaccine and wild type viruses by examining gene 62 alone must be interpreted with caution and by examining multiple sites of the gene [38]. These molecular studies have provided clues about what is responsible for the attenuation of the Oka strain, but require further investigation to identify more exactly the genetic basis for attenuation [39].

In addition to improved diagnostic methods, acyclovir (ACV) becomes available in the early 1980s, which enabled one to treat individuals who had varicella, especially those with highly aggressive disease. It is of interest, however, that despite ACV for treatment, leukemic children continued to succumb to natural varicella [40]. In general, though, ACV proved to be a highly effective drug for treating VZV infections.

Availability of ACV made it possible to carry out clinical trials of varicella vaccine in children with leukemia, in case it was necessary to control the Oka strain in these immunocompromised children. It was reasoned that these children had the most appropriate risk–benefit ratio for the initial studies of the vaccine in the United States, because the vaccine was potentially life saving. If there were any long-term adverse effects, the theoretical risk was worth taking under the circumstances. About 20 years previously, severe measles had resulted when vaccination of leukemic children against this disease was attempted; there was no antiviral therapy against measles, however, and the project had to be abandoned when severe reactions were noted [41]. Fortunately, for varicella vaccine testing, ACV was available if needed.

For varicella vaccine, thus, the appropriate diagnostic tools for identifying VZV and immune responses to it had been developed, and there was a potential treatment if VZV Oka caused adverse events in vaccinees. In addition, the availability of passive immunization with varicella-zoster immune globulin (VZIG) since the 1970s afforded another medical approach to control the virus if necessary. These available approaches might be characterized as the "Three Graces of VZV": passive immunization, antiviral therapy, and active immunization; availability of the first two made it possible to develop the third.

Now this is not the end. It is not even the beginning of the end. But it is, perhaps, the end of the beginning.

Churchill

In Japan, small numbers of children with varying degrees of immunocompromise were vaccinated after immunization of a larger group of healthy children against chickenpox [42]. It appeared that children with leukemia were capable of developing antibodies and CMI to VZV after vaccination, but the question of whether or how well this would protect them from developing chickenpox remained. Several studies were carried out in the United States, beginning in 1979, by Gershon in NY, Brunell in Texas and, Alan Arbeter, and Stanley Plotkin, at Children's Hospital of Philadelphia [33, 43, 44]. The largest was a collaborative study of immunization of children with leukemia in remission, which was organized and supported by the National Institute of Allergy and Infectious Diseases (NIAID); this study lasted roughly until 1990 [33]. Over 500 children with leukemia that was in remission for at least 1 year were immunized with the Oka vaccine in this study; most received 2 doses a month apart. Most vaccinees were receiving maintenance antileukemic therapy, which was withheld for 1 week before and 1–2 weeks after the first dose of vaccine. About 25% of these children developed a vaccine-associated rash with more than 50 skin lesions; usually this occurred about 1 month after the first vaccination. These vaccinees were given antiviral medication (usually ACV) to prevent further multiplication of the vaccine virus. The efficacy of the vaccine was found to be 85% in preventing chickenpox following a household exposure to VZV [33, 45]. When children developed what was termed "breakthrough" chickenpox after an exposure, however, the cases were uniformly mild and did not require antiviral therapy, as did the vaccine-associated rashes described above. The breakthrough illness was significantly less of a medical problem than the vaccination. There were no deaths in this study from either the Oka strain or the wild type strain of VZV; undoubtedly lives were saved in many children from this early clinical trial. The Oka vaccine proved to be both safe and effective.

This collaborative study was fortunate to have the input a number of highly distinguished pediatricians and/or vaccinologists of the time, including Drs. Saul Krugman, William S. Jordan, and Wolf Szmuness. Saul Krugman, educated at Ohio State University, was involved in World War II in the Pacific Theatre and eventually became the Professor and Chairman of Pediatrics at NYU for 20 years, beginning in the early 1960s. He was well known for his research in infectious diseases, particularly development of measles and hepatitis A and B vaccines. This first cousin of Albert Sabin cleverly distinguished between hepatitis A and B by observing and studying children who had two attacks of clinical hepatitis [46]. Krugman was very concerned about the potential of varicella as a significant illness and had observed severe and even fatal cases particularly in adults and immunocompromised children, when there was no available antiviral therapy. Wolf Szmuness, born in Poland, emigrated to the United States after World War II, and eventually came to head the Department of Epidemiology at the New York Blood Center. He conducted the first efficacy trials of hepatitis B, which was a scourge of the gay community. His trials indicated that hepatitis B vaccine was highly protective against hepatitis B in this population [47]. Jordan was the Director of the Microbiology and Infectious Diseases Program at NIAID; his mission was the advancement of research initiatives concerning vaccines within the National Institutes of Health (NIH), a position in which he could advocate for studies on varicella vaccine. He was highly respected in the medical community, having previously served in senior academic positions, at Western Reserve, the University of Virginia, and the University of Kentucky. He was awarded the Gold Medal of the Sabin Vaccine Foundation in 2004. He developed The Jordan Report, which is still issued annually by NIH, and is considered the most complete and up-to-date reference on vaccine research and development.

Krugman, Szmuness, and Jordan were extremely supportive of the development of varicella vaccine in the United States, which, as mentioned, was initially highly controversial. It was the opinion of Szmuness that merely measuring cellular and humoral immunity would be inadequate to indicate immunity to varicella in these high-risk leukemic children. Including data on protection against disease after an exposure to VZV would solidify evidence that the vaccine appeared to be protective. Without the interest, scientific perspective, and confidence of Krugman, Jordan, and Szmuness, this major project in leukemic children could not have been carried out, and the vaccine might not have eventually been studied and used in healthy children.

True clinical trials were still in developmental stages in the late 1970s, and computers were rarely utilized to facilitate them. The NIH collaborative study benefited vastly from a computer program developed by an enterprising high school student, Perry Gershon, who happened to be the son of the Principal Investigator of the study. Because of this program, regular follow up of vaccinees was highly likely to be assured. Antibody and CMI determinations could be requested regularly at local sites, and information such as household exposures to varicella, and clinical illnesses such as breakthrough varicella and zoster were automatically requested and recorded. The investigators carrying out the study knew on a regular basis which child was retaining antibodies and CMI to VZV; in that way children closely exposed to VZV could be passively immunized with varicella zoster immune globulin (VZIG) if necessary, to prevent severe varicella. VZIG was rarely required, however, because most children were known to be seropositive if and when an exposure to VZV occurred [33, 45]. There was no instance of any of these vaccinees developing severe wild type varicella. Many are alive and well even today, as long as 28 years later, long-term survivors of both leukemia and varicella vaccine.

It was noticed, however, that some vaccinees experienced loss of antibodies with time after 1 dose, and some failed to seroconvert after vaccination. At that time the common wisdom, which turned out to be naïve, was that development of a positive immune reaction to a virus meant immunity for many years if not for life. To deal with the problems of antibody loss and failure to seroconvert, a second dose of varicella vaccine was decided upon as a possible solution, during a coffee shop discussion between the PI of the study and the NIAID Program Officer, Dr. George Galasso. This was years before a second dose of measles vaccine was mandated, and required some coaxing of the investigators in the collaborative group, although most complied. A second dose increased the seroconversion rate and antibody persistence, and became the norm. Eventually, a two-dose schedule was recommended even for young healthy children [19, 48]. While the initial studies were performed in leukemic children in the United States, a number of varicella-susceptible health care workers were also immunized in early clinical trials. They too experienced loss of antibodies and failure to seroconvert after 1 dose, and were therefore also given 2 doses [48–50].

In the early 1980s, the dose of live vaccine virus was thought not to be important because the virus was expected to multiply and stimulate immunity in that manner. In a seminal double-blind, placebo-controlled study on varicella vaccine efficacy, published in 1984, in the *New England Journal of Medicine*, the dose of virus in the administered vaccine is not mentioned [51]. Eventually it became apparent that the immunizing dose or number of plaque forming units (pfus) of vaccine virus in the injection was critical and that lower doses stimulated weaker immune responses and higher doses caused stronger immune reactions.

In studies involving vaccination of leukemic children, it was not ethically possible to carry out a placebo-controlled study. Because vaccination in Japan strongly suggested a benefit to these children, a control group could not be included in the NIAID study. It was possible to determine vaccine efficacy in this study, however, because varicella is such a highly contagious disease [15]. The attack rate of breakthrough chickenpox in vaccinees after a household exposure to varicella could be compared with historical attack rates of susceptibles, which do not change, to determine efficacy. With this method, the vaccine was judged to be 85% effective in leukemic children, after 2 doses of vaccine [33].

The cited publication of Weibel and colleagues in the *New England Journal of Medicine* was a double-blind, placebo-controlled study, and it marked a milestone in the development of varicella vaccine because healthy children were successfully immunized and protected against the disease. The protective efficacy was 100% in the first year and 97% in the second year; however it later came to light that the dose of vaccine used (17,000 pfu) in this study was about 13 times higher than the dose which was licensed by the US FDA in 1995 (1,350 pfu) [2]. In clinical practice in the United States, as determined by a case–control study, the efficacy in young healthy children who had 1 dose of vaccine (containing 1,350 pfus) was 87% over an 8-year period [52, 53]. This degree of protection was less than originally anticipated, and is important, so we will return to this observation again.

Not only was it found that varicella vaccine prevented chickenpox in immunocompromised children, it also decreased the incidence of zoster. This was noted first by Japanese investigators [54], and was confirmed in American studies in children with leukemia and HIV infection, and also in French studies on children who underwent renal transplantation [4, 55–59]. It is clear that the Oka strain can establish latent infection and also reactivate, based on in vitro and in vivo animal models of VZV latency [60–62]. A correlation between the presence of skin lesions (either Oka or wild type virus) and zoster has been recognized for many years and in many populations [9, 57, 63, 64], which has led to the hypothesis that without obvious skin infection, latency of VZV may be less likely to develop than if there are skin lesions. Although on occasion latency may develop without the presence of an obvious rash, the presence of rash seems to increase the chance of development of latent VZV. This may be the mechanism by which varicella vaccination decreases the incidence of zoster, since skin lesions are unusual after it.

A pessimist sees the difficulty in every opportunity; an optimist sees the opportunity in every difficulty.

Churchill

The recommendation of universal vaccination was spearheaded by Jane Seward, MBBS, MPH, then deputy director of the Division of Viral Diseases in the National Center for Immunizations and Respiratory Diseases at the CDC [65–67]. One dose of varicella vaccine was recommended as part of the standard immunization schedule in the United States for children aged 1–12 years in 1995. Since that time, vaccine coverage of children has risen to 90%, and the epidemiology of VZV infections has changed. Varicella has become much less common an illness, and fatalities due to it have become rare. Both personal and herd immunity play roles in protection. A number of other countries have licensed varicella vaccine, which is manufactured by the Biken Institute at Osaka University, Merck and Company, and Glaxo SmithKline. Because there has never been a head-to-head comparison of these vaccines, it is presumed that they are equally effective. Routine universal immunization of infants is now mandated in Canada, Uruguay, Sicily, Germany, South Korea, Qatar, Taiwan, Germany, Israel, and Australia [2]. But not, amazingly, in the UK, home of Jenner, where varicella vaccination is used very little.

Not long after licensure in the United States, however, a number of outbreaks in day care facilities and schools where most children had been vaccinated began to be described [2]. Because most published immunologic studies suggested that the vaccine would be over 95% protective in children after 1 dose, the situation was confusing. Seward's CDC team that investigated the postlicensure effectiveness of the vaccine had cleverly identified three locations in the United States, in California, Texas, and Pennsylvania, where active surveillance of vaccination, varicella, and zoster were carried out beginning in 1995, as a substitute for national reporting of infections that were far too common to be reportable diseases. Because active surveillance was being performed, it was possible to generalize from these sentinel locations whether the vaccine was preventing varicella and to what degree. These observations indicated that while the incidence of varicella decreased dramatically after 1995, a low rate of disease remained from about 2002 onward [66, 68–70]. It was also noted that the seroconversion rate in children from three locations in New York, California, and Tennessee, measured by FAMA, was only 76% after 1 dose of vaccine [19]. Therefore it was decided that like leukemic children, healthy adults, and children who were over 13 years of age, two routine doses of varicella ought to be given to children aged 1–12 years. This approach was recommended by the CDC in 2006 [71]. It remains to be seen if two doses will result in a further decline in the incidence of varicella. Dramatic boosts in humoral and cellular immunity to VZV have been observed following a second dose of vaccine, which suggests that 2 doses will provide increased protection [71].

> **Every day you may make progress. Every step may be fruitful. Yet there will stretch out before you an ever-lengthening, ever-ascending, ever-improving path. You know you will never get to the end of the journey. But this, so far from discouraging, only adds to the joy and glory of the climb.**
>
> **Churchill**

At about the time that varicella vaccine was licensed in the United States, scientists began to hypothesize that it might be possible to develop a therapeutic vaccine against zoster, by immunizing varicella-immunes to boost their cellular immunity to VZV. It had been recognized in the early 1980s that older individuals, who were at high risk to develop zoster, had well-preserved humoral immunity to VZV, but lost their cellular immune responses [22, 72]. Therefore attempts were made to prevent zoster by vaccinating elderly individuals, who had latent infection with VZV, with various doses of varicella vaccine. After many open-label studies that suggested efficacy [73], a historic double-blind, controlled study was conducted in almost 40,000 individuals over the age of 60 years. At that time this was the largest vaccine clinical trial ever performed. It proved the efficacy of a VZV vaccine to prevent or modify zoster [74]. The vaccine, Zostavax™ contained on average 20,000 pfu of the Oka strain, as compared to 1,350 pfu in the Merck varicella vaccine, Varivax™. This dose was necessary to stimulate cellular immunity in the elderly. The vaccine was remarkably safe, even at these doses, in individuals with low cellular immune responses to VZV. Overall, it was 51% effective in preventing zoster in subjects aged 60–85, but there was better protection in those in the sixth decade (64%). Individuals over age 80 had less protection. There was also significant modification of zoster in those who developed it; in particular, the incidence of postherpetic neuralgia (PHN) was 67% lower in vaccinees than in placebo recipients. The effect against PHN was greatest in vaccinees in their seventh decade of life. This saga is not yet over; there are ongoing studies as to duration of immunity after 1 dose, whether booster doses are useful, and whether better protection occurs when 50-year olds are vaccinated, among others.

Another important aspect of the Oxman study [74] is that the annual incidence of zoster in the United States in healthy elderly people was found to be an astonishing one million cases. The burden of this disease had not previously been appreciated. Recall that the incidence of varicella in the prevaccine era was four million annual cases.

It is clear that boosting of immunity due to exposure to VZV can prevent zoster [74–76], but whether reduced circulation of wild type VZV will increase the incidence of zoster remains unclear, despite modeling studies that suggest it will [77]. The mechanism responsible for long-term immunity against VZV has not yet been elucidated, and it is possible that subclinical reactivation of VZV boosts immunity. Whether the prevalence of zoster is increasing in the United States since 1995 is controversial [69, 78–81], and, increases in zoster prevalence occurred in the twentieth century prior to licensure of varicella vaccine [80, 82, 83]. Additional data are necessary to answer the question as to whether increased prevalence of zoster will occur in the unimmunized when there is widespread vaccination against varicella. The use of the zoster vaccine could be helpful in managing this theoretical circumstance, should it develop and become a medical problem.

When the ACIP recommended that 2 doses of varicella vaccine be administered routinely in 2006, it stated that the newly licensed combination vaccine measles–mumps–rubella–varicella (MMRV) was the preferred vaccine. Use of this product would presumably increase vaccination rates by decreasing the number of injections for children. By 2008, however, there were reports from Merck and Co. and from the CDC that there were safety concerns regarding MMRV [84]. The rate of febrile seizures occurring in the 7–10 days after the first dose was 4/10,000 after MMR, but increased to 9/10,000 after MMRV. The reason for this doubling increase is not fully understood although it may be because the dose of varicella virus is much greater in MMRV (10,000 pfu) compared to Merck's monovalent product (1,350 pfu). More importantly, the significance of these unusual febrile seizures,

Anne Gershon and Phillip L. Russa

which rarely have sequelae, and which obviously also follow MMR itself albeit at a lower rate, is also not fully understood. Whether to recommend MMRV over MMR, therefore, is still under discussion at the CDC.

Wherever VZV vaccines are going, however, it must be appreciated that they have changed the profiles of varicella and zoster in the United States; varicella is becoming milder and less frequent than previously, as is zoster among vaccinees. Patrick Henry memorably pointed out that "The price of liberty is eternal vigilance." Diseases due to VZV are beginning to yield to us, but we need to continue to monitor their activity as well as the vaccines in order to safely continue this trend.

This manuscript is dedicated to pediatricians, virologists, and public health experts, especially those of Great Britain, in the hope that they will again heed the wisdom of Winston Churchill. Churchill's insights, leadership, and inspiration enabled Britain to successfully and, perhaps improbably, overcome a challenge to its survival that more than matched that of the Spanish Armada. These insights can surely be applied to lesser challenges, such as that of control of VZV, with great effect. Vaccines can prevent varicella as well as zoster. The time is now to grasp the nettle and protect the Island from VZV.

References

1. Cohen, J.I., S.E. Straus, and A. Arvin, *Varicella-Zoster Virus: Replication, Pathogenesis, and Management*, in *Fields Virology*, D.M. Knipe and P.M. Howley, Editors. 2007, Lippincott Williams and Wilkins: Philadelphia. p. 2773–2818.

2. Gershon, A., M. Takahashi, and J. Seward, *Live attenuated varicella vaccine*, in *Vaccines*, S. Plotkin, W. Orenstein, and P. Offit, Editors. 2008, WB Saunders: Philadelphia. p. 915–958.
3. Garland, J., *Varicella following exposure to herpes zoster.* N. Engl. J. Med., 1943. **228**: p. 336–337.
4. Hayakawa, Y., et al., *Analysis of varicella zoster virus (VZV) DNAs of clinical isolates by endonuclease HpaI.* J. Gen. Virol., 1986. **67**: p. 1817–1829.
5. Straus, S.E., et al., *Endonuclease analysis of viral DNA from varicella and subsequent zoster infections in the same patient.* N. Engl. J. Med., 1984. **311**: p. 1362–1364.
6. Williams, D.L., et al., *Herpes zoster following varicella vaccine in a child with acute lymphocytic leukemia.* J. Pediatr., 1985. **106**: p. 259–261.
7. Weller, T.H. and A.H. Coons, *Fluorescent antibody studies with agents of varicella and herpes zoster propagated in vitro.* Proc Soc Exp Biol Med, 1954. **86**(4): p. 789–94.
8. Cook, M.L. and J. Stevens, *Labile coat: reason for noninfectious cell-free varicella zoster virus in culture.* J Virol, 1968. **2**: p. 1458–1464.
9. Chen, J.J., et al., *Mannose 6-phosphate receptor dependence of varicella zoster virus infection in vitro and in the epidermis during varicella and zoster.* Cell, 2004. **119**(7): p. 915–926.
10. Brunell, P.A., *Transmission of chickenpox in a school setting prior to the observed exanthem.* Amer J Dis Child, 1989. **143**: p. 1451–1452.
11. Ku, C.C., et al., *Tropism of Varicella-Zoster Virus for Human Tonsillar CD4(+) T Lymphocytes That Express Activation, Memory, and Skin Homing Markers.* J Virol, 2002. **76**(22): p. 11425–33.
12. Guess, H.A., et al., *Population-based studies of varicella complications.* Pediatrics, 1986. **78**(4 Pt 2): p. 723–7.
13. Ku, C.C., et al., *Varicella-Zoster Virus Transfer to Skin by T Cells and Modulation of Viral Replication by Epidermal Cell Interferon-{alpha}.* J Exp Med, 2004. **200**(7): p. 917–925.
14. Ku, C.C., et al., *Varicella-Zoster virus pathogenesis and immunobiology: new concepts emerging from investigations with the SCIDhu mouse model.* J Virol, 2005. **79**(5): p. 2651–8.
15. Ross, A.H., E. Lencher, and G. Reitman, *Modification of chickenpox in family contacts by administration of gamma globulin.* N. Engl. J. Med., 1962. **267**: p. 369–376.
16. Brunell, P. and H. Casey, *Crude tissue culture antigen for determination of varicella-zoster complement fixing antibody.* Publ Heal Rep, 1964. **79**: p. 839–841.
17. Demmler, G., et al., *Rapid enzyme-linked immunosorbent assay for detecting antibody to varicella-zoster virus.* J. Infect. Dis., 1988. **157**: p. 211–212.
18. Williams, V., A. Gershon, and P. Brunell, *Serologic response to varicella-zoster membrane antigens measured by indirect immunofluorescence.* J. Infect. Dis., 1974. **130**: p. 669–672.
19. Michalik, D.E., et al., *Primary vaccine failure after 1 dose of varicella vaccine in healthy children.* J Infect Dis, 2008. **197**(7): p. 944–9.
20. Rand, K.H., et al., *Cellular immunity and herpesvirus infections in cardiac transplant patients.* N Engl J Med, 1977. **296**: p. 1372–1377.
21. Berger, R., G. Florent, and M. Just, *Decrease of the lympho-proliferative response to varicella-zoster virus antigen in the aged.* Infect. Imm., 1981. **32**: p. 24–27.
22. Burke, B.L., et al., *Immune responses to varicella-zoster in the aged.* Arch. Intern. Med., 1982. **142**: p. 291–293.
23. Takahashi, M., et al., *Live vaccine used to prevent the spread of varicella in children in hospital.* Lancet, 1974. **2**: p. 1288–1290.
24. Takahashi, M., et al., *Development of varicella vaccine.* J Infect Dis, 2008. **197 Suppl 2**: p. S41–4.
25. Arbeter, A.M., et al., *Live attenuated varicella vaccine: the KMcC strain in healthy children.* Pediatrics, 1983. **71**: p. 307–312.
26. Neff, B.J., et al., *Clinical and laboratory studies of KMcC strain of live attenuated varicella virus.* Proc Soc Exp Biol Med, 1981. **166**: p. 339–347.
27. Takahashi, M., et al., *Development of a live attenuated varicella vaccine.* Biken J, 1975. **18**: p. 25–33.

28. Takahashi, M., et al., *Active immunization for varicella zoster virus*, in *The Human Herpesviruses*, A.J. Nahmais, W.R. Dowdle, and R.E. Schnazi, Editors. 1981, Elsevier: New York. p. 414–431.
29. Takahashi, M., et al., *Active immunization for varicella-zoster virus*, in *The Human Herpesviruses*, A.J. Nahmias, W.R. Dowdle, and R.E. Schinazi, Editors. 1984, Elsevier: New York. p. 255.
30. Takahashi, M. and K. Baba, *A live varicella vaccine: its protective effect and immunological aspects of varicella-zoster virus infection*, in *Medical Virology*, L. de al Maza and E.M. Peterson, Editors. 1984, Elsevier: New York. p. 255.
31. Brunell, P.A., *Varicella vaccine: the crossroads is where we are not!* Pediatrics, 1978. **62**(5): p. 858–9.
32. Gelb, L.D., et al., *Molecular epidemiology of live, attenuated varicella virus vaccine in children and in normal adults.* J Infect Dis, 1987. **155**: p. 633–640.
33. Gershon, A.A., et al., *Live attenuated varicella vaccine: efficacy for children with leukemia in remission.* J.A.M.A., 1984. **252**: p. 355–362.
34. LaRussa, P., et al., *Restriction fragment length polymorphism of polymerase chain reaction products from vaccine and wild-type varicella-zoster virus isolates.* J. Virol., 1992. **66**: p. 1016–1020.
35. Loparev, V.N., et al., *Improved identification and differentiation of varicella-zoster virus (VZV) wild type strains and an attenuated varicella vaccine strain using a VZV open reading frame 62-based PCR.* J. Clin. Micro., 2000. **38**: p. 3156–3160.
36. Loparev, V.N., et al., *Rapid genotyping of varicella-zoster virus vaccine and wild type strains with fluorophore-labeled hybridization probes.* J. Clin. Micro., 2000. **38**: p. 4315–4319.
37. Loparev, V.N., Rubtcova, E., Seward, J.F., Levin, M.J., Schmid, D.S., *DNA sequence variability in isolates recovered from patients with postvaccination rash or herpes zoster caused by Oka varicella vaccine.* J Infect Dis, 2007. **196**: p. 801–2.
38. Lopez, A.S., et al., *Transmission of a Newly Characterized Strain of Varicella-Zoster Virus from a Patient with Herpes Zoster in a Long-Term-Care Facility, West Virginia, 2004.* J Infect Dis, 2008. **197**(5): p. 646–653.
39. Quinlivan, M.L., et al., *Natural selection for rash-forming genotypes of the varicella-zoster vaccine virus detected within immunized human hosts.* Proc Natl Acad Sci U S A, 2007. **104**(1): p. 208–12.
40. Feldman, S. and L. Lott, *Varicella in children with cancer: impact of antiviral therapy and prophylaxis.* Pediatrics, 1987. **80**: p. 465–472.
41. Mitus, A., et al., *Attenuated measles vaccine in children with acute leukemia.* Am J Dis Child, 1962. **103**: p. 413–8.
42. Izawa, T., et al., *Application of a live varicella vaccine in children with acute leukemia or other malignant diseases.* Pediatrics, 1977. **60**: p. 805–809.
43. Arbeter, A., et al., *Immunization of children with acute lymphoblastic leukemia with live attenuated varicella vaccine without complete suspension of chemotherapy.* Pediatrics, 1990. **85**: p. 338–344.
44. Brunell, P.A., et al., *Administration of live varicella vaccine to children with leukemia.* Lancet, 1982. **2**: p. 1069–1073.
45. Gershon, A., P. LaRussa, and S. Steinberg, *Varicella vaccine: use in immunocompromised patients*, in *Infectious Disease Clinics of North America*, R.E. J. White, Editor. 1996, Saunders: Philadelphia. p. 583–594.
46. Krugman, S., J.P. Giles, and J. Hammond, *Infectious hepatitis. Evidence for two distinctive clinical, epidemiological, and immunological types of infection.* JAMA, 1967. **200**(5): p. 365–73.
47. Szmuness, W., et al., *Hepatitis B vaccine: demonstration of efficacy in a controlled clinical trial in a high-risk population in the United States.* N Engl J Med, 1980. **303**(15): p. 833–41.
48. Centers-for-Disease-Control, *Prevention of varicella: recommendations of the Advisory Committee on Immunization Practices (ACIP).* MMWR, 2007. **56**: p. 1–40.

49. Ampofo, K., et al., *Persistence of immunity to live attenuated varicella vaccine in healthy adults.* Clin Infect Dis, 2002. **34**(6): p. 774–9.
50. Gershon, A.A., S. Steinberg, and NIAID-Collaborative-Varicella-Vaccine-Study-Group., *Live attenuated varicella vaccine: protection in healthy adults in comparison to leukemic children.* J. Infect. Dis., 1990. **161**: p. 661–666.
51. Weibel, R., et al., *Live attenuated varicella virus vaccine: efficacy trial in healthy children.* N Engl J Med, 1984. **310**: p. 1409–1415.
52. Vazquez, M., et al., *The effectiveness of the varicella vaccine in clinical practice.* N. Engl. J. Med., 2001. **344**: p. 955–960.
53. Vazquez, M., et al., *Effectiveness over time of varicella vaccine.* JAMA, 2004. **291**(7): p. 851–5.
54. Kamiya, H., et al., *Immunization of acute leukemic children with a live varicella vaccine (Oka strain).* Biken J, 1984. **27**: p. 99–102.
55. Broyer, M. and B. Boudailliez, *Prevention of varicella infection in renal transplanted children by previous immunization with a live attenuated varicella vaccine.* Transpl Proc, 1985. **17**: p. 151–152.
56. Broyer, M., et al., *Varicella and zoster in children after kidney transplantation: long term results of vaccination.* Pediatrics, 1997. **99**: p. 35–39.
57. Hardy, I.B., et al., *The incidence of zoster after immunization with live attenuated varicella vaccine. A study in children with leukemia.* N. Engl. J. Med., 1991. **325**: p. 1545–1550.
58. Hayakawa, Y., et al., *Biologic and biophysical markers of a live varicella vaccine strain (Oka): identification of clinical isolates from vaccine recipients.* J. Infect. Dis., 1984. **149**: p. 956–963.
59. Son, M., et al., *Vaccination of Children with Perinatal HIV Infection Protects against Varicella and Zoster*, in *Pediatric Academic Societies Annual Meeting*. 2008: Honolulu.
60. Chen, J., et al., *Latent and lytic infection of isolated guinea pig enteric and dorsal root ganglia by varicella zoster virus.* J. Med. Virol., 2003. **70**: p. S71–78.
61. Gershon, A.A., J. Chen, and M.D. Gershon, *A model of lytic, latent, and reactivating varicella-zoster virus infections in isolated enteric neurons.* J Infect Dis, 2008. **197 Suppl 2**: p. S61–5.
62. Rentier, B., et al. *Varicella-zoster virus latency in the nervous system of rats and humans is accompanied by the abundant expression of an immediate-early protein that is also present in acute infection.* in *Keystone Symposium on Virus entry, replication, and pathogenesis*. 1996. Santa Fe.
63. Head, H. and C. A.W., *The pathology of herpes zoster and its bearing on sensory localization.* Brain, 1900. **23**: p. 353–523.
64. Seiler, H.E., *A study of herpes zoster particularly in its relationship to chickenpox.* J Hyg (Lond), 1949. **47**(3): p. 253–62.
65. Seward, J., et al. *Decline of varicella disease: evidence of vaccine impact.* in *Society for Pediatric Research*. 2000. Boston.
66. Seward, J.F., et al., *Varicella disease after introduction of varicella vaccine in the United States, 1995–2000.* JAMA, 2002. **287**(5): p. 606–11.
67. Seward, J.F. and W.A. Orenstein, *Commentary: the case for universal varicella immunization.* Pediatr Infect Dis J, 2006. **25**(1): p. 45–6.
68. Nguyen, H.Q., A.O. Jumaan, and J.F. Seward, *Decline in mortality due to varicella after implementation of varicella vaccination in the United States.* N Engl J Med, 2005. **352**(5): p. 450–8.
69. Yih, W.K., et al., *The incidence of varicella and herpes zoster in Massachusetts as measured by the Behavioral Risk Factor Surveillance System (BRFSS) during a period of increasing varicella vaccine coverage, 1998–2003.* BMC Public Health, 2005. **5**(1): p. 68.
70. Zhou, F., et al., *Impact of varicella vaccination on health care utilization.* JAMA, 2005. **294**(7): p. 797–802.
71. Centers-for-Disease-Control, *Prevention of Varicella.* Morb. Mort. Wkly. Rep., 2007. **56**: p. 1–55.

72. Berger, R., D. Luescher, and M. Just, *Enhancement of varicella-zoster-specific immune responses in the elderly by boosting with varicella vaccine.* J Infect Dis, 1984. **149**: p. 647.
73. Levin, M.J., *Use of varicella vaccines to prevent herpes zoster in older individuals.* Arch Virol Suppl, 2001. **17**: p. 151–60.
74. Oxman, M.N., et al., *A vaccine to prevent herpes zoster and postherpetic neuralgia in older adults.* N Engl J Med, 2005. **352**(22): p. 2271–84.
75. Gershon, A., et al., *The protective effect of immunologic boosting against zoster: an analysis in leukemic children who were vaccinated against chickenpox.* J. Infect. Dis., 1996. **173**: p. 450–453.
76. Wood, S.M., et al., *Primary Varicella and Herpes Zoster Among HIV-Infected Children From 1989 to 2006.* Pediatrics, 2008. **121**(1): p. e150-e156.
77. Edmunds, W.J. and M. Brisson, *The effect of vaccination on the epidemiology of varicella zoster virus.* J Infect, 2002. **44**(4): p. 211–9.
78. Insinga, R.P., et al., *The incidence of herpes zoster in a United States administrative database.* J Gen Intern Med, 2005. **20**(8): p. 748–53.
79. Jumaan, A.O., et al., *Incidence of herpes zoster, before and after varicella-vaccination-associated decreases in the incidence of varicella, 1992–2002.* J Infect Dis, 2005. **191**(12): p. 2002–7.
80. Leung, J., N. Molinari, and R. Harpaz. *Trends in the incidence of herpes zoster using a National Insurance Database: United States 1993–2003.* in *43rd Annual Infectious Disease Society of America Meeting*. 2005. San Francisco.
81. Mullooly, J.P., et al., *Incidence of herpes zoster, 1997–2002.* Epidemiol Infect, 2005. **133**(2): p. 245–53.
82. Brisson, M., et al., *Epidemiology of varicella zoster virus infection in Canada and the United Kingdom.* Epidemiol Infect, 2001. **127**(2): p. 305–14.
83. Ragozzino, M.E., et al., *Population-based study of herpes zoster and its sequelae.* Medicine, 1982. **61**: p. 310–316.
84. Centers-for-Disease-Control, *Update: Recommendations from the Advisory Committee on Immunization Practices (ACIP) Regarding Administration of Combination MMRV Vaccine.* MMWR, 2008. **57**: p. 258–260.

Developmental History of HPV Prophylactic Vaccines

John T. Schiller and Douglas R. Lowy

Douglas Lowy and John Schiller

J.T. Schiller (✉)
Laboratory of Cellular Oncology, Center for Cancer Research, National Cancer Institute, Bethesda, MD, USA
e-mail: schillej@dc37a.nci.nih.gov

S.A. Plotkin (ed.), *History of Vaccine Development*,
DOI 10.1007/978-1-4419-1339-5_27,

Introduction

Prophylactic human papillomavirus (HPV) vaccines are a recent addition to our vaccine arsenal, first becoming commercially available in mid-2006. Although they have not yet made a public health impact, they likely represent an important addition to cancer prevention strategies. Their development is the result of a fortunate temporal convergence of two distinct lines of scientific enquiry. One produced the molecular biological and epidemiological evidence that established, beyond a reasonable doubt, that HPV infection is the central cause of cervical cancer. The second was the development of molecular technologies for production of the L1 virus-like particles (VLPs) vaccines and serological assays to measure their immunogenicity. HPV vaccine development and clinical testing was based on both critical lessons learned in the development of other vaccines and solutions to unique problems that arose because of the specific biology of HPVs and their relationship to neoplastic disease. This review offers our reflection on some of the milestones in the preclinical and clinical development of HPV prophylactic vaccine and includes some of our personal experiences in the enterprise. It is not meant to be a comprehensive review of the subject.

The Association of HPV and Cancer

It has been known for many decades that filterable transmissible agents, i.e. viruses, cause cutaneous and mucosal papillomas (warts) in animal models and in humans [1]. The oncogenic potential of papillomaviruses was first demonstrated by Payton Rous and J.W. Beard in the 1930s. They observed that epidermal papillomas induced in domestic rabbits by Shope cottontail rabbit papillomavirus (CRPV) could progress to squamous cell carcinomas, and that viral oncogenicity was enhanced by chemical cocarcinogens. However, the progression of human common and genital warts to cancer was virtually never observed, and so a link between HPV infection and human carcinogenesis was largely discounted. However, there were indications, first noted in the mid-1800s by the Italian physician, Rigoni-Stern, that cervical cancer had the epidemiological characteristics of a sexually transmitted infection. He astutely noted that cervical cancer was common among prostitutes, but rare in virgins or nuns [2]. Much later, in the mid-1970s, Alex Meisels and others observed that low-grade precancerous cervical lesions often contained areas of koilocytic atypia, characterized by vacuolated nuclei that histologically resembled nuclear abnormalities seen in cutaneous papillomas [3]. The dense bodies within the vacuolated nuclei were subsequently shown to contain typical HPV virion structures, reinforcing the idea that cervical neoplasia was virally induced. This led to the speculation that HPV might cause cervical cancer [4]. However, further investigations of the link between HPVs and cervical cancer were limited, in large measure by the inability to propagate HPVs in cultured cells or simple animal models. In the 1960s and 1970s, most of the attention was focused on the possibility that other STI agents, particularly herpes simplex virus 2, were the primary inducers of cervical carcinogenesis. However, the causal association with HSV was largely discounted by the mid-1980s, perhaps most convincingly in HSV2 serologic studies conducted by Vladimir Vonka [5].

In the mid-1970s, it was recognized that there were several HPV genotypes, based upon nucleic acid hybridization analysis of viral DNA extracted from wart-derived virions. However, the large number of distinct HPV genotypes (currently over 100), and the specific pathologies associated with them, were not appreciated. The advent of recombinant DNA technology and molecular cloning during this period proved to be the key technologic advances for further exploration of the biology of HPVs and their association with cancer. By the early 1980s, a number of distinct cutaneous and genital wart genotypes had been molecularly cloned into bacterial plasmids. Gerard Orth et al. identified an association between certain cutaneous types and skin cancer in a rare genetic disease, epidermodysplasia verruciformis [6]. Critically, in 1983, Harald zur Hausen et al. at the German Cancer Research Institute (DKFZ) reported the cloning of the sixteenth HPV genotype, in this case derived from a cervical cancer biopsy. Using this sequence as a probe, the group reported that HPV16 related sequences were detected by Southern blotting in approximately one-half of a set of cervical cancer tissue samples [7]. Shortly thereafter, the DKFZ group demonstrated that the E6 and E7 genes of HPV16 or HPV18 (another type they cloned soon after HPV16) were selectively retained and expressed in cervical cancers and cervical cancer-derived cell lines [8]. In 2008, Dr. zur Hausen received the Nobal Prize for Medicine for these groundbreaking discoveries. These findings launched studies of many groups that resulted in the classification of genital HPVs into two broad biological groups [9]. Low risk types, most often HPV6 and HPV11, cause genital warts and other benign genital neoplasia, but almost never cancer. The more than a dozen types designated high risk also normally cause benign neoplasia as part of their productive life cycle. However, these infections have an appreciable propensity for carcinogenic progression, particularly at the cervical transformation zone, where the columnar cells of the endocervix meet the stratified squamous cells of the ectocervix. Virtually all cervical cancers contain high-risk HPV DNA and a variable fraction of several other anogenital and oral cancer are also attributed to one of these viruses (Table 1).

Table 1 Estimate of annual incidence of cancers attributable to HPV infection

Site	% Attributable to HPV	Total cancers	Attributable to HPV	Attributable to HPV16/18
Cervix	100	492,800	492,800	333,900
Anus	90	30,400	27,300	25,100
Vulva, vagina	40	40,000	16,000	12,800
Penis	40	26,300	10,500	6,600
Mouth	3	274,300	8,200	7,800
Oro-pharynx	12	52,100	6,200	5,500
All sites		10,862,500	561,100	402,900

Adapted from [10]

It has been estimated that 5.2% of all human cancers are caused by HPV infection, with HPV16 alone accounting for 3.7% of all cancers [10].

Biological plausibility for a causative association between HPV infection and cancer was rapidly provided by laboratory studies. In the late 1980s/early 1990s, it was demonstrated that the E6 and E7 of high-risk types, but not low risk types, can independently transform certain rodent cell lines in vitro and cooperate to efficiently immortalize primary human epithelial cells, including cervical and foreskin keratinocytes [11]. These activities of high-risk E6 and E7 were attributed, at least in part, to their preferential ability to interact with and inactivate, respectively, p53 and pRb, the products of two tumor suppressor genes that are frequently inactivated by mutation in HPV-independent cancers [12, 13].

Surprisingly, the strong cancer association inferred from laboratory-based studies was not supported in the initial set of case/control epidemiological studies [14]. However, it was later determined that these studies were compromised by substantial misclassification of HPV status due to the use of relatively insensitive assays for measuring HPV DNA [15]. More sensitive and validated PCR-based assays that could simultaneously amplify the DNA of many HPV types were used in subsequent studies. These studies established a uniformly strong association between HPV infection and cervical cancer. Relative risks of greater than 100 were obtained in most studies conducted from the early 1990s onward, making high-risk HPV infection among the strongest risk factor ever observed for a specific cancer [16]. Several large prospective studies followed in the late 1990s. They demonstrated that the entire spectrum of cervical neoplasia from low grade dysplasias through carcinoma in situ, the accepted precursor of cervical cancer, arises from incident HPV infection [17]. By the beginning of the twenty-first century, there was essentially uniform agreement that sexually transmitted infection by high-risk HPVs was the central cause of cervical cancer [16]. However, HPV infection is considered a necessary, but not sufficient, cause of cervical cancer, since, although virtually all cases contain HPV DNA, most cervical HPV infections do not progress to cancer. The clear implication that could be drawn from this series of discoveries is that prevention of high-risk HPV infection would prevent cervical cancer, as well as a substantial proportion of other anogenital and oral cancers, thus providing the impetus for the development and commercialization of prophylactic HPV vaccines.

Preclinical Vaccine Development

Fortunately, vaccine developers did not wait for the conclusions of the prospective epidemiology studies before initiating work on prophylactic HPV vaccines. Perhaps because these investigators were primarily laboratory-based researchers, and even, as in our case, involved in the studies of the HPV oncogenes, they were not unduly inhibited by the weak correlation between HPV infection and cervical disease seen in the early epidemiologic studies. In any event, considerable activity aimed toward developing HPV vaccines was underway by 1990. A vaccine based on a live attenuated HPV strain was not considered a viable option for two reasons. First, HPVs could not be propagated in replicating cells in culture, so there was no reasonable means of virus production. Second, the virus was known to contain at least three oncogenes, E5, E6, and E7, and a vaccine that delivered oncogenes might not be considered safe for general use as a prophylactic vaccine. Therefore, most efforts involved subunit vaccine strategies. Since it was generally appreciated that most effective viral vaccines functioned primarily through the induction of neutralizing antibodies [18], attention was mainly focused on subunit vaccine strategies that might induce neutralizing antibodies to the papillomavirus virion proteins.

Almost equally important were the studies to develop in vitro neutralization assays that could critically evaluate the antibody responses to the vaccine candidates. In addition to protection from experimental challenge with wart-derived virions in rabbit, bovine, and canine models, there were only two assays for measuring neutralizing antibodies induced by papillomavirus vaccine candidates as of 1990. First, infectious events by bovine papillomavirus (BPV) types 1 and 2, but not HPVs, could be monitored in vitro by the induction of transformed foci in NIH3T3 and C127 mouse cell lines [19]. Neutralizing antibody titers were determined from the reduction in the number of foci induced by a standard stock of wart-derived virions (Fig. 1). We had developed this assay in the late 1970s, and it formed the basis for studying morphologic transformation by BPV. Second,

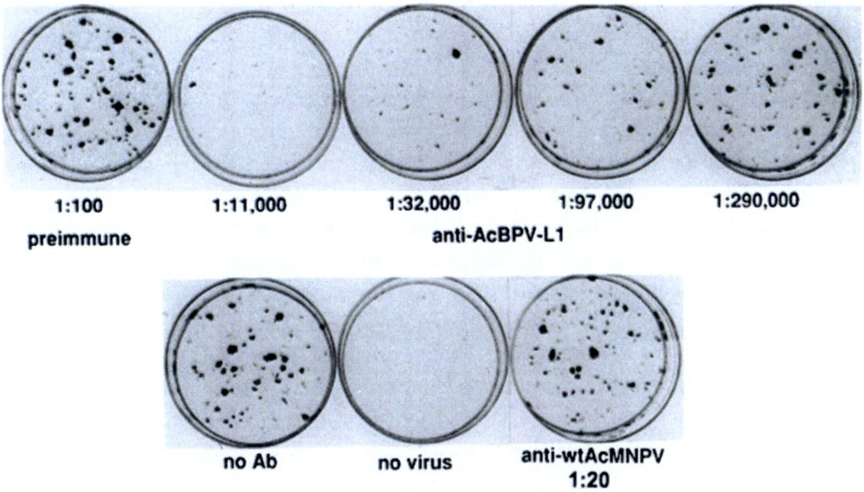

Fig. 1 In vitro neutralization of BPV1 virions by BPV1 L1 VLP antisera conducted in 1992. Foci of transformed cells induced by BPV1 infection of mouse C127 cells were visualized by staining with methylene blue/carbol fuchin. "Anti-AcBPV-L1" was sera from a rabbit vaccinated with BPV1 VLPs derived from L1 recombinant baculovirus infected insect cells. Anti-wt AcMNPV was serum from a rabbit vaccinated with an extract of wild type baculovirus infected insect cells. Numbers refer to the dilution of the sera used in the assay. "No Ab" demonstrates the number of focal transformation events induced by the BPV1 inoculum without added serum and "preimmune" demonstrates the number of foci induced in the presence of the rabbit's serum prior to VLP vaccination

wart-derived virions could be used to infect mucosal or cutaneous epithelial chips that were then placed under the renal capsule of athymic mice [20]. Infection was monitored by hyperproliferative changes in the transplanted chips and antibodies could be semiquantitatively evaluated for inhibition of in vitro infection prior to transplantation. Infection by BPV1, CRPV, HPV1, and HPV11 (but not other HPVs), and infection inhibition by type-specific antibodies, was evaluated by this relatively cumbersome xenograft assay [21].

Although papillomavirus virions were relatively poorly characterized at the time, it was known that the nonenveloped icosahedral capsid was composed of 360 copies of L1, the major capsid protein, and 12–72 copies of L2, the minor capsid protein [22]. Early studies employing animal papillomaviruses had established that interperitoneal or intramuscular injection of wart-derived virions, which does not induce an active infection, was effective at inducing neutralizing antibodies [19] and protecting from experimental infection [23, 24]. In contrast, IM injection of denatured virions or disordered L1 polypeptides derived from *Escherichia coli* inclusion bodies was ineffective at inducing neutralizing antibodies or protecting from experimental challenge [25, 26]. From these results, it was concluded that L1 needed to be in a "native" conformation to induce high titers of neutralizing antibodies. To this day, no relatively small peptide fragment of L1 has been shown to induce neutralizing antibodies. However, low levels of neutralizing antibodies, and partial protection from experimental challenge with homologous virus was induced using a bacterial fusion protein of either CRPV or BPV L1, suggesting that at least a minority of L1 molecules could display conformation-dependent neutralizing epitopes after bacterial expression [25, 27, 28].

At this point in time, it was appreciated that the major capsid protein of at least some naked icosahedral viruses had the intrinsic capacity to self-assemble into VLPs. For instance, *E. coli*-derived VP1 of several polyomaviruses, which are structurally similar to papillomaviruses, had been shown to self-assemble into morphologically correct VLPs from capsomeric subunits in in vitro reactions [29]. In addition, the commercial hepatitis B vaccine since the 1980s was based upon self-assembly of the viral S protein into lipid bilayer-containing VLPs [30]. Thus, by 1990 it was reasonable to consider a VLP-based vaccine displaying conformationally correct L1, or L1 and L2, as a potential candidate for an HPV vaccine.

In 1991, Zhou and Frazer reported that coexpression of HPV16 L1 and L2 in monkey CV-1 cells via a vaccinia virus vector resulted in the generation of "virus-like" particles that could be concentrated by sucrose gradient centrifugation [31]. These irregular particles had a mean diameter of 35–40 nm, compared to the 50–55 nm symmetrical particles reported for authentic virions, and were described as "incorrectly assembled arrays of HPV capsomeres." Particles were not detected when L1 was expressed separately. The vaccine potential of these L1/L2 capsomere arrays could not be critically evaluated because no HPV16 neutralizing assay was available at this time. Whether this study was the foundation for the subsequent development of the HPV prophylactic vaccines or taught against their development became the subject of much discussion, particularly among patent lawyers.

Shortly thereafter, Richard Schelgel and colleagues published the production of the L1 of HPV1 (a cutaneous wart type) in COS cells using an SV40 replicon vector [32]. The protein reportedly reacted with monoclonal antibodies that recognized native but not denatured HPV1 virions. However, no VLPs could be detected in cell extracts, and the L1 was neither purified nor evaluated for the induction of neutralizing antibodies. Thus it was unclear from these reports whether L1 alone, or L1 and L2 together, could assemble into conformationally correct VLPs or, if they could, whether they would efficiently induce neutralizing antibodies, and thus be attractive vaccine candidates. Because papillomavirus virions are only produced in terminally differentiated squamous epithelial cells, there was a concern that molecular chaperones unique to these cells might be required for correct folding and assembly of the virion proteins into capsids, and so it might be impossible for generation of morphologically correct VLPs in normal cell culture.

Our involvement in papillomavirus vaccine development began in early 1991, soon after Reinhard Kirnbauer, a dermatologist from the University of Vienna, arrived to begin a postdoctoral fellowship. As an initial project, we asked Reinhard to attempt to generate and characterize papillomavirus VLPs, both as vaccine candidates and as reagents for basic studies of virion/cell interactions, since there was no ready source of authentic virions. This project was a clear departure from the core activities of the laboratory, which for more than 10 years had centered on the molecular biology of the viral transforming genes and the regulation of viral gene expression. We have often reflected how fortunate we were to be in the intramural program of the National Cancer Institute, because our review was primarily retrospective. It is very doubtful that, given our lack of experience in virion structural proteins or vaccines, we would have convinced an extramural grant review committee to fund the project. After a review of potential production systems, we decide to express L1 in insect cells via recombinant baculovirus vectors, primarily for two reasons. First, they were known to produce exceptionally high levels of recombinant protein, and a critical capsid protein concentration might be needed to drive the VLP assembly reaction [33]. Second, the FDA had already approved clinical trials of other proteins produced in this system. Thus there was reason to expect that, should preclinical vaccine studies produce encouraging results, there would be a reasonable path to GMP production and human vaccination trials.

We focused initially on expressing BPV1 L1, rather than an HPV L1, because we had a stock of cow wart-derived infectious BPV virions in the laboratory and expertise in the in vitro focal transformation assay that could be used to evaluate neutralizing antibodies elicited by any vaccine candidate that would be generated in the study. In relatively short order, particularly considering his limited expertise in molecular biology, Reinhard was able to generate the L1 recombinant baculovirus, infect insect cells, and demonstrate robust L1 expression in the infected cells. We were thrilled when the first set of electron photomicrographs of thin sections from the infected insect cells revealed approximately 50 nm particles, the size expected for authentic papillomavirus virions. In addition, particles purified by CsCl gradient centrifugation displayed a regular array of capsomeric structures in transmission electron photomicrographs (Fig. 2). Most importantly, rabbits injected with partially

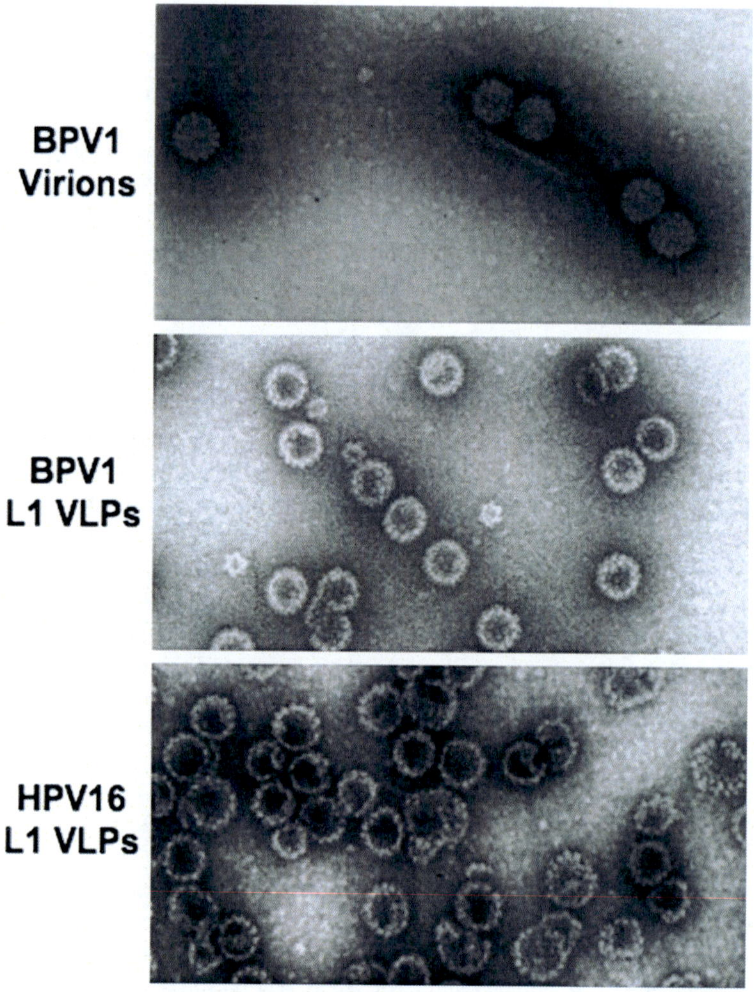

Fig. 2 Comparison of wart-derived BPV1 virions with recombinant baculovirus infected insect cell-derived BPV1 and HPV16 L1 VLPs, purified in 1992–1993. Electron photomicrographs after negative staining with 1% uranyl acetate are shown

purified VLPs, or even crude extracts of the infected cells, generated very high titers of BPV neutralizing antibodies (Fig. 1). The titers were so unexpectedly high that Reinhard had to conduct three consecutive neutralizing antibody titration experiments, each involving successively higher sera dilutions, until he finally reaching an endpoint titer when the sera was diluted more than 100,000-fold. In contrast to the titers of 10^5 induced by our BPV1 VLPs, the highest titer reported for bacterially derived BPV1 L1 had been 36 [25]. Consistent with the concept that neutralizing antibodies recognize conformationally dependent L1 epitopes, no neutralizing activity was detected in the sera of rabbits inoculated with denatured VLPs. The above results established that BPV L1 without any other viral proteins had the intrinsic capacity to assemble into VLPs that were able to induce high titers of neutralizing antibodies [34]. It is worth noting that assembly of a major capsid protein in VLPs did not necessary predict the ability of the VLPs to efficiently induce neutralizing antibodies. Neil Young's laboratory had recently published the production of B19 parvovirus VLPs [35]. That study demonstrated that the major capsid protein was sufficient to generate VLPs, but significant neutralizing antibodies were induced only when the minor capsid was coassembled into the particles.

Since the prevention of cow warts was not our ultimate goal, the obvious question became whether these findings could be translated into HPV vaccines. There were several possible explanations for why L1 VLPs were not detected in the HPV16 and HPV1 studies described earlier. One possibility was that the levels of L1 expression were insufficient to drive the assembly reaction. Another possibility was that L1s of different types had different intrinsic self-assembly capabilities. The latter possibility was raised since BPV1 belongs to a group of animal papillomavirus that uniquely induces cutaneous fibropapillomas and is distantly related to HPV16 and HPV1. To address these possibilities, Reinhard expressed HPV16 via an analogous recombinant baculovirus and was able to demonstrate abundant L1 protein expression in the infected insect cells. However, to our surprise and consternation, it was exceedingly difficult to find VLPs in the extracts. We estimated that the efficiency of HPV16 VLP formation was 1,000-fold lower than for BPV L1 [34]. Fortunately, Reinhard had also begun work on generating L1 VLPs of rhesus papillomavirus (RhPV1), since we were considering the possibility of future vaccine studies in nonhuman primates. RhPV1 L1 produced excellent yields of VLPs in our production system, much like BPV1 L1. This was an important observation because RhPV1 is closely related to HPV16, in fact more closely related to HPV16 than HPV16 is to some other high-risk HPVs, such as HPV18. We therefore felt that it was unlikely that the inefficient assembly of HPV16 was attributable to its phylogeny. It seemed more likely that the L1 of the prototype HPV16 clone used in our study was an assembly defective mutant. This widely distributed clone was the initial HPV16 DNA isolated by Harald zur Hausen's group [7]. The fact that it was isolated from a cervical cancer, and cancer cells are genetically unstable, supported the mutant hypothesis. To test this possibility, we obtained from the DKFZ's Mathias Durst and Lutz Gissmann two HPV16 clones that they had isolated from low-grade virus-producing lesions. We were much relieved to find that the L1 of these clones efficiently produced VLPs in our baculovirus expression system [36] (Fig. 2). Sequencing of the clones revealed that a single aspartate to histidine change in the prototype strain was responsible for the assembly defect of the prototype L1. Whether the inability to demonstrate assembly of HPV16 L1 into morphologically correct VLPs in the 1991 study [31] was due to the use of the prototype gene, the production system used, or some other technical difference was never firmly established.

In the same year as we published the efficient assembly of HPV16 L1 VLPs, the laboratory of Denise Galloway published the production of HPV1 L1 VLPs using a vaccinia virus vector, indicating that recombinant vaccinia virus infected cells could be permissive for L1 VLP self-assembly [37]. In addition, Bob Rose et al. published the production of HPV11 L1 VLPs using a baculovirus vector [38]. Thus, by the end of 1993, there was considerable evidence that the ability to self-assemble into VLPs is a general property of wild type papillomavirus L1 proteins. These findings were substantiated in many subsequent studies involving the L1s of other human and animal papillomavirus types.

Armed with the data demonstrating induction of high titer neutralizing antibodies by L1 VLPs, and a provisional patent application, we began a series of visits to commercial vaccine manufactures. We were received with interest, but also some reservations, by most of the companies. No doubt this was in part because we lacked credentials in vaccine development. However, there was also considerable skepticism in general about the prospects of developing an effective vaccine against a sexually transmitted mucosal pathogen. A cautious view at that time was understandable, given the conspicuous failure, despite extensive efforts in both academic and commercial sectors, to develop effective vaccines against HIV, HSV, and other STIs. The exceptional response came from Maurice Hilleman, one of the godfathers of modern vaccinology, who was then an emeritus employee at Merck. After a short private presentation of our data in his office, he unequivocally stated that the vaccine was going to work and Merck was going to make it. He turned out be correct on both accounts. Shortly after our discussion with Hilleman, we were approached by MedImmune, a biotechnology company headquartered just a few miles north of our laboratory in Bethesda, with an expression of interest in our vaccine concept. We were enthusiastic about the prospects of more than one company undertaking the commercial development of the vaccine. In our view, public health interest would most likely be served by competition during the development phase and hopefully, eventually in the marketplace. In keeping with its general policy, the NIH ultimately granted nonexclusive licenses to both Merck and Medimmune. Merck also exclusively licensed competing patent applications from Zhou and Frazer, and MedImmune exclusively licensed competing patent applications from the Schlegel and Rose groups. This led to a long series of patent disputes that for practical purposes was settled in 2005 when Merck and GlaxoSmithKline (which by this time had sublicensed Medimmune's HPV vaccine patent portfolio) agreed to a financial settlement. This agreement gave the two companies unrestricted and coexclusive access to the papillomavirus VLP vaccine patent claims of all four parties. This exclusivity solidified the sustained commercial investment in the vaccines.

Further insights into papillomavirus VLP vaccines were provided by the publication in 1995–1996 of several proof-of-concept trials in animal models. Intramuscular injection of low microgram amounts of L1 VLPs of COPV, CRPV, or BPV4, even without adjuvant, was shown to induce strong protection from high dose experimental challenge in dogs, rabbits, and calves, respectively [39–42]. Protection was type specific and could be passively transferred in immune sera or purified IgG, indicating that neutralizing antibodies were sufficient to confer protection. However, VLP vaccination did not induce regression of established lesions, suggesting that HPV VLP vaccines would not be therapeutic. These overall encouraging results strengthened commercial and academic interest in the vaccines.

The animal challenge studies noted earlier assessed cross-type protection against distantly related animal papillomavirus types. It also seemed important to assess the potential for cross-protection of HPV VLP vaccines against genital HPV types, which form relatively closely related clusters around HPV16, HPV18, and HPV6, respectively. Such an assessment would help in making decisions concerning the valency of an HPV VLP vaccine aimed at preventing cervical cancer and/or genital warts. Richard Roden, then a postdoctoral fellow in the laboratory, initially addressed the question by investigating the ability of homologous and heterologous L1 VLP rabbit sera to inhibit VLP agglutination of mouse red blood cells. Hemagglutination inhibition (HAI) has been used for a number of viruses as a surrogate for a true virus neutralization assay. Homologous HAI titers of several thousand or more were obtained with the VLP sera. However, only low HAI titers were seen across types, and then only for closely related pairs such as HPV6 and HPV11 or HPV18 and HPV45, arguing that VLPs would induce type-restricted protection against HPV infection [43]. However, we discovered that only a subset of the monoclonal antibodies that neutralized BPV1 in our focal transformation assay exhibited HAI activity. Therefore, we considered HAI to be a somewhat imperfect surrogate assay for assessing HPV infection inhibition.

The limitations of the HAI assay led Richard to develop an in vitro neutralization assay based on HPV pseudoviruses [44]. The pseudovirions were generated by coexpression of L1 and L2 via Semliki Forest Virus vectors in a hamster cell line that contained a relatively high copy number of autonomous replicating BPV1 genomes. Expression of L1 and L2 in these cells led to assembly of capsids that had incorporated the BPV genome. Infectious events could be scored by counting transformed foci on C127 cells, as in the case of the authentic BPV1 (Fig. 1). Using this rather laborious neutralization assay, or easier later versions in which a marker gene-expressing plasmid rather than the BPV genome was encapsidated [45], we were able to demonstrate that wild type HPV16 L1 VLPs, but not the prototype HPV16 L1 protein, generated high titers of HPV16 neutralizing antibodies. In agreement with the HAI data, VLP-induced neutralizing antibodies were clearly type restricted, with low levels of cross-neutralization detected only for closely related types. However, antibodies raised to VLPs of one HPV16 L1 variant were equally effective at neutralizing pseudovirions of other HPV16 variants [46]. From these studies, we concluded that HPV genotypes represent distinct serotypes, but that it is unlikely that distinct serotypes exist within a given genotype. The results supported the prediction that immunoprophylaxis by HPV VLP vaccines would be type restricted, which implied that multivalent vaccines would be needed for broad-spectrum protection against HPV-induced diseases.

Clinical Trials

Three groups, Merck, MedImmune, and the U.S. National Cancer Institute independently moved forward with GMP production of VLPs for phase I clinical trials. The MedImmune and the NCI (under a contract to Novavax) chose to continue producing their VLPs in recombinant baculovirus infected insect cells. However,

Merck decided to manufacture their vaccine in *Saccharomyces cerevisiae*, presumably because they had extensive experience in yeast production of their HBV vaccine. GMP process development and scale-up proved to be challenging. For example, the VLPs have a propensity to interact with solid surfaces, leading to aggregation and flocculent precipitation. However, Merck scientists determined that this problem could be overcome by the addition of small amounts of certain nonionic detergents [47]. Both companies eventually settled on purification schemes involving the dissociation of the VLPs into pentameric capsomeric subunits followed by in vitro reassembly into VLPs [48, 49]. The conformation dependency of the neutralizing epitopes also presented difficulties. For instance, MedImmune/GSK added thimerosal to their phase I product as a preservative. This formulation retained the VLP structure and induced VLP-reactive antibodies in vaccinees, but none of the antibodies were neutralizing [50]. Although the problem was solved by formulating the VLPs without thimerosal, this experience proved to be a substantial set back in GSK's development program. Additional insights into the commercial product development of these vaccines are provided in a recent review [50].

Results of the phase I clinical trials of HPV VLP vaccines were first published in 2001. The NCI collaborated with Johns Hopkins University to demonstrate that their HPV16 L1 VLPs were safe and highly immunogenic after three intramuscular doses of 10 or 50 μg when formulated in alum, MF59 or unadjuvanted [51]. At the higher dose, there was no significant difference in the titers induced by the three formulations. VLP binding and HPV16 pseudovirus neutralizing titers were both high and highly correlated. Encouragingly, the mean antibody titers in vaccinees were approximately 40-fold higher than those seen after natural HPV16 infection and similar to the titers that induced strong protection from experimental infection in the rabbit and bovine challenge models. MedImmune collaborated with the University of Rochester to conduct a trial of their HPV11 L1 VLPs. HPV11 was chosen for the initial study because the investigators had access to a recently developed semiquantitative in vitro HPV11 neutralizing assay. It was based on infection of an immortalized human keratinocyte line with SCID mouse-derived HPV11 virions and monitoring infection by production of HPV-specific mRNA [52]. Three intramuscular doses of 3, 9, 30, or 100 μg doses in aluminum hydroxide adjuvant were well tolerated and induced high titers of HPV11 VLP binding and HPV11 neutralizing antibodies [53]. Similar responses were seen with the three highest doses. The encouraging results of these trials provided the impetus for sponsorship of the larger phase II and III trials by GSK.

In 2002, the results of the first proof of concept efficacy trial of an HPV VLP vaccine were published. Three 40 μg doses of Merck's HPV16 VLP in an aluminum adjuvant were administered over a 6-month period. Over the course of 1.5 years of follow up, 41 of the approximately 1,000 placebo controls developed persistent HPV16 infection, while none of the 1,000 VLP vaccines became persistently infected. Laura Koutsky's unannounced presentation of these results at the 2002 International Papillomavirus conference is etched in our memories. It was electrifying to hear that simple intramuscular injection of a VLP vaccine could induce 100% protection against persistent cervical infection, even in the short term.

Table 2 Comparison of commercial HPV VLP vaccines

	Gardasil	Cervarix
Manufacturer	Merck & Co.	GlaxoSmithKline
VLP types	6/11/16/18	16/18
L1 protein dose	20/40/40/20 μg	20/20 μg
Production	*Saccharomyces cerevisiae* expressing L1	L1 recombinant baculovirus infection of *Trichoplusia ni* (Hi 5) insect cells
Adjuvant	225 μg aluminum hydroxyphosphate sulfate	500 μg aluminum hydroxide, 50 μg 3-O-deacylated-4′-monophosphoryl lipid A
Injection schedule	0, 2, 6 months	0, 1, 6 months

Two important decisions needed to be made before proceeding to large-scale phase III studies aimed at generating the data required for licensure. The first was the composition of the vaccine. In this instance, the development paths diverged for Merck and GSK, which, as noted above, had taken over the vaccine development from MedImmune after their phase I trial. GSK decided to concentrate exclusively on cancer prevention and therefore included L1 VLPs of HPV16 and HPV18, the two types responsible for approximately 70% of cancer worldwide. They chose to use their proprietary adjuvant, AS04, which, in addition to an aluminum salt contains monophosphoryl lipid A (a detoxified form of LPS). AS04 tends to induce a Th1 type T helper response, compared to the Th2 type response generally associated with aluminum salt vaccines. Merck decided to target genital warts, in addition to cervical cancer, and therefore included VLPs of types 6 and 11, which induce about 90% of genital warts, in addition to VLPs of types 16 and 18. Merck chose to use a conventional aluminum salt adjuvant (Table 2).

The second critical decision was what primary endpoint to use in the pivotal efficacy trials. Although the main goal of both vaccines is to prevent cervical cancer, there was consensus that cancer was not a reasonable clinical trial endpoint. First, it would take an extremely large trial, several decades in duration, to accumulate a sufficient number of cervical cancer cases from incident HPV infection. Second, the trial would be unethical, because careful active follow up of participating women would identify virtually all high grade precancerous lesions, and standard of care would require their excision before they progressed to invasive cancer. At the opposite end of the spectrum, there was general agreement that incident HPV infection was not a sufficiently stringent endpoint. Most genital HPV infections regress spontaneously and so are not on the causal pathway to cancer. Also, HPV infection is identified by the presence of HPV DNA in genital swabs or scrapes and so transient infection cannot be unequivocally distinguished from contamination, as might be acquired during sexual activity with an infected partner. A strong argument was put forth that persistent HPV infection was an appropriate endpoint for efficacy trials. Essentially all cervical cancers arise from persistent infection and HPV DNA measurements using PCR-based technologies are sensitive and reproducible. However, objections were raised that persistent

HPV infection by itself is not an indication for therapeutic intervention. In addition, the duration of infection that distinguishes harmless transient infections from persistent infections with a high probability of progression had not been established. Therefore, advisory groups to national regulatory agencies, such as the U.S. FDA, recommended a histologically confirmed neoplastic disease endpoint, in particular intermediate or high grade cervical intraepithelial neoplasia (CIN2/3). These dysplasias are routinely treated by ablative therapy when they are diagnosed in Pap screening programs, and at least CIN3 is widely accepted as an obligate cervical cancer precursor. Low grade cervical dysplasia (CIN1) was not considered an appropriate endpoint, since it is a normal manifestation of productive HPV infection and in most cases spontaneously regresses.

The selection of CIN2/3 from incident infection by the vaccine-targeted types as the primary endpoint meant that the pivotal phase III trials had to be large and of relatively long duration. Merck, GSK, and the NCI independently initiated randomized, controlled and double-blind phase III trials of 5,000–19,000 women, with a duration of 4 years (Table 3). The NCI chose to test the GSK bivalent vaccine, because of difficulties in generating sufficient quantities of GMP vaccine in a timely fashion under contract. The Merck and GSK trials were multinational involving more than 100 individual sites each in Europe, North America, South America, Asia, and Australia. In contrast, the NCI chose to conduct, in collaboration with the Costa Rican government, a community-based study centered in the Guanacaste province of Costa Rica, in part because of the infrastructure established during a long-standing natural history study of HPV and cervical cancer in the province [54]. The trials enrolled nonpregnant young women, ages 15–26, generally with fewer than five or seven lifetime sexual partners. This exclusion criterion was used to limit the number of women with prior exposure to the vaccine types of HPV, since the primarily intent of the trials was to test prophylactic efficacy.

In 2004 and 2005 publications, phase IIb trials of the GSK bivalent vaccine, designated Cervarix, and the Merck quadrivalent vaccine, designated Gardasil, provided encouraging preliminary results [55, 56]. They demonstrated high levels of protection against persistent infection and cervical dysplasia (of any grade) by the vaccine targeted types. These publications were followed, in 2007, by the publication of interim analyses of the phase III trials, which were triggered 1.5–3.0 years postvaccination by the accumulation of a predetermined number of disease events [57, 58]. The according to protocol (ATP) analysis revealed that, in women without infection by the vaccine types at enrollment, three doses of either vaccines induced virtually complete protection from incident CIN2/3 lesions in which the vaccine types were detected (Table 3). Gardasil also demonstrated close to 100% protection from external genital lesions, including genital warts, induced by the vaccine types. As expected from the understanding of HPV carcinogenesis obtained in the prospective epidemiology studies, high levels of protection against persistent infection and CIN1 by the vaccine-targeted types were also observed for both vaccines. The high type-specific efficacy of these vaccines certainly exceeded our expectations and those of others in the field. However, their efficacy was clearly limited in that they did not induce clearance of preexisting infections or prevent

Table 3 VLP vaccine efficacy trials in young women: ATP analyses for vaccine-specific HPV types in women negative for vaccine type infections at enrollment

Characteristic	GSK 001/07	Merck 007	Patricia	Future I	Future II	Costa rica
Vaccine	Cervarix	Gardasil	Cervarix	Gardasil	Gardasil	Cervarix
Phase	II	II	III	III	III	III
Control	500 μg aluminum hydroxide	225 μg aluminum hydroxy-phosphate sulfate	Hepatitis A vaccine	225 μg aluminum hydroxy-phosphate sulfate	225 μg aluminum hydroxy-phosphate sulfate	Hepatitis A vaccine
No. participants	1,113	552	18,644	5,455	12,167	7,466
Age range	15–25	16–23	15–25	16–24	15–26	18–25
Lifetime no. of sex partners	≤6	≤4	≤6	≤4	≤4	Any no.
Screening frequency	6 months	6 months	12 months	6 months	12 months	12 months
Mean duration of follow-up	48 months	60 months	15 months[a]	36 months[a]	36 months[a]	48 months[b]
Endpoint: vaccine efficacy (95% CI)	Persistent HPV DNA: 96 (75–100)	Persistent HPV DNA: 96 (83–100) CIN1+, AIS: 100 (<0–100)	CIN2+: 90 (53–99)[c]	CIN1+, AIS: 100 (94–100) EGL: 100 (94–100)	CIN2+, AIS: 98 (92–100)	NA

ATP according to protocol; *CIN1+* cervical intraepithelial neoplasia grade 1 or worse; *CIN2+* cervical intraepithelial neoplasia grade 2 or worse; *AIS* adenocarcinoma in situ; *EGL* external genital lesions; *NA* not available

[a] Interim analysis of 4-year trial

[b] In progress

[c] Modified intention to treat analysis: received at least one dose, case counting started at first dose. Confidence intervals are 97.9%. A post hoc analysis including HPV-specific causal attribution of CIN2+ with multiple type infections generated efficacy estimates of 100% (97.5% CI: 74.2–100)

their progression [57, 59]. Also, the two vaccines have limited ability to prevent infections by other high-risk genital HPV types, as predicted by the in vitro neutralization studies. Submission of these interim analyses led to regulatory approval of Gardasil in the United States, European Union, and elsewhere, starting in mid-2006. Cervarix was approved in the European Union and elsewhere (in 2010 in the United States), starting in mid-2007, making it the first licensed produced to be produced via recombinant baculoviruses. Analysis of the full data sets accumulated over the complete 4 years of these trials, and also of the Costa Rican trial, will likely be published soon.

Perspectives

It is interesting to speculate why the efficacy of this vaccine has proven to be so high, while that of other STI vaccines have not. First, the vaccine is based on the production of neutralizing antibodies, not T cell effector responses, and antibody-mediated protection is a well-established principle on which to base a prophylactic viral vaccine. Second, strong neutralizing antibody responses are induced in greater than 99% of vaccinees. The mammalian immune system has clearly evolved to induce exceptionally strong antibody responses against the highly ordered and repetitive epitopes characteristically displayed on the surface of a VLP, presumably because virion neutralizing antibodies are so critical for defense against most viral infections. Third, papillomaviruses have a unique lifecycle that involves limited exposure of the virions to the systemic immune system. Therefore, there has been limited pressure on the virus to evolve mechanisms to escape systemic antibody responses induced by intramuscular vaccination. The fact that papillomaviruses have DNA genomes and therefore evolve very slowly also makes it unlikely that the viruses will rapidly generate escape variants under the selective pressure of vaccine-induced antibodies. Fourth, the site and mechanism of the mucosal infection appear to make the virus exceptionally susceptible to neutralization by systemic antibodies. Fortunately, IgG, presumably transudated from serum, comprises a large proportion of the antibody in cervical mucus, in contrast to most mucosal surfaces. The levels of VLP-specific IgG in the cervical mucus of parenterally vaccinated women are approximately 10% the levels in their sera [60]. In addition, we have determined in animal models that HPV infection of the female genital tract requires trauma or permeabilization sufficient to expose the epithelial basement membrane to virus binding [61]. This requirement would expose the virus to direct exudation of serum antibodies at the site of trauma. The virtually complete protection against genital warts, many of which are on genital skin that is not bathed by mucus, argues that the latter mechanism is sufficient to prevent infection. Further, papillomavirus infection is a remarkably slow process [62], and the virus is susceptible to antibody-mediated neutralization for several hours, even after attachment to cell surfaces [63]. Therefore, there is an exceptionally long window of opportunity for neutralizing antibodies to act. Taken together these factors may explain the truly remarkable finding that the HPV vaccines appear to induce sterilizing immunity in most women, in that the viral DNA is never detected at the cervix, despite the use of sensitive PCR-based assays.

Some of the principles established with the HPV vaccines, e.g. the high intrinsic immunogenicity of VLPs, might be applicable to vaccines targeting other local mucosal STIs. However one is left with the impression that several of the unusual characteristics of papillomavirus infection make them exceptionally susceptible to a prophylactic vaccine. We consider ourselves lucky to have been involved in developing a vaccine against a virus that, in retrospect, turned out to be a easy target. However, in the beginning not even the most optimistic of vaccinologists would have predicted the remarkable level of efficacy achieved by these vaccines. In 1990, the prospects of developing HPV vaccines seemed no better than the prospects are today for developing effective HIV or malaria vaccines. Hopefully, the story of the development of the HPV VLP vaccines will encourage continued investment in the development of vaccines against other challenging targets.

References

1. Lowy DR. History of Papillomavirus Research. In: Garcea RL, Dimaio D, editors. The Papillomaviruses. New York, Springer, 2007: 13–28.
2. zur Hausen H. Papillomaviruses in the causation of human cancers – a brief historical account. Virology 2009;384(2):260–5.
3. Meisels A, Fortin R. Condylomatous lesions of the cervix and vagina. I. Cytologic patterns. Acta Cytol 1976;20(6):505–9.
4. zur Hausen H. Condylomata acuminata and human genital cancer. Cancer Res 1976;36 (2 pt 2):794.
5. Krcmar M, Suchankova A, Kanka J, Vonka V. Prospective study on the relationship between cervical neoplasia and herpes simplex type 2 virus. III. Presence of herpes simplex type-2 antibody in sera of subjects who developed cervical neoplasia later in the study. Int J Cancer 1986;38(2):161–5.
6. Orth G, Jablonska S, Favre M, Croissant O, Jarzabek-Chorzelska M, Rzesa G. Characterization of two types of human papillomaviruses in lesions of epidermodysplasia verruciformis. Proc Natl Acad Sci USA 1978;75:1537–41.
7. Dürst M, Gissmann L, Ikenberg H, zur Hausen H. A papillomavirus DNA from a cervical carcinoma and its prevalence in cancer biopsy samples from different geographic regions. Proc Natl Acad Sci USA 1983;80:3812–15.
8. Schwarz E, Freese UK, Gissmann L, Mayer W, Roggenbuck B, Stremlau A, et al. Structure and transcription of human papillomavirus sequences in cervical carcinoma cells. Nature 1985;314(6006):111–4.
9. zur Hausen H. Human papillomaviruses in the pathogenesis of anogenital cancer. Virology 1991;184:9–13.
10. Parkin DM, Bray F. Chapter 2: The burden of HPV-related cancers. Vaccine 2006;24 Suppl 3:S11–25.
11. Hawley-Nelson P, Vousden KH, Hubbert NL, Lowy DR, Schiller JT. HPV16 E6 and E7 proteins cooperate to immortalize primary human foreskin keratinocytes. EMBO J 1989;8:3905–10.
12. Werness BA, Levine AJ, Howley PM. Association of human papillomavirus types 16 and 18 E6 proteins with p53. Science 1990;248:76–79.

13. Dyson N, Howley PM, Munger K, Harlow E. The human papillomavirus-16 E7 oncoprotein is able to bind the retinoblastoma gene product. Science 1989;243:934–37.
14. Munoz N, Bosch X, Kaldor JM. Does human papillomavirus cause cervical cancer? The state of the epidemiological evidence. Br J Cancer 1988;57(1):1–5.
15. Franco EL. Measurement errors in epidemiological studies of human papillomavirus and cervical cancer. IARC Sci Publ 1992;119:181–97.
16. Bosch FX, Lorincz A, Munoz N, Meijer CJ, Shah KV. The causal relation between human papillomavirus and cervical cancer. J Clin Pathol 2002;55(4):244–65.
17. Moscicki AB, Schiffman M, Kjaer S, Villa LL. Chapter 5: Updating the natural history of HPV and anogenital cancer. Vaccine 2006;24 Suppl 3:S3/42–51.
18. Plotkin SA. Vaccines: correlates of vaccine-induced immunity. Clin Infect Dis 2008;47(3):401-9.
19. Dvoretzky I, Shober R, Chattopadhyay SK, Lowy DR. Focus assay in mouse cells for bovine papillomavirus. Virology 1980;103:369–75.
20. Kreider JW, Howett MK, Leure DA, Zaino RJ, Weber JA. Laboratory production in vivo of infectious human papillomavirus type 11. J Virol 1987;61(2):590–3.
21. Christensen ND, Kreider JW. Antibody-mediated neutralization in vivo of infectious papillomaviruses. J Virol 1990;64:3151–56.
22. Baker TS, Newcomb WW, Olson NH, Cowsert LM, Olson C, Brown JC. Structures of bovine and human papillomaviruses. Analysis by cryoelectron microscopy and three-dimensional image reconstruction. Biophys J 1991;60(6):1445–56.
23. Shope RE. Immunization of rabbits to infectious papillomatosis. J Exp Med 1935;65:219–31.
24. Olson C, Skidmore LV. Immunity to bovine cutaneous papillomatosis produced by vaccine homologous to the challenge agent. J Am Vet Med Assoc 1959;135:499–502.
25. Pilacinski WP, Glassman DL, Glassman KF, Reed DE, Lum MA, Marshall RF, et al. Immunization against bovine papillomavirus infection. In: Papillomaviruses: Ciba Foundation Symposium 120. Chichester, Wiley, 1986: 136–56.
26. Ghim S, Christensen ND, Kreider JW, Jenson AB. Comparison of neutralization of BPV-1 infection of C127 cells and bovine fetal skin xenografts. Int J Cancer 1991;49(2):285–9.
27. Lin Y-L, Borenstein LA, Selvakumar R, Ahmed R, Wettstein FO. Effective vaccination against papilloma development by immunization with L1 or L2 structural protein of cottontail rabbit papillomavirus. Virology 1992;187:612–19.
28. Jarrett WF, Smith KT, O'Neil BW, Gaukroger JM, Chandrachud LM, Grindlay GJ, et al. Studies on vaccination against papillomaviruses: prophylactic and therapeutic vaccination with recombinant structural proteins. Virology 1991;184:33–42.
29. Salunke D, Caspar DLD, Garcea RL. Self-assembly of purified polyomavirus capsid protein VP1. Cell 1986;46:895–904.
30. Valenzuela P, Medina A, Rutter WJ, Ammerer G, Hall BD. Synthesis and assembly of hepatitis B virus surface antigen particles in yeast. Nature 1982;298:347–50.
31. Zhou J, Sun XY, Stenzel DJ, Frazer IH. Expression of vaccinia recombinant HPV 16 L1 and L2 ORF proteins in epithelial cells is sufficient for assembly of HPV virion-like particles. Virology 1991;185(1):251-57.
32. Ghim SJ, Jenson AB, Schlegel R. HPV-1 L1 protein expressed in cos cells displays conformational epitopes found on intact virions. Virology 1992;190(1):548–52.
33. Miller LK. Baculoviruses: high-level expression in insect cells. Curr Opin Genet Dev 1993;3(1):97–101.
34. Kirnbauer R, Booy F, Cheng N, Lowy DR, Schiller JT. Papillomavirus L1 major capsid protein self-assembles into virus-like particles that are highly immunogenic. Proc Natl Acad Sci U S A 1992;89(24):12180–84.
35. Kajigaya S, Fujii H, Field A, Anderson S, Rosenfeld S, Anderson LJ, et al. Self-assembled B19 parvovirus capsids, produced in a baculovirus system, are antigenically and immunogenically similar to native virions. Proc Natl Acad Sci USA 1991;88:4646–50.
36. Kirnbauer R, Taub J, Greenstone H, Roden RBS, Durst M, Gissmann L, et al. Efficient self-assembly of human papillomavirus type 16 L1 and L1-L2 into virus-like particles. J Virol 1993;67(12):6929–36.

37. Hagensee ME, Yaegashi N, Galloway DA. Self-assembly of human papillomavirus type 1 capsids by expression of the L1 protein alone or by coexpression of the L1 and L2 capsid proteins. J Virol 1993;67(1):315–22.
38. Rose RC, Bonnez W, Reichman RC, Garcea RL. Expression of human papillomavirus type 11 L1 protein in insect cells: in vivo and in vitro assembly of viruslike particles. J Virol 1993;67(4):1936–44.
39. Breitburd F, Kirnbauer R, Hubbert NL, Nonnenmacher B, Trin-Dinh-Desmarquet C, Orth G, et al. Immunization with virus-like particles from cottontail rabbit papillomavirus (CRPV) can protect against experimental CRPV infection. J Virol 1995;69(6):3959–63.
40. Jansen KU, Rosolowsky M, Schultz LD, Markus HZ, Cook JC, Donnelly JJ, et al. Vaccination with yeast-expressed cottontail rabbit papillomavirus (CRPV) virus-like particles protects rabbits from CRPV-induced papilloma formation. Vaccine 1995;13(16):1509–14.
41. Kirnbauer R, Chandrachud L, O'Neil B, Wagner E, Grindlay G, Armstrong A, et al. Virus-like particles of Bovine Papillomavirus type 4 in prophylactic and therapeutic immunization. Virology 1996;219:37–44.
42. Suzich JA, Ghim S, Palmer-Hill FJ, White WI, Tamura JK, Bell J, et al. Systemic immunization with papillomavirus L1 protein completely prevents the development of viral mucosal papillomas. Proc Natl Acad Sci USA 1995;92:11553–57.
43. Roden RBS, Hubbert NL, Kirnbauer R, Christensen ND, Lowy DR, Schiller JT. Assessment of the serological relatedness of genital human papillomaviruses by hemagglutination inhibition. J Virol 1996;70:3298–301.
44. Roden RBS, Greenstone HL, Kirnbauer R, Booy FP, Jessie J, Lowy DR, et al. In vitro generation and type-specific neutralization of a human papillomavirus type 16 virion pseudotype. J Virol 1996;70:5875–83.
45. Pastrana DV, Buck CB, Pang YY, Thompson CD, Castle PE, FitzGerald PC, et al. Reactivity of human sera in a sensitive, high-throughput pseudovirus-based papillomavirus neutralization assay for HPV16 and HPV18. Virology 2004;321(2):205–16.
46. Pastrana DV, Vass WC, Lowy DR, Schiller JT. An HPV16 VLP vaccine induces human antibodies that neutralize divergent variants of HPV16. Virology 2001;279(1):361–9.
47. Shi L, Sanyal G, Ni A, Luo Z, Doshna S, Wang B, et al. Stabilization of human papillomavirus virus-like particles by non-ionic surfactants. J Pharm Sci 2005;94(7):1538–51.
48. Mach H, Volkin DB, Troutman RD, Wang B, Luo Z, Jansen KU, et al. Disassembly and reassembly of yeast-derived recombinant human papillomavirus virus-like particles (HPV VLPs). J Pharm Sci 2006;95(10):2195–206.
49. McCarthy MP, White WI, Palmer-Hill F, Koenig S, Suzich JA. Quantitative disassembly and reassembly of human papillomavirus type 11 virus-like particles in vitro. J Virol 1998;72:32–41.
50. Inglis S, Shaw A, Koenig S. Chapter 11: HPV vaccines: Commercial Research & Development. Vaccine 2006;24 Suppl 3:S99–S105.
51. Harro CD, Pang YY, Roden RB, Hildesheim A, Wang Z, Reynolds MJ, et al. Safety and immunogenicity trial in adult volunteers of a human papillomavirus 16 L1 virus-like particle vaccine. J Natl Cancer Inst 2001;93(4):284–92.
52. Smith LH, Foster C, Hitchcock ME, Leiserowitz GS, Hall K, Isseroff R, et al. Titration of HPV-11 infectivity and antibody neutralization can be measured in vitro. J Invest Dermatol 1995;105:438–44.
53. Evans TG, Bonnez W, Rose RC, Koenig S, Demeter L, Suzich JA, et al. A Phase 1 Study of a Recombinant Viruslike Particle Vaccine against Human Papillomavirus Type 11 in Healthy Adult Volunteers. J Infect Dis 2001;183(10):1485–93.
54. Herrero R, Hildesheim A, Rodriguez AC, Wacholder S, Bratti C, Solomon D, et al. Rationale and design of a community-based double-blind randomized clinical trial of an HPV 16 and 18 vaccine in Guanacaste, Costa Rica. Vaccine 2008;26(37):4795–808.
55. Harper DM, Franco EL, Wheeler C, Ferris DG, Jenkins D, Schuind A, et al. Efficacy of a bivalent L1 virus-like particle vaccine in prevention of infection with human papillomavirus types 16 and 18 in young women: a randomised controlled trial. Lancet 2004; 364(9447):1757–65.

56. Villa LL, Costa RL, Petta CA, Andrade RP, Ault KA, Giuliano AR, et al. Prophylactic quadrivalent human papillomavirus (types 6, 11, 16, and 18) L1 virus-like particle vaccine in young women: a randomised double-blind placebo-controlled multicentre phase II efficacy trial. Lancet Oncol 2005;6(5):271–8.
57. Garland SM, Hernandez-Avila M, Wheeler CM, Perez G, Harper DM, Leodolter S, et al. Quadrivalent vaccine against human papillomavirus to prevent anogenital diseases. N Engl J Med 2007;356(19):1928–43.
58. Paavonen J, Jenkins D, Bosch FX, Naud P, Salmeron J, Wheeler CM, et al. Efficacy of a prophylactic adjuvanted bivalent L1 virus-like-particle vaccine against infection with human papillomavirus types 16 and 18 in young women: an interim analysis of a phase III double-blind, randomised controlled trial. Lancet 2007;369(9580):2161–70.
59. Hildesheim A, Herrero R, Wacholder S, Rodriguez AC, Solomon D, Bratti MC, et al. Effect of human papillomavirus 16/18 L1 viruslike particle vaccine among young women with pre-existing infection: a randomized trial. JAMA 2007;298(7):743–53.
60. Nardelli-Haefliger D, Wirthner D, Schiller JT, Lowy DR, Hildesheim A, Ponci F, et al. Specific antibody levels at the cervix during the menstrual cycle of women vaccinated with human papillomavirus 16 virus-like particles. J Natl Cancer Inst 2003;95(15):1128–37.
61. Roberts JN, Buck CB, Thompson CD, Kines R, Bernardo M, Choyke PL, et al. Genital transmission of HPV in a mouse model is potentiated by nonoxynol-9 and inhibited by carrageenan. Nat Med 2007;13(7):857–61.
62. Day PM, Lowy DR, Schiller JT. Papillomaviruses infect cells via a clathrin-dependent pathway. Virology 2003;307(1):1–11.
63. Christensen ND, Cladel NM, Reed CA. Postattachment neutralization of papillomaviruses by monoclonal and polyclonal antibodies. Virology 1995;207(1):136–42.

History of Rotavirus Vaccines Part I: RotaShield

Albert Z. Kapikian

Discovery of the Noroviruses and Rotaviruses

According to hieroglyphic evidence, diarrhea is one of the oldest recorded illnesses, dating as far back as 3300 BC [1]. However, despite major discoveries in microbiology during the past century, the etiology of most diarrheal illnesses remained elusive until relatively recently [2, 3]. Although volunteer studies in the 1940s and 1950s showed that oral administration of bacteria-free stool filtrates derived from patients with diarrhea could induce illness, a suspected viral etiologic agent could not be identified.

A.Z. Kapikian (✉)
Laboratory of Infectious Diseases, National Institute of Allergy and Infectious Diseases, National Institutes of Health, DHHS, Bethesda, MD, USA
e-mail: akapikian@niaid.nih.gov

S.A. Plotkin (ed.), *History of Vaccine Development*,
DOI 10.1007/978-1-4419-1339-5_28, © Springer Science+Business Media, LLC 2011

This frustration intensified following the advent of the "golden age" of virology in the 1950s and 1960s when tissue culture technology led to the discovery of hundreds of new viruses, but none of them could be established as an important cause of diarrhea [2, 3].

A second generation of volunteer studies was initiated in the 1970s utilizing bacteria-free filtrates from gastroenteritis outbreaks, anticipating that newer techniques such as organ culture technology might enable the detection of a viral agent, but again without success [4]. However, in 1972 the discovery of the 27 nm Norwalk virus and its association with epidemic gastroenteritis in older children and adults, followed in 1973 by the discovery of the 70 nm rotavirus and its association with severe gastroenteritis in infants and young children ushered in a new era in the etiology of diarrheal diseases (Figs. 1 and 2) [5, 6].

1975	**<u>Diarrheal Disease Viruses:</u>** **Coronavirus Astrovirus Calicivirus** **Enteric Adenoviruses** **Rotavirus** **Norwalk Agent**
1950	**Herpes** **Influenza**
1925	**Bacteriophage**
	Poliomyelitis **Yellow Fever**
1900	**Foot and Mouth Disease** **Tobacco Mosaic**
	Bacteria **Protozoa** **Fungi**
1800	
1700	
	Arthropods
1600	
1379	**Helminths**

Fig. 1 Cumulative discovery of infectious agents (from ref. [135])

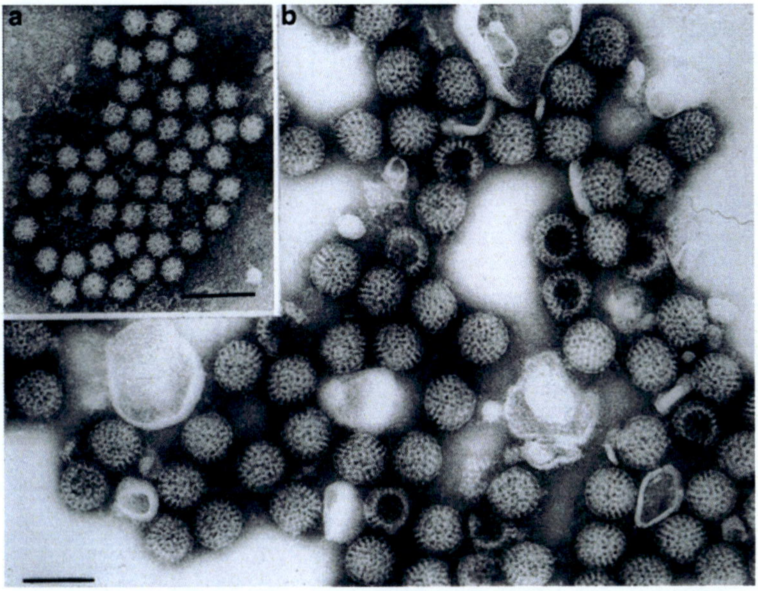

Fig. 2 (**a**) 27-nm Norwalk virus particles observed in a stool filtrate by immune electron microscopy from a volunteer with acute gastroenteritis (adapted from ref. [5]); (**b**) 70-nm rotavirus particles observed in a stool suspension by electron microscopy from an infant with acute gastroenteritis (adapted from ref. [136])

These fastidious new viruses were discovered without the benefit of tissue culture, but instead relied on the electron microscope: (a) the Norwalk virus by immune electron microscopic examination of a stool specimen of a volunteer who had developed illness following administration of a stool filtrate derived from a gastroenteritis outbreak at an elementary school in Norwalk, Ohio [5]; and (b) the human rotavirus, by electron microscopic examination of thin sections of duodenal mucosa of hospitalized infants with gastroenteritis in Australia [6]. It became apparent that the Norwalk virus and related viruses (now comprising a genus called norovirus in the family Caliciviridae) were an important cause of community or institutional outbreaks of nonbacterial gastroenteritis whereas the rotavirus was recognized as the long sought after single most important cause of severe diarrheal illnesses in infants and young children [7, 8]. It is noteworthy that because of newer sensitive assays, noroviruses have now emerged as the second most important cause of diarrhea in infants and young children accounting for up to 200,000 deaths in children <5 years of age in developing countries [9].

It did take a while, however, to establish that these new agents had an important impact on public health. For example, shortly after these and related discoveries had been reported, an editorial in the *BMJ* in 1974 noted that "Acute transient attacks of diarrhea and vomiting are so common that they can almost be regarded as part of the normal way of life.... In bygone days the family doctor would be able to reassure the patient that he had merely eaten something which disagreed with

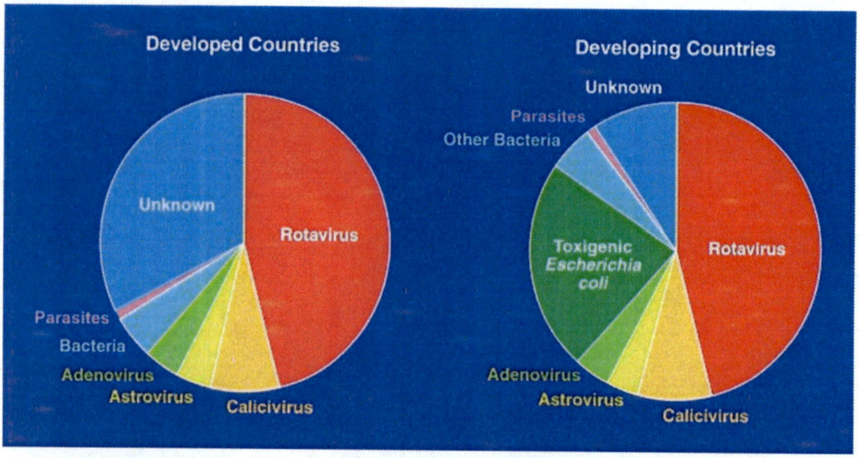

Fig. 3 Estimates of the role of etiologic agents of severe diarrheal illnesses requiring hospitalization of infants and young children in developed and developing countries (from ref. [8])

him-perhaps an indiscretion with green apples. However, modern patients expect a more fashionable label such as virus gastroenteritis or one of its many synonyms. This diagnosis appears to be ideal since, until recently it could neither be confirmed nor refuted…. For the moment, the general practitioner can be reassured that a disease which he regularly diagnoses actually exists" [10].

Relatively quickly, studies from all over the world established that the rotaviruses were the single most important etiologic agents of severe diarrhea in infants and young children accounting for one third to one half of these illnesses (Fig. 3) [8]. They were found to be an important cause of morbidity in the developed countries and of morbidity and mortality in the developing countries. The major role of diarrheal diseases, in general, in the developing world was highlighted by this UNICEF poster announcing the World Summit for Children at the United Nations in New York in September 1990 (Fig. 4).

For the first time, a virus, the rotavirus had been implicated as a major cause of severe diarrhea in infants and young children. The Institute of Medicine in the mid-1980s estimated that in the developing countries rotaviruses were responsible for over 870,000 deaths of infants and young children under 5 years of age annually and that in the United States they caused over one million cases of severe diarrhea and 150 deaths in the under 5-year age group [11, 12]. Although these estimates have declined somewhat in recent years, they are still very substantial – for example the number of deaths attributable to rotavirus diarrhea is estimated currently to be over 500,000 in the <5-year age group with about 85% occurring in the developing countries [13].

Rotaviruses are a major cause of morbidity in the developing countries and of morbidity and mortality in the developing countries (Fig. 5) [14]. They are egalitarian viruses as they infect almost all infants and young children during the first

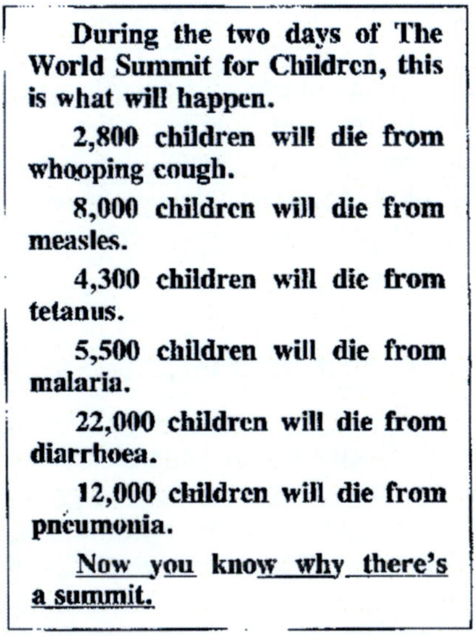
During the two days of The World Summit for Children, this is what will happen.

2,800 children will die from whooping cough.

8,000 children will die from measles.

4,300 children will die from tetanus.

5,500 children will die from malaria.

22,000 children will die from diarrhoea.

12,000 children will die from pneumonia.

Now you know why there's a summit.

Fig. 4 Adaptation of UNICEF Poster: World Summit for Children, United Nations, New York, 29–30 September, 1990

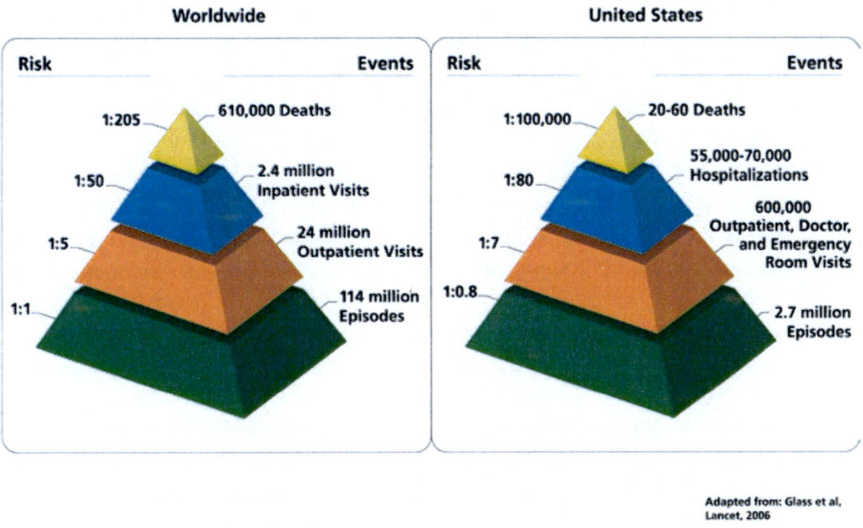

Fig. 5 Annual burden of rotavirus disease worldwide and in the United States in infants and young children <5 years of age (adapted from ref. [14])

few years of life, peaking at 6–24 months of age, regardless of socio-economic conditions. However, the consequences of these infections are strikingly different. As shown in Fig. 5, the risk of rotavirus disease in the US is quite similar to that observed worldwide (the latter composed predominantly of developing countries), but the risk of death from rotavirus is strikingly different in these settings [14]. However, hospitalizations and visits to a medical practitioner are substantial. The difference in mortality may be attributed, in part, to the widespread availability and use of fluid replacement therapy in developed countries.

It became clear from the morbidity and mortality statistics that a rotavirus vaccine was needed globally for infants and young children under 2 years of age who were most susceptible to the severe complications of dehydrating diarrhea.

Characteristics of Rotaviruses Related to Vaccine Development

Rotaviruses are 70 nm in diameter and are composed of three layers: outer, intermediate, and inner. The morphologic appearance by negative stain electron microscopy was responsible for the name "rotavirus," because it had the appearance of a sharply defined rim of a wheel (rota [Latin] = wheel) placed on short spokes radiating outward from a wide hub (Fig. 2) [15]. Within the intermediate layer is the third layer (the inner core layer [VP2]), which contains the 11 segmented double-stranded RNA genome [8].

Rotaviruses have three important antigenic specificities: group, subgroup, and serotype [8]. VP6, located between the outer and intermediate layers, is encoded by the sixth gene and determines group and subgroup specificities. There are six groups (serogroups) each designated by a letter (A–G), with group A being the epidemiologically most important by far, and therefore vaccines are aimed solely at preventing illnesses with this group. Two major subgroups (1 and 2) have been defined. Serotypes are determined by both VP7 (encoded by gene 7, 8, or 9 [depending on the strain]) and VP4 (encoded by gene 4). VP7 is located on the outer layer whereas VP4 protrudes from the outer layer as 60 discrete spikes 10–12 nm in length. The VP4 spikes are not visible by routine negative stain electron microscopy. The geography of the triple-layered particle is shown in the schematic reconstruction in Fig. 6 [16]. Both VP7 and VP4 induce neutralizing antibodies that in animal studies are independently associated with protection [8].

The Group A rotavirus serotypes are defined by a binary classification system according to VP4 and VP7 specificities established by neutralization but augmented by genotyping for practical purposes [8]. The VP7 serotype designation is used most frequently, but the VP4 serotype is also essential in delineating circulating strains. There are 11 human rotavirus VP7 serotypes (also designated as "G" serotypes because VP7 is a glycoprotein) and 11 human rotavirus VP4 serotypes or genotypes (also designated as "P" serotypes or genotypes because VP4 is protease sensitive). The most prevalent serotype globally is the G1 P1A [8] strain (the bracket indicates the P genotype, which was not described contemporaneously with the P neutralization specificity). G 1, 2, 3, and 4 serotypes are considered to be the epidemiologically most important serotypes but G9 has emerged more recently and appears to be of at least equal or greater importance in various parts of the world. In addition, the G8 serotype is important in parts of Africa. The G 1, 2, and 3 serotypes

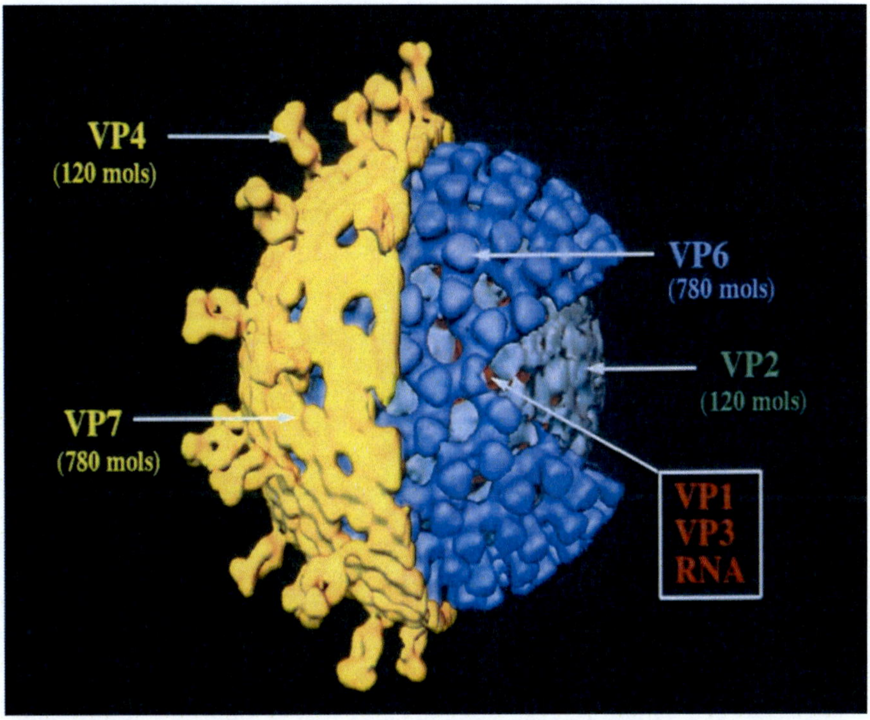

Fig. 6 Three dimensional reconstruction of a rotavirus particle (adapted from ref. [16] and further adaptation by BVV Prasad)

characteristically have P1A [8] specificity whereas the G2 serotype characteristically has P1B [4] specificity. A G serotype may have various P specificities.

Human to human transmission via the fecal-oral route is the major source of infection. The incubation period is approximately 24–48 h.

Rotaviruses also infect almost all animal species studied as well as several avian species [8]. There are many serotypes or genotypes of group A animal rotaviruses, most of which share VP4 or VP7 specificities with human strains. Animal to human transmission of rotaviruses has not proven to be an important epidemiologic mode of transmission.

Strategies for a Rotavirus Vaccine

The need for a rotavirus vaccine received strong international endorsement from the World Health Organization. Approaches to vaccine development ranged from conventional growth in cell culture of human or animal rotavirus strains, to the use of molecular biological techniques with the aim being the prevention of serious illness caused by the four epidemiologically important rotavirus serotypes.

The road to the development of RotaShield was rather circuitous and emerged after several detours involving various strategies. The following description will try to place this journey in perspective by tracing the various steps, hurdles, and detours.

Attenuated Human Rotavirus Vaccine Candidates: A Monovalent Cell Culture Passaged Virus, A Neonatal Strain, and Cold-Adapted Viruses

Volunteer studies were carried out with a human G1 rotavirus (the D strain) stool filtrate derived from an infant with diarrhea [17]. Oral administration of the inoculum induced a diarrheal illness in certain volunteers, allowing us to determine correlates of susceptibility to challenge [17, 18]. This gave us confidence that a vaccine could be evaluated in this model, an important prelude to the development of a vaccine for infants and young children.

Human rotaviruses were considered to be fastidious agents as none had been propagated efficiently in cell cultures until the Wa human rotavirus strain (a G1, P1A [8] virus) derived from the stool of an infant with diarrhea was cultivated in primary African green monkey kidney (AGMK) cells after 11 passages in gnotobiotic piglets [19]. This adapted mutant virus was triply plaque purified, and the 16th AGMK cell culture passage was evaluated first in animals and then adult volunteers as a potential oral vaccine candidate [18, 20, 21]. The oral route was selected for this enteric illness because animal studies affirmed that antibodies in the intestinal lumen were of major importance in preventing the disease [22–28]. The volunteer studies were carried out in stepwise fashion, initially in two volunteers with high levels of rotavirus serum antibody. Since neither volunteer became ill, ten additional volunteers with lower levels of serum antibody consistent with those observed in susceptible volunteers in the challenge study were given this virus orally. None developed illness, but six developed serologic evidence of infection, suggesting that the Wa strain was less virulent than the D strain and that it was attenuated, and might be a suitable vaccine candidate [18, 20, 21].

However, studies with this live attenuated vaccine strain were suspended because three volunteers developed low level serum transaminase elevations 10 days after vaccination with the values returning to the normal range 11 days later. Moreover, these three volunteers had mild transaminase and/or LDH levels within 4 weeks of the return to the normal transaminase levels. The volunteers did not develop overt liver disease. In addition, the seed virus used to prepare the vaccine but not the vaccine itself was found to be contaminated with simian foamy virus type 2, but none of the volunteers developed a serologic response to the contaminant. However, administration of the rotavirus to nonhuman primates orally and intravenously induced antibodies to simian foamy virus type 2 and to simian cytomegalovirus as well as transaminase elevations. Because of the difficulty in securing adventitious agent-free AGMK cells, attempts were made to grow the Wa virus in rhesus monkey diploid FRhL-2 cells and human diploid (WI38 and MRC-5) cells but without success. An ether-treated Wa virus suspension was prepared in AGMK cells that was free of adventitious agents, but further studies with this preparation were not pursued.

The use of a neonatal human rotavirus strain, M37, recovered from an asymptomatic neonate in Venezuela was also evaluated for safety, immunogenicity, and efficacy in 2–5 or 2–6-month-old infants [29, 30]. This was prompted by the finding in various settings that neonatal rotavirus strains induced a predominantly asymptomatic neonatal infection and, therefore, may have been naturally attenuated. There was also compelling evidence from Australia that an asymptomatic neonatal infection protected against postneonatal gastroenteritis [31]. The M37 candidate vaccine efficacy trial failed to show clinical protection against rotavirus diarrhea [30].

Finally, cold-adapted mutants for the Wa (G1:P1A), DS-1 (G2: P1B) viruses and Wa (P 1A)×DS-1 (G2) and Wa (G1)×P (P1A) reassortants that exhibited different degrees of growth restriction in vitro were generated as possible candidate vaccines [32]. Limited safety and immunogenicity studies were carried out with the Wa cold-adapted strain in different age groups in stepwise fashion [33].

Further studies with the various attenuated viruses were suspended in favor of the Jennerian approach, which eventually led to the RotaShield vaccine.

The Jennerian Approach

The most extensively studied approach to rotavirus vaccination was the concept pioneered at the end of the eighteenth century by Edward Jenner for human smallpox vaccination, in which a related, live, attenuated agent from a nonhuman host was used as the immunogen in humans [34]. Early serologic and animal studies were instrumental in suggesting the feasibility of the "Jennerian" approach to rotavirus vaccination [15, 35–37]. These studies demonstrated that human and animal rotaviruses shared a common group antigen and children infected with a human rotavirus developed a seroresponse to both human and animal rotaviruses, including those of bovine or simian origin.

Moreover, an important study in 1979 demonstrated the feasibility of this strategy: colostrum-deprived gnotobiotic calves infected in utero experimentally by injection of bovine rotavirus NCDV into the amniotic sac 2–14 weeks before delivery were protected against challenge with a heterologous human rotavirus shortly after birth [38]. Furthermore, it was found later that most of the calves developed serum neutralizing antibodies to heterotypic G1, G2 or G3 specificity [39]. This study established the rationale for adopting the "Jennerian" approach for a rotavirus vaccine. Later, in 1983, this concept was confirmed in colostrum-deprived piglets with the RIT 4237 bovine rotavirus candidate vaccine [40].

The first immunogenicity and safety trials evaluating this strategy in humans were reported in 1983 by Finnish investigators who evaluated the bovine rotavirus NCDV (G6) strain (designated as RIT 4237) in adults, children, and infants [41].

They also showed that this heterologous vaccine protected against naturally occurring rotavirus diarrhea in the target population of infants [42, 43]. We pursued this strategy with another strain of bovine rotavirus, the UK strain, by initiating studies in two adults with high levels of rotavirus antibodies and then in nine adult volunteers with low levels [21, 44]. Although none of the volunteers developed confirmed rotavirus illness (two had single bouts of nonrotavirus diarrhea), none of them developed a seroresponse or shed the virus. In addition, 3 of the 11 volunteers developed transient serum transaminase elevations.

At this time, our studies with the UK strain were suspended in favor of a rhesus monkey rotavirus (RRV) strain (MMU 18006) described by investigators in California in 1980 [45]. It was recovered from a 3.5-month-old rhesus monkey with acute diarrhea. We favored this strain as a vaccine candidate because (1) it appeared to be relatively safe as it had not been detected under natural conditions in humans. In addition, under stringent hybridization conditions, RRV failed to hybridize with G 1–4 human rotavirus serotypes and with various human rotavirus field strains; (2) it shared VP7 serotype specificity with the human rotavirus serotype G3; (3) it grew to high titer in primary simian tissue culture and was readily adapted to $FRhL_2$ cells, a semi-continuous simian diploid cell strain developed by the FDA as a potential substrate for vaccines [46, 47]. This was advantageous because of the frequency of adventitious agents found in primary monkey kidney cells.

Our clinical studies with RRV, which involved numerous collaborators, were initiated on January 20, 1984 by administering it orally first to two young adult volunteers with high levels of rotavirus serum antibody and proceeding 4 days later in nine adult volunteers with lower levels of serum antibody [46, 48]. Once again the transaminase issue arose as 4 of the 11 volunteers developed transient serum transaminase elevations. The transaminase issue needed to be reconciled before we could proceed further. Therefore over 3 months later, a study was carried out in which RRV or control material was administered orally to 25 young adults [38, 49]. Three of 13 vaccinees and 5 of 12 controls developed transiently elevated serum transaminase levels beginning 3–7 days after oral administration. In a follow-up study 10 days later, none of eight vaccinees and 1 (who had a naturally occurring rotavirus infection) of 7 controls developed transiently elevated serum transaminase levels beginning 1 week after oral administration [38, 49]. It was concluded from these events that the transaminase elevations were random events unrelated to the vaccine.

Six months after the initiation of the RRV adult volunteer studies, the RRV vaccine studies were extended in stepwise fashion beginning with children 5–12 years of age, and finally reached the target population of 4-month-old infants over 3 months later [48]. These clinical studies were carried out exclusively in the US for the first 11 months and then extended to Sweden and Finland [48–57]. The RRV vaccine was also evaluated in nine placebo-controlled efficacy trials (with RIT 4237 vaccine as a third arm in one trial) in the US, Venezuela, Sweden, and Finland, which included over 1,600 infants and young children [58–69]. The vaccine efficacy against moderately severe or severe diarrhea ranged from 85% to nil [67]. This inconsistency was attributed to a mismatch of the vaccine with prevailing serotypes in young infants without prior exposure to rotavirus resulting in a failure to develop heterotypic antibodies. Further development of a monovalent RRV vaccine was discontinued in favor of developing a multivalent vaccine.

It should be noted that although other investigators showed initially that the monovalent bovine rotavirus vaccines NCDV and WC3 protected against rotavirus illness, additional trials in several developing countries showed only limited protection. Therefore, clinical studies with these vaccines were also discontinued (see later chapter).

Modified Jennerian Approach

The failure of the RRV strain to induce consistent protection against nonserotype 3 strains led us to change our strategy from an exclusively "Jennerian" strategy to a combined "Jennerian-modified Jennerian" approach whereby a quadrivalent vaccine was formulated in order to broaden the protection to include the other epidemiologically important VP7 serotypes 1, 2, and 4, in addition to VP7 serotype 3, which was represented by RRV [67]. In 1985 and 1986, by taking advantage of the ability of rotaviruses to undergo genetic reassortment, three human rotavirus-RRV reassortants were generated, each possessing ten RRV genes and a single human rotavirus gene encoding VP7 serotype 1, 2, or 4 (modified "Jennerian" strategy) [70, 71] (Fig. 7). These reassortants, along with RRV, which represented serotype 3 (the "Jennerian" approach), were formulated into a quadrivalent candidate vaccine [67]. We considered serotype-specific immunity to be important for optimal protection in infants who had never been infected previously with rotavirus, a conclusion that remains controversial [72].

Development of the First Licensed Rotavirus Vaccine (RotaShield)

We conducted stepwise safety and efficacy studies of the G1, G2, and G4 monovalent components of the individual components of the quadrivalent candidate reassortant vaccine in similar fashion as we did for the monovalent RRV strain [67, 73]. Our studies with the G1 component of the reassortant vaccine began in April 1986 proceeding from two adults with high levels of antibodies, to eight adults with the lowest available levels of serum antibodies and proceeded in stepwise fashion to the 2–4-month-age group over 6 months later [48]. Subsequently, the four reassortants were also combined into a quadrivalent vaccine. Twelve separate studies were carried out in 10–20-week-old infants to determine the most efficient and immunogenic dose. A regimen of three doses of the quadrivalent vaccine containing 1×10^5 plaque forming units (PFU) of each serotype was adopted for further testing [74–82].

Extensive placebo-controlled studies were carried out in over 10,000 infants and young children (neonates to 7 months old) with individual reassortants or the

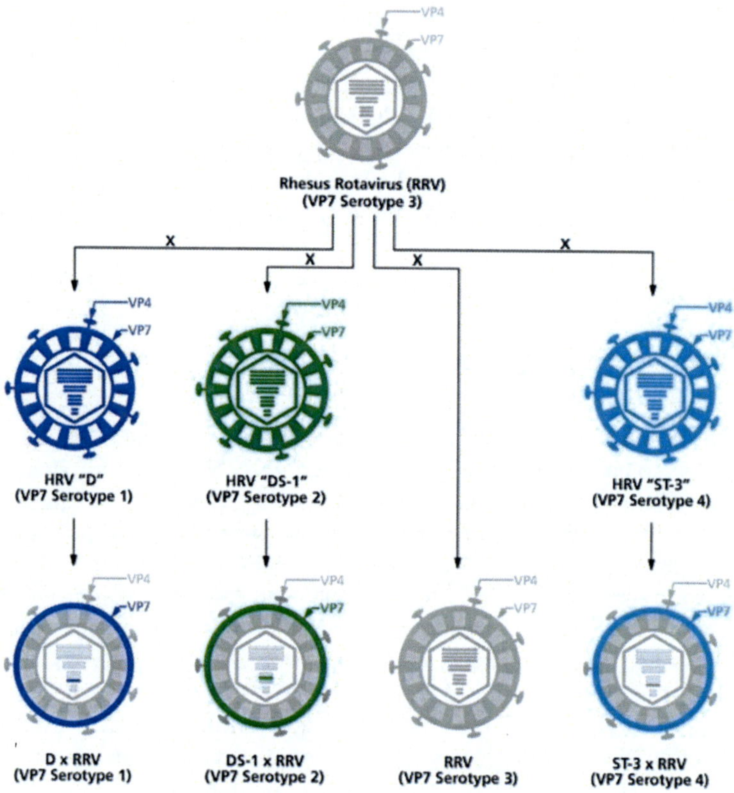

Fig. 7 Human rotavirus (HRV) × Rhesus RV reassortant quadrivalent vaccine with VP7 serotype 1, 2, 3, and 4 specificities (RotaShield) (adapted from ref. [73])

quadrivalent vaccine by various investigators in the US, Finland, Peru, Myanmar, Brazil, and Venezuela [74–84]. We also entered into a collaborative agreement with Wyeth-Ayerst, which led the studies for the development of the quadrivalent vaccine for commercial use.

Four large field trials with three oral doses of 4×10^5 PFU of quadrivalent RotaShield precursor (RRV-TV) yielded important information regarding the vaccine's efficacy: (1) a multicenter trial at 24 US sites in which over 1,200 infants completed surveillance; about one third received RRV-TV, the G1 reassortant (Wa x RRV) or a placebo. RRV-TV had an efficacy of 80% against severe diarrhea and 100% efficacy against dehydrating rotavirus illness [85]; (2) over 2,300 Finnish infants received either RRV-TV or placebo. The vaccine was 91% effective against severe rotavirus gastroenteritis, 100% effective against hospital admission and 97% against dehydration [86]; (3) in a catchment study in Venezuela over 2,200 infants received either RRV-TV or placebo; the vaccine was 88% effective against severe rotavirus diarrhea, 75% against dehydration and 70% against hospital admission [87]; and (4) over 1,000 Native American infants received RRV-TV, the G1 reassortant or placebo; RRV-TV had an efficacy of 69% against severe rotavirus diarrhea [88]. There were very few cases of dehydration in the latter study, likely due to "...aggressive use of oral rehydration therapy...." [88]. In these four trials, the vaccine was safe but was associated with a self-limited febrile response, but not diarrhea, in up to 25% of the vaccinees, characteristically after the first dose during the first week after vaccination.

The US FDA gave a license to the Wyeth vaccine (named "RotaShield") in August 1988 after demonstration of safety and efficacy. The vaccine was recommended for routine use in healthy infants in a 3-dose oral schedule at 2, 4, and 6 months of age by the ACIP. It became available for general use in October 1998 and suspended for use in July 1999 and withdrawn about 3 months later [89–91].

The "effectiveness" of this vaccine – (i.e., how would this vaccine work in practice in the real world rather than in the efficacy trials conducted under optimal conditions [delivery, timing with regard to rotavirus seasonality, age, exclusions for medical conditions, etc.]) was examined retrospectively in three epidemiologic studies, 5, 9, and 12 years after RotaShield's withdrawal.

The first study included children who received medical care at a large urban pediatric practice and who because of their date of birth would have been eligible to receive RotaShield. In addition, if hospitalization was needed, they would have been admitted to a single Children's Hospital [92]. The investigators reviewed both their office records to determine vaccination history and the hospital records to determine inpatient (hospitalization) and outpatient (emergency department) treatment for acute gastroenteritis. They enrolled 1,099 children of whom 513 had not received RotaShield and 586 who had received at least 1–3 doses. They found that the attack rate for rotavirus hospitalization was 0.52 for unvaccinated children (no doses), 0.20 for children who had received 1 or 2 doses, and 0 for children who had received all 3 doses. This was translated to an overall protective vaccine effectiveness against rotavirus acute gastroenteritis requiring hospitalization of 70% (i.e., 61% among partially vaccinated and 100% among fully vaccinated children). The protection observed in the previous efficacy trials was strikingly similar to that found in this clinical assessment.

The second study was a case-control investigation that included children who were eligible to receive RotaShield because of their date of birth and who were admitted to one of three hospitals [93]. Cases were hospitalized for acute gastroenteritis. Controls not admitted with acute gastroenteritis were matched to cases by date of birth with the aim of having 3 controls for each case: 136 cases and 440 controls were available for analysis. RotaShield had an efficacy of 100% against severe rotavirus gastroenteritis requiring hospitalization when 3 or 2 doses were compared with 0 doses. One dose of RotaShield had a protective efficacy against severe gastroenteritis of 89% vs. 0 doses. Once again, the effectiveness of RotaShield was similar to that observed in the efficacy trials.

The third study was a retrospective cohort study analyzing children who were eligible to receive RotaShield and were enrolled in any of the six Managed Care Organizations participating in the Vaccine Safety Datalink project in collaboration with the CDC to monitor vaccine safety [94]. Children (64,599) who had been eligible to receive RotaShield were continuously enrolled for the 1-year period (August 1999 through July 31, 2000). Of the cohort 13,339 (21%) had received at least 1 dose of RotaShield. The outcomes of interest were hospitalization and Emergency Department discharges for all-cause gastroenteritis (not rotavirus gastroenteritis). The full 3-dose schedule of RotaShield had an effectiveness of 83% against all-cause gastroenteritis hospitalizations and 43% vs. all-cause gastroenteritis requiring emergency department visits when compared to children who did not receive the vaccine. In addition, 2 doses of RotaShield prevented 70% of all-cause gastroenteritis and 1 dose prevented 52%. Rotashield's effectiveness was noted to be "substantially greater than the 48–53% of year-round hospitalizations and 33% of emergency department visits estimated to result from rotavirus by indirect methods." This finding gave the authors a cause for optimism noting that "Thus, the potential health benefits of the new rotavirus vaccination program may be greater than previously anticipated."

Withdrawal of RotaShield: The Intussusception Issue

After over one million doses were given to approximately 600,000 infants, the ACIP-CDC recommendation for its use was suspended in July 1999 and it was withdrawn in October 1999 as CDC reported a link with intussusception. This decision led to considerable attention and controversy (Fig. 8).

This suspension was based heavily on the initial interpretation of CDC's case-control, case series, and cohort studies [95–97]. CDC had projected that the excess risk of intussusception attributable to RotaShield was as high as 1.8 (or 1 in 2,500 children), 1.7 (or 1 in 3,333 children), and 1.6 (or 1 in 4,323 children), for these three respective studies. These risks led to projections that there would be up to 1,600, 1,400, or 1,200 excess cases of intussusception if the entire birth cohort of the US had been vaccinated. This would have represented a 60, 70, or 80% increase over the background rate of intussusception of ~1 in 2,000 children. In addition, further analyses presented later described sharp peaks in the odds ratio (OR) for developing intussusception 3–7 days after the first dose for the case-control study (37.2), in the incidence rate ratio for case-series study (58.9) and the relative risk in the expanded MCO cohort study (30.4) [98–100].

Efforts Made to Reconsider the Withdrawal Decision

Despite concerted efforts, it was not possible to resuscitate the vaccine with such projections as noted above even though: (a) later the CDC presented lower excess risk values of up to 1 in 11,073 in their enlarged managed case cohort study [100]; (b) NIH population-based studies suggested a risk of excess intussusception of nil to 1 in 32,000 vaccinees [101–105]; and (c) NIH studies found not only no apparent

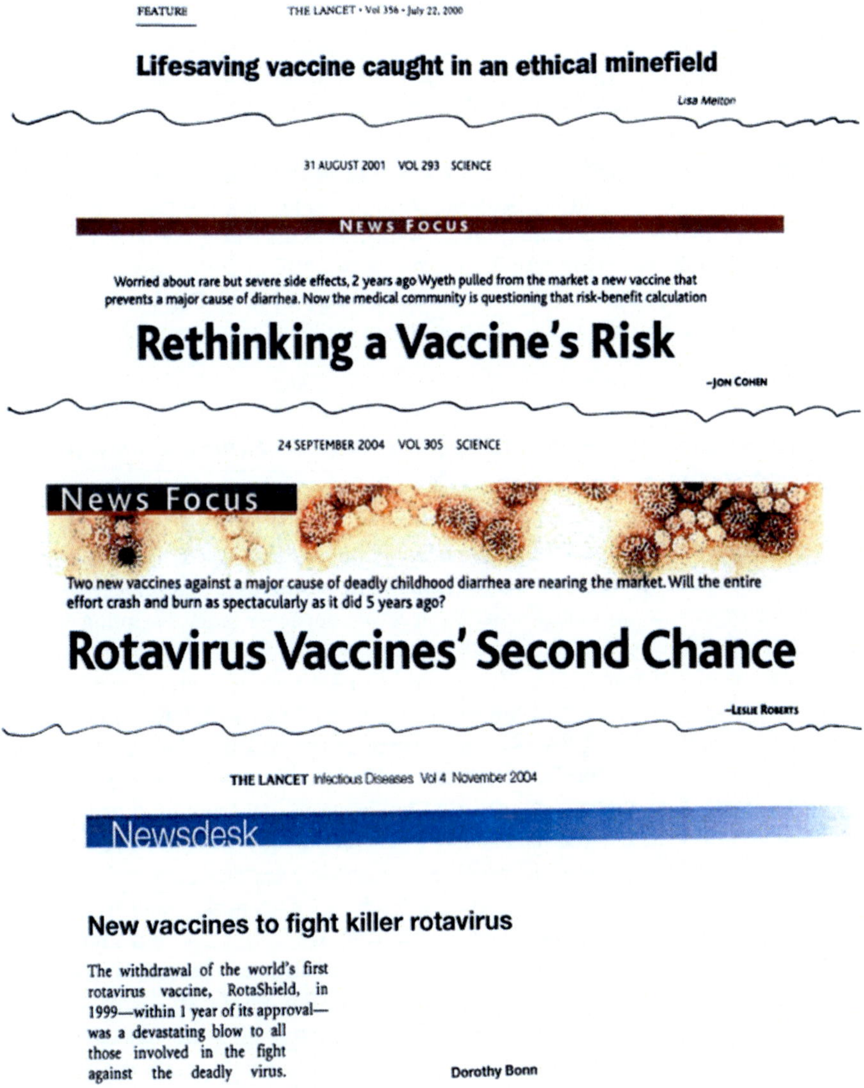
FEATURE THE LANCET • Vol 356 • July 22, 2000

Lifesaving vaccine caught in an ethical minefield

Lisa Melton

31 AUGUST 2001 VOL 293 SCIENCE

NEWS FOCUS

Worried about rare but severe side effects, 2 years ago Wyeth pulled from the market a new vaccine that prevents a major cause of diarrhea. Now the medical community is questioning that risk-benefit calculation

Rethinking a Vaccine's Risk

–JON COHEN

24 SEPTEMBER 2004 VOL 305 SCIENCE

News Focus

Two new vaccines against a major cause of deadly childhood diarrhea are nearing the market. Will the entire effort crash and burn as spectacularly as it did 5 years ago?

Rotavirus Vaccines' Second Chance

–LESLIE ROBERTS

THE LANCET Infectious Diseases Vol 4 November 2004

Newsdesk

New vaccines to fight killer rotavirus

The withdrawal of the world's first rotavirus vaccine, RotaShield, in 1999—within 1 year of its approval—was a devastating blow to all those involved in the fight against the deadly virus.

Dorothy Bonn

Fig. 8 Various aspects of the RotaShield withdrawal controversy in various publications

overall risk of intussusception linked to RotaShield in an extended analysis of CDC's case-control study but also paradoxically found a decrease in intussusception in vaccinees in comparison to controls (OR of 0.3) after the third week of vaccination, thus proposing a compensatory decrease following the initial increase [97]. This provided an explanation for the failure to find an overall increase in intussusception in vaccinees.

The CDC in a similar additional analysis of their case-control study also reported an OR of 0.3 among vaccinated vs. unvaccinated infants during a risk period extending >3 weeks after RotaShield administration [106]. They explained this unexpected finding in their case-control study in which 4 controls were matched carefully with each case by suggesting "...that confounding bias from socioeconomic status may be responsible for the low value of this OR. Infants with higher socioeconomic status were more likely to receive RRV-TV and had a lower risk of intussusception than were infants with lower socioeconomic status." They later noted that the NIH hypothesis of a compensatory decrease with a protective effect of vaccination with ORs of <1 "...while intriguing, are not proven." There were insurmountable differences among the various groups regarding the actual attributable risk of RotaShield with intussusception, these differences were very apparent at a subsequent reappraisal of the vaccine in February 2002 by CDC and ACIP and the withdrawal recommendation was reaffirmed [107].

Catch-Up Vaccination Found to be Related to Intussusception

However, there was an important association that for the most part had eluded the various investigating groups, a relationship if known and acted upon may have rescued the vaccine. It was clear that routine administration of RotaShield in the US was doomed because no matter what the true incidence of intussusception was in the US, it was likely not small enough in the age schedule that had been used for RotaShield's administration. It is interesting to note that in September 1999, the interlude between suspension of the vaccine in July and its withdrawal in October, a communication regarding RotaShield's suspension provided an important clue on the effect of age on naturally occurring intussusception that was not pursued (Fig. 9a) [108]. The authors showed the incidence of intussusception by age in months in 833 infants hospitalized over a 2-year period at National Health Service Hospitals in England, where the vaccine was not in use. The authors noted that "The highest incidence was in those aged 3–6 months. This peak coincides with the age at which rotavirus vaccination was scheduled in the USA (2, 4, and 6 months)." Another study reported in 1989 showed a relatively refractory period for intussusception in the under 3 months age group in Taiwan (Fig. 9b) [109].

An important analysis was made by NIH investigators after the vaccine's withdrawal. They reported that age of vaccination with RotaShield was a major factor in its link with intussusception in the first 2–3 weeks after the first dose [110, 111]. In an analysis of the CDC case-control study, they found that vaccinees who were 90 days of age or greater at the time of the first dose developed 81% (35/43) of all intussusception cases occurring within 2 weeks after vaccination even though this age group had received only 38% of all first doses according to the CDC National Immunization Survey (Fig. 10) [111]. In addition, no cases of intussusception were detected in the ~70,000 vaccinees who were <60 days of age when they received the first dose of RotaShield [110]. It thus appeared that "catch-up" vaccination of older infants in whom the first dose was given during the age period of highest vulnerability for intussusception (rather than the recommended first dose at 2 months of age) contributed disproportionately to the number of cases [110, 111]. It should be noted that it had been reported by CDC investigators that in the

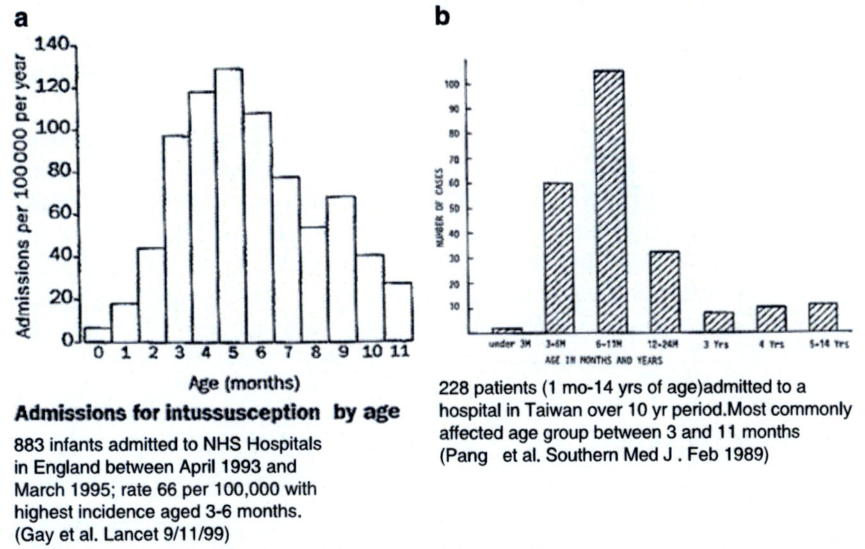

Fig. 9 (**a**) Incidence rates by age for 888 hospitalizations of infants for intussusception to NHS hospitals in England between April 1993 and March 1995 (from ref. [108]); (**b**) age distribution of 228 patients admitted to a hospital for intussusception in Taiwan over a 10-year period in Taiwan (from ref. [109])

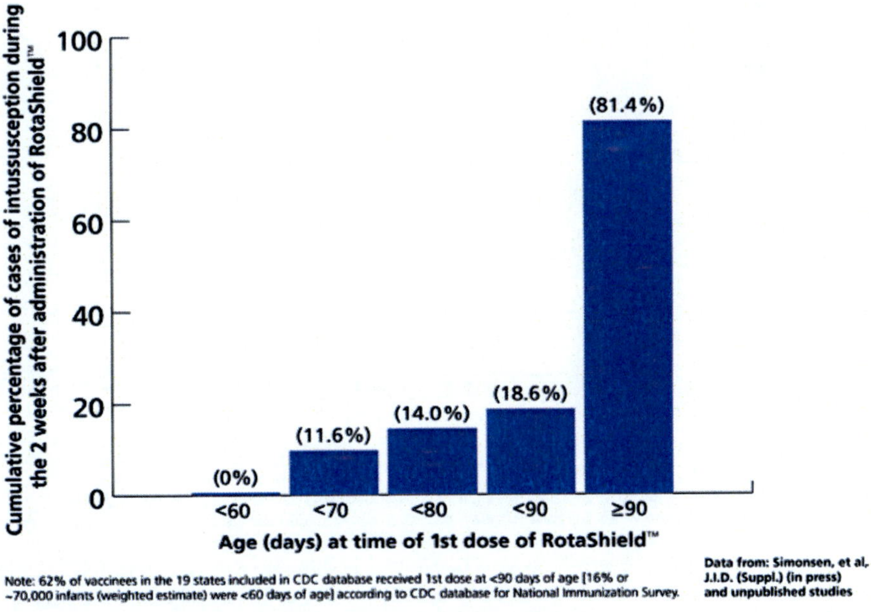

Fig. 10 Age distribution of 43 cases of intussusception with onset during the 2 weeks after the first dose of RotaShield in CDC case-control study (from ref. [111])

case-control study "We found no evidence that age or other variables, except for feeding with breast milk, modified the risk of intussusception among infants given RRV-TV" [98]. The role of age and its relationship to intussusception after RotaShield has remained a controversial issue [112–117].

Further analysis by NIH investigators indicated that the early cluster of intussusception cases in the CDC case-control study during the first 2 weeks after the first dose was derived from a population of ~435,000 vaccinated infants with a mean age of 123 days (about 4 months of age) (Fig. 11a) [98, 118]. However, the early cluster of cases disappeared when the analysis was limited to a population derived from ~135,000 children who had a mean age of <70 days at the time of the first dose of RotaShield (Fig. 11b) [118].

Age Factor Important for Current Rotavirus Vaccines

The age factor was not totally ignored by the WHO which, in a position paper in 2007 on newly available rotavirus vaccines, issued a firm admonition in a WHO position paper on the current rotavirus vaccines stating that the first dose of Rotarix

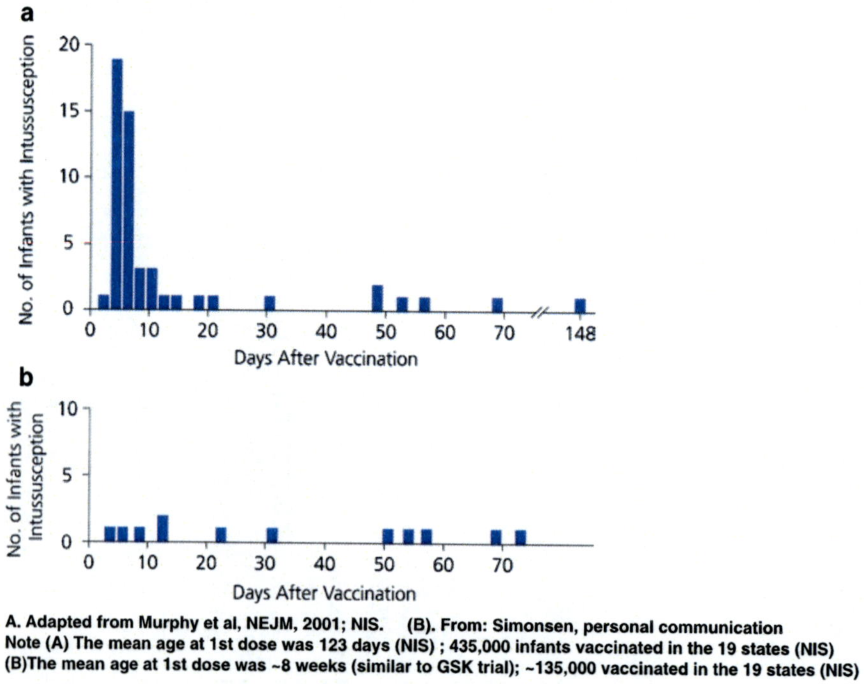

Fig. 11 (**a**) Occurrence of intussusception in infants who received the first dose of RotaShield at 60–290 days (2–6 months) of age (from ref. [98]); (**b**) occurrence of intussusception in infants who received the first dose of RotaShield at <70 days of age (from ref. [118])

"...should be given no later than at the age of 12 weeks" and for RotaTeq that "Vaccination should not be initiated for infants aged >12 weeks" [119]. They concluded by noting that "There is a potentially higher risk of intussusception when the first dose of these vaccines is given to infants aged >12 weeks; consequently, current rotavirus vaccines should not be used in catch-up vaccination campaigns, where the exact age of the vaccinees may be difficult to ascertain." It was apparent that the RotaShield experience with "catch-up" vaccination provided an important lesson on the age of administration of the new vaccines.

Age Factor in Naturally Occurring Intussusception

Further observations of the age distribution of naturally occurring intussusception are particularly instructive and give additional biologically plausible support for the age-related link of RotaShield with intussusception resulting from "catch-up" vaccination. Studies from various parts of the world indicate that there is a relatively refractory period for developing intussusception during the first 60–90 days of life with a gradual rise until peaking at ~6 months of age [108, 109, 120–123]. For example, a comprehensive US study of intussusception hospitalization rates per 100,000 infants <12 months of age in 39 states from 1993 to 2004 (1999 excluded) reported that intussusception rates were remarkably low in <8-week-old infants ranging from ~2/100,000 at birth to ~5/100,000 at 8 weeks but then increasing rapidly to a peak of ~52/100,000 for infants aged 26–29 weeks (Fig. 12) [123]. The "catch-up" vaccination of RotaShield coincided with this heightened susceptibility phase of naturally occurring intussusception, and we believe that this phenomenon was not unique to RotaShield but would likely apply to any rotavirus vaccine if the age relationship is ignored. It is clear that it was important to limit the first dose of any rotavirus vaccine to the age period of low vulnerability to developing intussusception and to not exceed about 12 weeks of age; even greater safety would be achieved if vaccination were limited to those less than 8 weeks of age [124].

Currently Licensed Rotavirus Vaccines and Intussusception

The recently (February 2006) licensed pentavalent rotavirus vaccine Rotateq described later in this book underwent extremely large prelicensure safety trials involving almost 70,000 infants because of the RotaShield experience [125]. Fortunately, intussusception was not associated with the vaccine. The age factor described above for RotaShield was controlled by stringent age restrictions in the Rotateq studies. The mean and median ages of the vaccinees and controls in this large safety trial were identical (<3 months [mean 68.6 days and median 70 days]). The package insert after licensure specified that the vaccine should be given "... starting at 6–12 weeks of age."

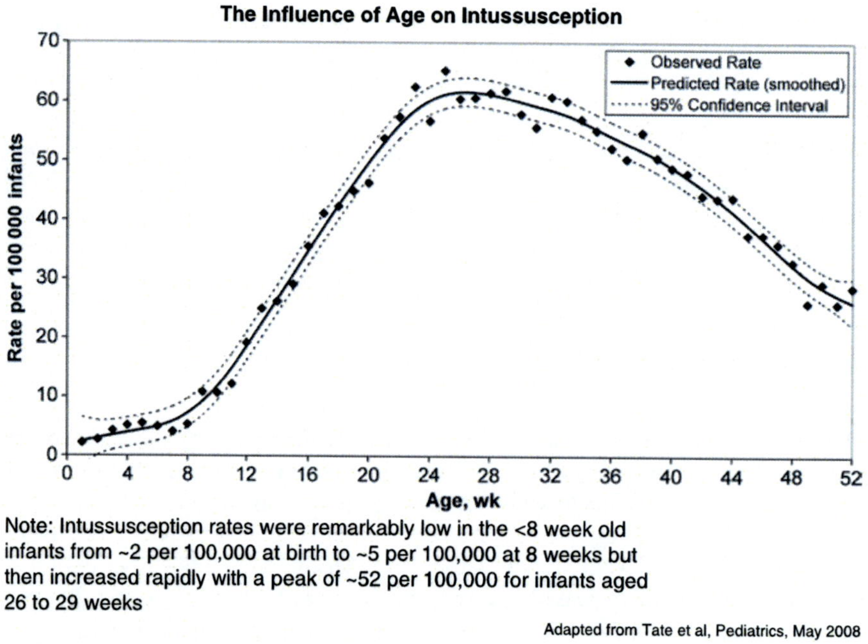

Fig. 12 Intussusception rates per 100,000 12-month-old infants by week of age at time of hospitalization in 39 states that participated at least 1 year from 1993 to 2004 (1999 excluded) (from ref. [123])

The intussusception "cloud" emerged for Rotateq in February 2007 about 1-year postlicensure after about 3.5 million doses had been distributed, when the FDA released a "Public Health Notification" with "Information on RotaTeq and Intussusception" in which the FDA notified "health care providers and consumers about 28 postmarketing reports of intussusception following administration" of RotaTeq [126]. They also noted that about half of the cases had occurred 1–21 days following vaccination. The FDA cautioned that "Of the 28 cases of intussusception, the number that may have been caused by the vaccine, or occurred by coincidence, is unknown." The Notification also indicated that "The RotaTeq label and Patient Product Information have been updated to include postmarketing reports of intussusception."

Moreover, in a later postmarketing analysis in which it was estimated that over nine million doses of Rotateq had been distributed, it was noted that with 100% reporting completeness of doses given and of VAERS cases reported, there was a dramatic decrease in intussusception following vaccination [127]. But if reporting completeness was 50% and administration of vaccine was 50%, there would be a statistically significant association (RR 2.01) between the vaccine and intussusception for all doses combined. Moreover, using the same 50% assumptions, the authors note that "Limiting the sensitivity analysis to the first dose with the same conservative reporting and dose administration assumptions would yield a RR of 3.7 (96% CI 1.9–6.9)." Finally, they state "…Direct comparison of the risks of intussusception associated with RotaShield and RotaTeq requires cautious interpretation, because of their schedules of administration (i.e., catch-up vaccination with the first dose of RotaShield was permitted for infants up to 6 months of age, whereas the first dose of RotaTeq has to be administered by 12 weeks of age)..."

Similarly, the Rotarix vaccine, which was licensed in the US in April 2008, underwent a very large prelicensure trial for safety regarding intussusception, which included over 60,000 infants [128]. Fortunately, intussusception was not found to be associated with the vaccine. However, it should be noted that the mean age of the vaccinees and controls at dose 1 were identical being <2 months (57.4 days). As noted earlier, RotaShield was not associated with intussusception in infants who received the first dose at <70 days of age. The package insert for Rotarix also set age limitations stating that the first dose should be given "...beginning at 6 weeks of age" and the "second dose after an interval of at least 4 weeks and prior to 24 weeks of age."

A Permissive Recommendation for RotaShield Was Needed

The fate of RotaShield may have been very different if the review process was more deliberative without the hasty withdrawal decision in October 1999, only 3 months after the suspension in July. Later at a 2002 ACIP meeting where the withdrawal decision was reviewed, Wyeth indicated that it would not reintroduce RotaShield unless there was a routine recommendation [107]. A plea was made that despite Wyeth's decision, a permissive recommendation by the ACIP could save millions of children's lives in the developing countries as it would send a powerful message that RotaShield could have been used on a permissive basis in the US if it were still available [107]. It was also noted that with Wyeth's withdrawal the RotaShield strains would be available from NIH for manufacturers in these areas of the world where the vaccine was needed most [107]. The plea fell on "deaf ears." There was a sentiment that the ACIP role was to form policy for the US and not the world. However, this decision has had a profound negative impact on the infants and young children in the developing countries, because of the irretrievable loss of time for vaccine development – over a decade – and the continued high infant mortality from rotavirus diarrhea in these areas of the world. It had been noted previously by CDC in the 1999 withdrawal statement regarding RotaShield that "... the ACIP's decision may not be applicable to other settings, where the burden of disease is substantially higher and where the risks and benefits of rotavirus vaccination could be different" [91]. This conditional statement failed to alleviate the ethical anxieties of the developing countries despite an international WHO meeting in 2000, which had concluded that it was ethical to test RotaShield in the developing world in spite of its withdrawal in the US because of the risk-benefit equation. However, the developing countries rejected this idea because of the ethical-political fall-out of using a vaccine that was not fit for children in the US. One can readily understand this position. A permissive recommendation for the US could have overcome this ethical dilemma for the developing countries.

Ironically, Wyeth, prior to its withdrawal, had already initiated studies with RotaShield in Asia and Africa where most of the 500,000 deaths of children under 5 years of age occur annually. In this regard, a clinical study of RotaShield was initiated in 1998 for safety and immunogenicity in Bangladesh [129]. The authors concluded that "In this population RRV-tetravalent vaccine was comparably immunogenic and safe as in trials conducted in developed countries, where this vaccine has been proved effective in preventing severe diarrhea." Immunogenicity studies of RotaShield had also been initiated in South Africa and Ghana. Such studies were not pursued further because of the withdrawal of the vaccine in the US At this writing in the last trimester of 2009, there is still no rotavirus vaccine for routine administration for the poorest children of the world where most of the 500,000 deaths occur from rotavirus diarrhea annually.*

Finally, if the age factor described above had been known, there may have been a striking change in the fate of RotaShield in the US as well.

Ethicist's View Regarding the Withdrawal of RotaShield on the Developing Countries

With regard to this quandary of risk vs. benefit and RotaShield, a physician-ethicist, who participated in the WHO meeting on RotaShield's use in developing countries in 2000 cited above, wrote an editorial soon after the withdrawal of RotaShield entitled "The future of research into rotavirus vaccine. Benefits of vaccine may outweigh risks for children in developing countries" [130]. He concluded the editorial as follows: "…If the next vaccine in development takes 3–5 years to get to the stage where tetravalent rhesus rotavirus is now, the choice to wait must be weighed against the cost of waiting: 1.4–3.2 million preventable deaths. Some have falsely assumed that inaction is a morally neutral state. But if one is culpable for vaccine-related deaths, then one is also culpable for deaths caused by withholding the vaccine."

> "Is there a moral difference between a treatment that may cause a sick child to die and a vaccine that may cause a healthy child to die? Because public health doctors treat unhealthy populations rather than unhealthy patients the risk of death or serious disability must be lower with vaccines than with clinical treatments. The risks of tetravalent rhesus rotavirus vaccine seem comparable to the risks associated with measles, mumps, and rubella vaccine. The moral yardstick for the public health physician is ultimately the same as for clinicians: do the benefits of vaccination exceed the risks? In a developing country in which a child's risk of death from rotavirus diarrhea is 1 in 200 or greater the answer may well be yes" [130].

**Editor's note*: As of 2011, both new vaccines have demonstrated efficacy in developing countries and Rotarix is widely used in Latin America.

RotaShield in a Field Trial in Ghana

There is an unanticipated postscript to this saga. A nonprofit organization, the International Medica Foundation, has acquired the license to the RotaShield technology from the NIH Office of Technology transfer. With this license came the master virus seeds and other reagents developed by Wyeth for its US studies [131]. The goal of this effort by the IMF was to produce RotaShield vaccine at a low cost for the developing countries. RotaShield has been produced in Germany under IMF auspices, and a phase 2 coded placebo-controlled double blind study was initiated on August 28, 2009 in Ghana. This study will include approximately 1,000 children who will receive the first dose of RotaShield at 0–4 weeks of age and the second dose at 4–8 weeks of age with a minimum of 3 weeks between doses.

Benefits of Neonatal Vaccination

The schedule of beginning rotavirus vaccination in the neonatal period has several advantages: (a) it may prove to be the safest time to administer an oral, live, attenuated vaccine; (b) it may be a period of low reactogenicity; RRV-TV did not induce a febrile response in infants vaccinated during the neonatal period and moreover, the neonatal dose gave significant protection against a febrile response at 2 months of age when compared to neonates who received a placebo neonatally [132]; (c) RotaShield did not induce intussusception in ~70,000 infants under the age of 60 months [110]; (d) it is a relatively refractory period for naturally occurring intussusception (this may also apply to a vaccine); (e) a single dose may yield adequate protection vs. severe diarrhea; (f) it may afford protection of vulnerable infants during the first 2 months of life who are now excluded by the conventional schedule starting at ~2 months of age; (g) there is more likely exposure to a health-care provider during the neonatal period, thereby facilitating vaccine delivery.

Why Is Another Vaccine Needed When There Are Already Two Licensed Rotavirus Vaccines?

The advantage of RotaShield (and other vaccines being produced in developing countries) is that it is made under the auspices of a nonprofit foundation with later transfer to a developing country where it will be produced at an affordable price. This vaccine can be "expanded" and thereby designed as necessary in a formulation that will correspond to circulating newer serotypes such as G8, 9, or 10, which are available as single VP7 gene substitution reassortants [133]. It will also have sustainability because of manufacture by nonprofit sponsorship and ultimately by a developing country manufacturer.

Sustainability is crucial for public health globally, and its need was stated eloquently recently as follows [134]: "A single entity cannot address the complex issues of global health; the confluence of many is required. Long term success in global health requires building a sustainable infrastructure in developing nations and, importantly, solve their own problems through the establishment of economic stability and self-sufficiency. Unfortunately, popular Western culture tends to have a short attention span, and today's latest trend can quickly become yesterday's news."

References

1. Kumate J, Isibasi A. Pediatric diarrheal diseases: a global perspective. Pediatr. Infect Dis. 1986; 5(Suppl): S21–28
2. Connor JD, Barrett-Connor E. Infectious Diarrheas. Pediatr Clin North Am 1967;14:197–221
3. Yow MD, Melnick JL, Blattner RJ, Stephenson NB, Robinson NM, Burkhardt MA. The association of viruses and bacteria with infantile diarrhea. Am J Epidemiol 1970;92:33–39
4. Kapikian AZ. The discovery of the 27-nm Norwalk Virus: an historic perspective. J. Infect Dis 2000;181 (Suppl 2) 295–302
5. Kapikian AZ, Wyatt RG, Dolin R, Thornhill TS, Kalica AR, Chanock RM. Visualization by immune electron microscopy of a 27 nm particle associated with acute nonbacterial gastroenteritis. J. Virol 1972;10:1075–1081
6. Bishop RF, Davidson GP, Holmes IH, Ruck BJ. Virus particles in epithelial cells of duodenal mucosa from children with acute non-bacterial gastroenteritis. Lancet 1973;302: 1281–1283
7. Green KY. Caliciviridae: The Noroviruses. In Fields Virology, Knipe DM, Howley PM (eds) (Fifth Ed) 2007; pp 949–97 Lippincott Williams & Wilkins, Philadelphia
8. Estes MK, Kapikian. Rotaviruses. in Fields Virology, Knipe DM, Howley PM (eds) (Fifth Ed) pp 1917–1974 (Lippincott Williams & Wilkins, Philadelphia)
9. Patel MM, Widdowson M-A, Glass RI, Akazawa K, Vinje J, Parashar UD. Systematic literature review of role of noroviruses in sporadic gastroenteritis. Emerg Infect Dis 2008;14:1224–1231
10. Editorial. More about D and V. Brit. Med J 1974; 5935;1–2
11. Institute of Medicine. The prospects for immunizing against rotavirus. In New Vaccine Development. Establishing Priorities. Diseases of importance in Developing Countries. (1986). Vol 2: pp 308–316. National Academy Press, Washington
12. Institute of Medicine. The prospects for immunizing against rotavirus. In New Vaccine Development. Establishing Priorities. Diseases of importance in the United States. (1986). Vol 1: pp 410–423. National Academy Press, Washington
13. Parashar UD, Burton A, Lanata C, Boschi-Pinto C, Shibuya K, Steele D, Birmingham M, Glass RI. Global mortality associated with rotavirus disease among children in 2004. J Infect Dis 2009;200 (Suppl 1):9–15
14. Glass RI, Parashar UD, Bresee JS, Turcios R, Fischer TK, Widdowson M-A, Jiang B, Gentsch JR. Rotavirus vaccines: current prospects and future challenges. Lancet 2006;368:323–332
15. Flewett TH, Bryden AS, Davies H, Woode GN, Bridger JC, Derrick JM. Relation between viruses from acute gastroenteritis of children and newborn calves. Lancet 1974; 304: 61–63
16. Prasad BVV. In Rotaviruses. Current Topics in Microbiology and Immunology 1994; 185:Cover (adapted from). Ramig RF (ed) Springer, Berlin
17. Kapikian AZ, Wyatt RG, Levine MM, Yolken RH, Van Kirk DH, Dolin R, Greenberg HB, Chanock RM. Oral administration of a human rotavirus to volunteers: induction of illness and correlates of resistance. J. Infect Dis 1983; 147:95–106

18. Kapikian AZ, Wyatt RG, Levine MM, Black RE, Greenberg HB, Flores J, Kalica AR, Hoshino Y, Chanock RM. Studies in volunteers with human rotaviruses. Develop. Biol. Standard 1983;53;209–2218
19. Wyatt RG, James WD, Bohl EH, Theil KW, Saif LJ, Kalica AR, Greenberg HB, Kapikian AZ, Chanock RM. Human rotavirus type 2: cultivation in vitro. Science 1980;207: 189–191
20. Wyatt RG, Kapikian AZ, Greenberg HB, Kalica AR, Flores J, Hoshino Y, Chanock RM, Levine MM. Development of vaccines against rotavirus disease. Prog Fd Sci 1983;7: 189–192
21. Wyatt RG, Kapikian AZ, Hoshino Y, Flores J, Midthun K, Greenberg HB, Glass RI, Askaa J, Levine MM, Black RE, Clements ML, Potash L, London WT. Development of rotavirus vaccines. Control and eradication of infectious diseases an international d symposium PAHO copublication series No. 1 1985; pp 17–38
22. Kapikian AZ, Wyatt RG, Greenberg HB, Kalica AR, Kim HW, Brandt CD, Rodriguez WJ, Parrott RH, Chanock RM. Approaches to immunization of infants and young children against gastroenteritis due to rotaviruses. Rev Infect Dis 1980;2;459–469
23. Woode GN, Jones J, Bridger J. Levels of colostral antibodies against neonatal calf diarrhea virus. Vet Rec 1975;96:148–149
24. Bridger JC, Woode GN. Neonatal calf diarrhea: Identification of a reovirus-like (rotavirus) agent in faeces by immunofluorescence and immune electron microscopy. Br Vet J 1975; 131;528–535
25. Snodgrass DR, Wells PW. Rotavirus infection in lambs: studies on passive protection. Arch Virol 1976; 52:201–205
26. Snodgrass DR, Madeley CR, Wells PW, Angus KW. Human rotavirus in lambs: infection and passive protection. Infect Immun 1977;16:268–270
27. Snodgrass DR, Wells PW. Passive immunity in rotavirus infections. J Am Vet Med Assoc 1978;173:L565–569
28. Lecce JG, King MW, Mock R. Reovirus-like agent associated with fatal diarrhea in neonatal pigs. Infect Immun 1976;14:816–825
29. Flores J, Perez-Schael I, Blanco M, Garcia D, Vilar M, Cunto W, Gonzalez R, Ubina C, Boher J, Mendez M, Kapikian AZ. Comparison of the reactogenicity and antigenicity of the M37 rotavirus vaccine and the rhesus-rotavirus-based quadrivalent vaccine. Lancet 1980; 336:330–334
30. Vesikari T, Ruuska T, Koivu HP, Green KY, Flores J, Kapikian AZ. Evaluation of the M37 human rotavirus vaccine in 2-to 6-month-old infants. Pediatr Infect Dis J 1991;10:912–917
31. Bishop RF, Barnes GL, Cipriani E, Lund JS. Clinical immunity after neonatal rotavirus infection. N. Eng J. Med 1983;309:72–76
32. Hoshino Y, Kapikian AZ, Chanock RM. Selection of cold-adapted mutants of human rotaviruses that exhibit various degrees of growth restriction in vitro. J. Virol 1994;68:7598–7602
33. Treanor JJ. Unpublished studies
34. Henderson DA. Edward Jenner's vaccine. Publ. Hlth Reports 1997:112: 116–121. (Reprinted from The Jordan Report: Accelerated Developments of Vaccines 1996, NIAID, NIH)
35. Woode GN, Bridger JC, Jones JM, Flewett TH, Bryden AS, Davies HA, White GBB. Morphological and antigenic relationships between viruses (rotaviruses) from acute gastroenteritis of children, calves, piglets, mice and foals. Infec Immun 1976;14:804–810
36. Kapikian AZ, Cline WL, Kim HE, Kalica AR, Wyatt RG, Van Kirk DH, Chanock RM, James HD Jr, Vaughn AL. Antigenic relationships among the five reovirus-like (RVL) agents by complement fixation (CF) and development of a new substitute CF antigen for the human RVL agent of infantile gastroenteritis. Proc Soc Exp Biol Med 1976;152:535–539
37. Matsuno S, Inouye S, Kono R. Plaque assay of neonatal calf diarrhea virus and the neutralizing antibody in human sera. J. Clin Microbiol 1977;5:1–4
38. Wyatt RG, Mebus CA, Yolken RH, Kalica AR, James HD Jr, Kapikian AZ, Chanock RM. Rotaviral immunity in gnotobiotic calves: heterologous resistance to human virus induced by bovine virus. Science 1979;203:548–550

39. Wyatt RG, Kapikian AZ, Mebus CA. Induction of cross-reactive serum neutralizing antibody to human rotavirus in calves after in utero administration of bovine rotavirus. J Clin Microbiol 1983;18;505–508
40. Zissis G, Lambert JP, Marbehant P, Marissens D, Lobman M, Charlier P, Delem A, Zygraich N. Protection studies in colostrum-deprived piglets of a bovine rotavirus vaccine candidate using human rotavirus strains for challenge. J Infect Dis 1983;148:1061–1068
41. Vesikari T, Isolauri E, Delem A, D'Hondt E, Andre FE, Zissis G. Immunogenicity and safety of live oral attenuated bovine rotavirus vaccine strain RIT 4237 in adults and young children. Lancet 1983;322;807–811
42. Vesikari T, Isolauri E, D'Hondt E, Delem A, Andre FE, Zissis G. Protection of infants against rotavirus diarrhoea by RIT 4237 attenuated bovine rotavirus strain vaccine. Lancet 1984;323:977–981
43. Vesikari T, Isolauri E, Delem A, D'Hondt E, Andre A, Beards GM, Flewett TH. Clinical efficacy of the RIT 4237 live attenuated bovine rotavirus vaccine in infants vaccinated before a rotavirus epidemic. J. Pediatr 1985;107:189–194
44. Kapikian AZ, Hoshino Y, Flores J, Midthun K, Glass RI, Nakagomi O, Nakagomi T, Chanock RM, Potash L, Levine MM, Dolin R, Wright PF, Belshe RE, Anderson EL, Vesikari T, Gothefors L, Wadell G, Perez-Schael I. Alternative approaches to the development of a rotavirus vaccine. In: Development of Vaccines and Drugs against Diarrhea. 11th Nobel Conf., Stockholm 1985, pp 192–214 (eds J. Holmgren, A Lindbergh & R Mollby) Studentlitteratur, Lund, Sweden, 1986
45. Stuker G, Oshiro L, Schmidt NL. Antigenic comparisons of two new rotaviruses from rhesus monkeys. J. Clin. Microbiol 1980; 11:202–203
46. Kapikian AZ, Midthun, K, Hoshino Y, Flores J, Wyatt RG, Glass RI, Askaa J, Nakagomi O, Nakagomi T, Chanock RM, Levine MM, Clements ML, Dolin R, Wright PF, Belshe RB, Anderson EL, Potash L. Rhesus rotavirus: a candidate vaccine for prevention of human rotavirus disease, In: Vaccines 85 pp 357–367. Cold Spring Harbor Laboratory
47. Wallace RE, Vasington PJ, Petricciani JC, Hopps HE, Lorenz DE, Kadanka Z. Development of a diploid cell line from fetal rhesus monkey lung for virus vaccine production. In Vitro 1973;8:323–332
48. Kapikian AZ, Flores J, Hoshino Y, Midthun K, Green KY, Gorziglia M, Chanock RM, Potash L, Perez-Schael I, Gonzalez M, Vesikari T, Gothefors L, Wadell G, Glass RI, Levine MM, Rennels MB, Losonsky G, Christy C, Dolin R, Anderson EL, Belshe RB, Wright PF, Santosham M, Halsey NA, Clements ML, Sears SD, Steinhoff MC, Black RE. Rationale for the development of a rotavirus vaccine for infants and young children. In: Progress in Vaccinology 1989 Vol 2 pp 151–180 (GP Talwar ed). Springer, New York
49. Christy C, Madore HP, Treanor JJ, Pray K, Kapikian AZ, Chanock RM, Dolin R. Safety and immunogenicity of live attenuated rhesus monkey rotavirus vaccine. J. Infect Dis. 1986;154:1045–1047(correspondence)
50. Losonsky GA, Rennels MB, Kapikian AZ, Midthun K, Ferra PJ, Fortier DN, Hoffman KM, Baig A, Levine MM. Safety, infectivity, transmissibility and immunogenicity of rhesus rotavirus vaccine (MMU 18006) in infants. Pediatr Infect Dis 1986; 5:25–29
51. Anderson EL, Belshe RB, Bartram J, Crookshanks-Newman F, Chanock RM, Kapikian AZ. Evaluation of rhesus rotavirus vaccine (MMU 18006) in infants and young children. J. Infect Dis 1986;153:823–831
52. Perez-Schael I, Gonzalez M, Daoud N, Perez M, Soto I, Garcia D, Daoud G, Kapikian AZ, Flores J. Reactogenicity and antigenicity of the rhesus rotavirus vaccine in Venezuelan children. J. Infect Dis 1987;155;334–338
53. Rennels MB, Losonsky GA, Schindledecker CL, Hughes TP, Kapikian AZ, Levine MM, the Clinical Study Group (Ferra PJ, Fortier DN, Sutton JM). Immunogenicity and reactogenicity of lowered doses of rhesus rotavirus vaccine strain MMU 18006 in young children. Pediatr Infect Dis J. 1987; 6:260–267
54. Wright PF, Tajima T, Thompson J, Kokubun K, Kapikian A, Karzon DT. Candidate rotavirus vaccine (rhesus rotavirus strain) in children: an evaluation. Pediatrics 1987;80:473–480

55. Losonsky GA, Rennels MB, Lim Y, Krall G, Kapikian KZ, Levine MM. Systemic and mucosal responses to rhesus roavirus vaccine MMU 18006. Pediatr Infect Dis J 1988;7:388–393
56. Vesikari T, Kapikian AZ, Delem A, Zissis. A comparative trial of rhesus monkey (RRV-1) and bovine (RIT 4237) oral rotavirus vaccines in young children. J. Infect Dis 1986; 153:832–839
57. Kapikian AZ, Flores J, Midthun K, Hoahino Y, Green KY, Gorziglia M, Nishikawa K, Chanock RM, Potash L, Perez-Schael I. Strategies for the development of a rotavirus vaccine against infantile diarrhea with an update on clinical trials of rotavirus vaccine. Adv Exp Med Biol 1987;257:67–89
58. Rennels MB, Losonsky GA, Levine MM, Kapikian AZ, and the Clinical Study Group (Fortier DN, Sutton JM, Ferra PJ and Hoffman KM) Preliminary evaluation of the efficacy of rhesus rotavirus vaccine strain MMU 18006 in young children. Pediatr Infect Dis 1986;5:587–588
59. Flores J, Perez-Schael I, Gonzalez M, Garcia D, Perez M, Daoud N, Cunto W, Chanock RM, Kapikian AZ. Protection against severe rotavirus diarrhoea by rhesus rotavirus vaccine in Venezuelan infants. Lancet 1987;329; 882–884
60. Gothefors L, Wadell G, Juto P, Taniguchi K, Kapikian AZ, Glass RI. Prolonged efficacy of rhesis rotavirus vaccine in Swedish children. J Infect Dis 1989;159:753–757
61. Rennels MB, Losonsky GA, Young AE, Shindledecker CL, Kapikian AZ, Levine MM and the Clinical Study Group. An efficacy trial of rhesus rotavirus vaccine in Maryland. Am J Dis Child 1990;144:601–604
62. Christy C, Madore HP, Pichichero ME, Gala C, Pincus P, Vosefski D, Hoshino Y, Kapikian AZ, Dolin R. The Elmwood and Panorama pediatric groups: field trial of rhesus rotavirus vaccine in infants. Pedaitr Infect Dis J 1988;7:645–650
63. Perez-Schael I, Garcia D, Gonzalez M, Gonzalez R, Daoud N, Perez M, Cunto W, Kapikian AZ, Flores J. Prospective study of diarrheal diseases in Venezuelan children to evaluate the efficacy of rhesus rotavirus vaccine. J. Med Virol 1990;30:219–229
64. Santosham M, Letson GW, Wolff M, Reid R, Gahagan S, Adams R, Callahan C, Sack RB, Kapikian AZ. A field study of the safety and efficacy of two candidate rotavirus vaccines in a Native American population. J. Infect Dis 1991;163: 483–487
65. Vesikari T, Rautanen T, Varis T, Beards GM, Kapikian AZ. Rhesus rotavirus candidate vaccine. Clinical trial in children vaccinated between 2 and 5 months of age. Am J Dis Child 1990;144:285–289
66. Madore, HP, Christy C, Pichichero M, Long C, Pincus P, Vosefsky P, Kapikian AZ, Dolin R, Elmwood, Panorama, and Westfall Pediatric Groups. Field trial of rhesus rotavirus or human-rhesus reassortant vaccine of VP7 serotype 3 or 1 specificity in infants. J. Infect Dis 1992;166:235–243
67. Kapikian AZ, Flores J, Midthun K, Hoshino Y, Green KY, Gorziglia M, Taniguchi K, Nishikawa K, Chanock RM, Potash L, Perea-schael I, Dolin R, Christy C, Santosham M, Halsey N, Clements ML, Sears SD, Black RE, Levine MM, Losonsky GA, Rennels MB, Gothefors L, Wadell G, Glass RI, Vesikari T, Anderson EL, Belshe RB. Wright PF, Urasawa. Development of a rotavirus vaccine by a "Jennerian" and a modified "Jennerian" approach. In: Modern Approaches to New Vaccines including Prevention of AIDS: Cold Spring Harbor Laboratory. Vaccines 88; 1988 pp 151–159
68. Lanata C, Black RE, Flores J, Lazo F, Butron B, Linhares A, Huapaya A, Ventura G, Gil A, Kapikian AZ. Immunogenicity, safety and protective efficacy of one dose of the rhesus rotavirus vaccine and serotype 1 and 2 human-rhesus rotavirus reassortants in children from Lima, Peru. Vaccine 1996;14:237–243
69. Kapikian AZ, Flores J, Hoshino Y, Midthun K, Gorziglia M, Green KY, Chanock RM, Potash L, Sears SD, Clements ML, Halsey NA, Black RE, Perez-Schael I. Prospects for development of a rotavirus vaccine against rotavirus diarrhea in infants and young children. Rev. Infect Dis (Suppl) 1989;3:539–546
70. Midthun K, Greenberg HB, Hoshino Y, Kapikian AZ, Wyatt RG, Chanock RM, Reassortant rotaviruses as potential live roavirus vaccine candidates. J Virol 1985;53: 949–954

71. Midthun K, Hoshino Y, Kapikian AZ Chanock RM. Single gene substitution rotavirus reassortants containing the major neutralization protein (VP7) of human rotavirus serotype 4. J Clin Microbiol 1986;24: 822–836
72. Kapikian AZ, Hoshino Y. To serotype or not to serotype: that is still the question. J. Infect. Dis 2007;195:611–614
73. Kapikian AZ, Hoshino Y, Chanock RM, Perez-Schael I. Efficacy of a quadrivalent rhesus rotavirus-based human rotavirus vaccine aimed at preventing severe rotavirus diarrhea in infants and young children. J. Infect Dis 1996;174 (suppl):65–72
74. Kapikian AZ, Vesikari T, Ruuska T, Madore HP, Christy C, Dolin R, Flores J, Green KY, Davidson BL, Gorziglia M, Hoshino Y, Chanock RM, Midthun K, Perez-Schael I. An update on the "Jennerian" and modified "Jennerian" approach to vaccination of infants and young children against rotavirus diarrhea. Adv Exp Med Biol 1992; 327:59–69
75. Kapikian AZ. Rotavirus vaccine: the clinical experience with the rhesus rotavirus-based vaccines. In: Mucosal Vaccines New Trends in Immunization 1996 pp 345–356 (Kiyono H, Ogra PL, McGhee JR eds)
76. Flores J, Perez-Schael I, Blanco M, Vilar M, Daoud N, Midthun K, Kapikian AZ. Reactions to and antigenicity of two human-rhesus rotavirus reassortant vaccine candidates of serotypes 1 and 2 in Venezelan infants. J. Clin. Microbiol 1989: 27; 512–518
77. Halsey NA, Anderson EL, Sears SD, Steinhoff M, Wilson M, Belshe RB, Midthun K, Kapikian AZ, Chanock RM, Samorodin R, Burns B, Clements ML. Human-rhesus reassortant vaccines: safety and immunogenicity in adults, infants and children. J. Infect Dis 1988;158:1261–1267
78. Perez-Schael I, Blanco M, Vilar M,Garcia D,Gonzalez R, Kapikian AZ, Flores J. Clinical studies of a quadrivalent rotavirus vaccine in Venezuelan infants J. Clin Microbiol 1990; 28: 553–558
79. Flores J, Perez-Schael I, Blanco M, Rojas AM, Alfonzo E, Cespo I, Cunto W, Pittman AL, Kapikian AZ. Reactogenicity and immunogenicity of a high-titer rhesus rotavirus-based quadrivalent rotavirus vaccine. J. Clin Microbiol 1993;31:2439–2445
80. Linhares AC, Gabbay YB, Mascarenhas JDP, de Freitas RB, Oliveira CS, Bellesi N, Monteiro TA, Lins-Lainson Z, Ramos FLP, Valente SA. Immunogenicity, safety and efficacy of tetravalent rhesus-human, reassortant rotavirus vaccine in Belem, Brazil. Bull WHO 1996:74;491–500
81. Linhares AC, Lanata CF, Hausdorff WP, Gabbay YB, Black RE. Reappraisal of the Peruvian and Brazilian lower titer tetravalent rhesus-human reassortant rotavirus vaccine efficacy trials: analysis by severity of diarrhea. Pediatr Infect Dis J 1999;18:1001–1006
82. Kapikian AZ, Flores J, Vesikari T, Ruuska T, Madore HP, Green KY, Gorziglia M, Hoshino Y, Chanock RM, Midthun K, Perez-Shael I. Recent advances in development of a rotavirus vaccine for prevention of severe diarrheal illness of infants and young children. In: Immunology of milk and the neonate 1991 pp 255–264 (Mestecky J et al eds). Plenum Press, New York
83. Bernstein DI, Glass RI, Rodgers G, Davidson BL, Sack DA, US Rotavirus Vaccine Efficacy Group. Evaluation of rhesus rotavirus monovalent and tetravalent reassortant vaccines in US children. JAMA 1985; 273:1191–1196
84. Vesikari T, Ruuska T, Green KY, Flores J, Kapikian AZ. Protective efficacy against serotype 1 rotavirus diarrhea by live oral rhesus-human reassortant rotavirus vaccines with human rotavirus serotype 1 or 2 specificity. Pediatr Infect Dis J 1992;11:535–542
85. Rennels MB, Glass RI, Dennehy PH, Bernstein DI, Pichichero ME, Zito ET, Mack ME, Davidson BL, Kapikian AZ for the United States otavirus Efficacy Group. Safety and efficacy of high-dose rhesus-human reassortant rotavirus vaccines-report of the national multicenter trial. Pediatrics 1996; 97: 7–13
86. Joensuu J, Koskenniemi E, Pang X-L, Vesikari T. Randomised placebo-controlled trial of rhesus-human reassortant rotavirus vaccine for prevention of severe gastroenteritis. Lancet 1997;350:1205–1209
87. Perez-Schael I, Guntinas MJ, Perez M, Pagone V, Rojas AM, Gonzalez R, Cunto W, Hoshino Y, Kapikian AZ. Efficacy of the rhesus rotavirus-based quadrivalent vaccine in infants and young children in Venezuela N. Eng. J. Med 1997;337: 1181–1187

88. Santosham M, Moulton LH, Reid R, Croll J, Weatherbolt R, Ward R, Forro J, Zito E, Mack M, Brenneman G, Davidson BL. Efficacy and safety of high-dose rhesus-human reassortant vaccine in Native American populations. J Pediatr 1997;131: 632–638
89. Centers for Disease Control and Prevention. Rotavirus vaccine for the prevention of rotavirus gastroenteritis among children: recommendation of the Advisory Committee on Immunization Practices (ACIP). MMWR 1999;48:577–581
90. Centers for Disease Control and Prevention. Intussusception among recipients of rotavirus vaccine: United States, 1998–1999. MMWR 1999;48:577–581
91. Centers for Disease Control and Prevention. Withdrawal of rotavirus vaccine recommendation. MMWR 1999;48:1007
92. Mato SP, Perrin K, Scardino D, Begue RE. Evaluation of rotavirus vaccine effectiveness in a pediatric group practice. Am J. Epidem 2002;156:1049–1055
93. Staat MA, Cortese MM, Bresee JS, Begue RE, Vitek C, Rhodes P, Zhang R, Gentsch J, Roberts NE, Jaeger JL, Wars R, Bernstein DI, Dennehy PH. Rhesus rotavirus vaccine effectiveness and factors associated with receipt of vaccine. Pediatr Infect Dis 2006; 25:1013–1016
94. Tate JE, Curns AT, Cortese MM, Weintraub ES, Hambidge S, Zangwill KM, Patel MM, Baggs JM, Parashar UD. Burden of acute gastroenteritis hospitalizations and emergency department visits in US children that is potentially preventable by rotavirus vaccination: a probe study using the now withdrawn RotaShield vaccine. Pediatrics 2009;123:744–749
95. Kapikian AZ. A rotavirus vaccine for preventing severe diarrhoea of infants and young children: development and withdrawal. In: Chadwick D, Goode JA, eds. Gastroenteritis viruses. Novartis Foundation Symposium 238. Chichester, England. John Wiley and Sons. 1001; 153–171
96. ACIP. Verbatim transcript of the ACIP meeting (Atlanta GA), 22 October 1999 Nancy Lee and Associates, Certified Cour Reporters 1999;3:1–174
97. Murphy BM, Morens DM, Simonsen L, Chanock RM, La Montagne JR, Kapikian AZ. Reappraisal of the association of intussusception with the licensed live rotavirus vaccine challenges initial conclusions J. Infect Dis 2003;187: 1301–1308
98. Murphy TV, Gargiullo PM, Massoudi MS et al. Intussusception among infants given an oral rotavirus vaccine. N Engl J Med 2001;344:564–572
99. Murphy TV, Gargiullo PM, Massoudi MS, et al. Corrcctions: intussusception among infants given an oral rotavirus vaccine. N Engl Med 2001;344:1564
100. Kramarz P, France EK, Destefano F, et al. Population-based study of rotavirus vaccination and intussusception. Pediatr Infect Dis J 2001;20:410–416
101. Simonsen L, Morens D, Elixhauser A, Gerber M,Van Raden M, Blackwelder W. Effect of rotavirus vaccination programme on trends in admission of patients to hospital for intussusception. Lancet 2001;358;1224–1229
102. Hall AJ. Ecological studies and debate on rotavirus vaccine and intussusception. Lancet 2001;358:1197–1198 (Commentary)
103. Kapikian AZ. Ecological studies, rotavirus vaccination, and intussusception. Lancet 2002;359:1065–1066 (letter)
104. Hall AJ. Ecological studies, rotavirus vaccination, and intussusception. Lancet 2002;359:1066 (letter)
105. Simonsen L, Morens DM, Blackwelder WC. Ecological studies, rotavirus vaccination, and intussusception. Lancet 2002;359:1066–1067 (letter)
106. Murphy TV, Gargiullo PM, Wharton M. More on Rotavirus Vaccination and Inussusception. N. Engl. J Med 2002;346: 211–212 (letter)
107. Centers for Disease Control and Prevention Advisory Committee on Immunization Practice. Records of the meeting held on February 20–21, 2002, Atlanta Marriott North Central Hotel, Atlanta
108. Gay N, Ramsay M, Waight P. Rotavirus vaccination and intussusception. Lancet 1999;354: 956 (Letter)
109. Pang LC. Intussusception revisited: clinicopathologic analysis of 261 cases, with emphasis on pathogenesis. South. Med J 1989; 82; 215–228

110. Simonsen L, Viboud C, Elixhauser A, Taylor RJ, Kapikian AZ. More on RotaShield and intussusception: the role of age at the time of vaccination. L. Infect Dis 2005;192 (Suppl 1): 36–42
111. Kapikian AZ, Simonsen L, Vesikaari T, Hoshino Y, Morens DM, Chanock RM, La Montagne JR, Murphy BM. A hexavalent human rotavirus-bovine rotavirus (UK) reassortant vaccine designed for use in developing countries and delivered in a schedule with the potential to eliminate the risk of intussusception J. Infect Dis 2005;192 (suppl):22–29
112. Rothman KJ, Young-Xu Y, Arellano F. Age dependence of the relation between reassortant rotavirus vaccine (RotaShield) and intussusception. J. Infect Dis 2006: 193:898 (Correspondence)
113. Rothman KJ, Lee L. Rotashield and intussusception: reanalysis of data demonstrates reduced risk. Chinese J Vaccines and Immunization 2005;11:71–73 (Correspondence)
114. Simonsen L, Taylor RJ, Elixhauser A, Viboud C, Kapikian AZ. Reply to Rothman et al. J. Infect Dis 2006; 193:898–899 (Correspondence)
115. Garguillo PM, Murphy TV, Davis RL. Is there a safe age for vaccinating infants with tetravalent rhesus-human reassortant rotavirus vaccine? J. Infect Dis 2006;194:1793–1794 (Correspondence)
116. Rothman KJ, Arellano FM, Young-Xu Y, Simonsen L, Gargiullo et al. J infect Dis 2006; 194;1794–1795
117. WHO Global Advisory Committee on Vaccine Safety 1-2 December 2005. Rotavirus vaccine safety. Wkly Epid Rec 2006; No 2 vol 81;16–17
118. Simonsen, L. Unpublished Studies
119. WHO Position paper Rotavirus Vaccines. Wkly Epid Rep 2007; no 32 vol 82;285–296
120. Perrin PS, Lindsay FC. Intussusception: a monograph based on 400 cases. Brit J. Surg 1921; 9:46–71
121. Chen YE, Beasley S, Grimwood K. Intussusception and rotavirus associated hospitalization in New Zealand. Arch Dis Cils. 2005; 10:1077–1081
122. Justice FA, Auldist AW, Bines JE, Intussusception: trends in clinical presentation and management. J. Gastroenterol Hepatol 2006;21:842–846
123. Tate JE, Simonsen L, Viboud C, Steiner C, Patel M, Curns AT, Parashar UD. Trends in intussusception hospitalizations among Us infants, 1993–2004: implications for monitoring the safety of the new rotavirus vaccination program. Pediatrics 2008;121: e1125–1132
124. Vesikari T, Matson DO, Dennehy P et al. Safety and efficacy of a pentavalent human-bovine (WC2) reassortant rotavirus vaccine N. Eng. J. Med 2006;354: 23–33
125. Simonsen L, Taylor RJ, Kapikian AZ. Rotavirus vaccines. N Engl J Med 2006;354:1748 (letter)
126. US FDA: FDA public Health Notification: Information on RotaTeq and Intussusception February 13, 2007
127. Haber P, Patel M, Izurieta HS, Baggs J, Gargiullo P, Weintraub E, Cortese M, Braun MM, Belongia EA, Miller E, Ball R, Iskander J, Parashr U. Pediatrics 2008;121:1206–1212
128. Ruiz-Palacios GM, Perez-Schael I, Velazquez FR et al. Safety and efficacy of an attenuated vaccine against rotavirus gastroenteritis. No Engl J Med 2006;354:11–22
129. Bresee JS, El Arifeen S, Azim T, Chakraborty J, Mounts AW, Podder G, Gentsch JR, Ward RL, Black R, Glass RI, Yunus M. Safety and immunogenicity of tetravalent rhesus-based roavirus vaccine in Bangladesh. Ped Inf DIs J 2001;20:1136–1143
130. Weijer C. The future of research into rotavirus vaccine. Benefits of vaccine may outweigh risks for children in developing countries. BMJ 2000;321:525–526
131. Ruiz L, Personal Communication
132. Vesikari T, Karvonen A, Forrest BD, Hoshino Y, Chanock RM, Kapikian AZ. Neonata administration of rhesus rotavirus tetravalent vaccine. Pediatr Infect Dis J 2006;25:118–122
133. Hoshino Y, Jones RW, Ross J, Kapikian AZ. Construction and characterization of rhesus monkey rotavirus (MMU18006)- or bovine rotavirus (UK)-based serotype G5, G8, G9 or G10 single VP7 gene substitution reassortant candidate vaccines. Vaccine 2003;21:3003–3010
134. Fauci AS. The expanding global health agenda: a welcome development. Nature Med 2007;13;1169–1171

Rotavirus Vaccines Part II: Raising the Bar for Vaccine Safety Studies

Paul A. Offit and H. Fred Clark

Fred Clark and Paul Offit

Rotaviruses are one of the important causes of disease and death worldwide. For this reason, there has been a great deal of public and private interest in developing a vaccine. Unfortunately, because of a rare adverse event associated with the first marketed rotavirus vaccine, subsequent vaccines have been difficult to develop.

In 1998, a rotavirus vaccine (RotaShield, Wyeth) was licensed and distributed in the USA. Ten months after licensure, it was found to be a rare cause of intestinal blockage (intussusception) affecting 1 per 10,000 vaccine recipients. When this rare adverse event was discovered, the Centers for Disease Control and Prevention (CDC) withdrew its recommendation for use. The problem with the first rotavirus

P.A. Offit (✉)
Children's Hospital of Philadelphia, Philadelphia, PA 19104, USA
e-mail: offit@email.chop.edu

S.A. Plotkin (ed.), *History of Vaccine Development*,
DOI 10.1007/978-1-4419-1339-5_29, © Springer Science+Business Media, LLC 2011

vaccine did not end subsequent attempts to make a safer vaccine; but it did dramatically increased the size of subsequent pre-licensure trials. Two new rotavirus vaccines (RotaTeq [Merck] and Rotarix [GlaxoSmithKline]) have each been tested for safety and efficacy in clinical trials of more than 60,000 infants. In February 2006, RotaTeq was licensed by the Food and Drug Administration (FDA) for use in infants and subsequently recommended by the Advisory Committee on Immunization Practices (ACIP) to the CDC. In June 2008, Rotarix was also licensed by the FDA and recommended by the ACIP. This review provides a description of these three vaccines.

The Burden of Rotavirus

The impetus to develop rotavirus vaccines was based on the enormous impact of the virus. Rotaviruses are the leading cause of severe dehydrating diarrhea in infants and young children throughout the world (Table 1). Virtually all children are infected by the time they are 2–3 years old [1, 2]. In developed nations, where standards of hygiene and sanitation are high, rotavirus is the most common cause of severe infant diarrhea [3].

Prior to the development of rotavirus vaccines, in the USA, rotavirus accounted for 2.7 million illness episodes, 500,000 physician visits, 55,000–70,000 hospitalizations, and 20–60 deaths every year [4, 5]. As a consequence, the economic burden of disease was high, estimated at approximately $1 billion each year in direct medical costs as well as indirect costs, such as time lost from work [4, 6, 7].

In less developed countries, rotaviruses are also a common cause of severe gastroenteritis in children and a common cause of mortality [8–10]. Prior to the worldwide use of current rotavirus vaccines, the virus caused 450,000–600,000 deaths in children each year – about 2,000 deaths every day [11–15].

Table 1 Burden of rotavirus disease in the USA and worldwide before the introduction of rotavirus vaccines

	United States		Worldwide	
Parameter	Total	Risk per child	Total	Risk per child
Births	3.9 million	–	130 million	–
Rotavirus gastroenteritis	2.7 million	1 in 1.4[a]	111 million	1 in 1.2
Physician, ED visits	600,000	1 in 6.5	2.5 million	1 in 5
Hospitalizations	55,000–70,000	1 in 70	2 million	1 in 6.5
Moderate-to-severe disease	–	–	16 million	1 in 8
Deaths	20–60	1 in 100,000	450,000–600,000	1 in 250
Medical costs	$400 million	–	–	–
Indirect and direct costs	$ 1 billion	–	–	–

ED emergency department

[a]Meaning that 1 of every 1.4 children born in the USA will have at least one episode of symptomatic rotavirus infection

The Disease

Rotavirus causes the sudden onset of watery diarrhea, fever, and vomiting [16–19]. Most rotavirus diseases are mild, but about 1 of every 75 children develops severe dehydration [4, 5]. In children admitted to the hospital with dehydration, fever and vomiting usually persist for 2–3 days and diarrhea persists for 4–5 days [16–19].

The Virus

Rotaviruses are a genus within the family Reoviridae [20]. The virus is composed of three shells (an outer and inner capsid and a core) that surround 11 segments of double-stranded RNA [20]. For the most part, each gene segment codes for a single protein.

Rotavirus contains two proteins, vp4 and vp7, on its outer capsid. Both surface proteins induce serotype-specific neutralizing antibodies as well as cross-reactive neutralizing antibodies [21–27]. VP7 (viral protein 7) is glycosylated and serotypes determined by this protein are termed G types [20]. VP4 is cleaved by the protease trypsin and serotypes determined by this protein are termed P types [20]. The major human rotavirus P types have been characterized by both serotype and genotype [28]. A complete rotavirus strain is described using a number (or a number and letter) for the P serotype, followed by a number in brackets that represents the P genotype. The most common human serotypes are listed in Table 2.

Hope for a Vaccine: Natural Infection Protects Against Disease Following Reinfection

In 1983, Ruth Bishop et al. showed the importance of immunity in protection against subsequent rotavirus disease [29]. Bishop found that neonates infected during the first month of life were not protected against rotavirus reinfection but were protected against moderate-to-severe disease following reinfection. Conversely, neonates not infected with rotavirus during the first month of life were fully susceptible to diarrheal disease associated with the first rotavirus infection. Since then, studies of neonates have been extended to infants and young children, which have also shown that the first infections protect against severe disease following reinfection [30–32].

Table 2 Common human rotavirus serotypes worldwide, 2009

VP4 serotype [genotypes]	Associated VP7 types
P1A[8]	G1, G3, G4, G9
P1B[4]	G2
P2[6]	G9

Although natural rotavirus infection protects against moderate-to-severe disease caused by reinfection, some children experience repeated episodes of diarrhea with the same serotype during the following rotavirus season [29, 33–43], and a small number of children develop symptomatic rotavirus infection twice within the same season [44]. These observations are consistent with the fact that effector functions at mucosal surfaces, such as production of virus-specific secretory IgA (sIgA), are usually short lived and that rotavirus-specific sIgA often is not detected at the intestinal mucosal surface 1 year after symptomatic infection [44, 45]. While large quantities of virus-specific sIgA at the intestinal mucosal surface at the time or reexposure can completely protect against the disease, modification of the severity of rotavirus disease caused by reinfection is most likely mediated by production of virus-specific sIgA by memory rotavirus-specific B cells in the intestinal lamina propria [46]. Because activation and differentiation of memory B cells to antibody-producing plasma cells take several days, modification of disease severity, not complete protection against disease, is the outcome.

The first rotavirus vaccines were made using the same approach as that used by Edward Jenner. Jenner had found that an animal strain of smallpox (cowpox) was similar enough to human smallpox to induce protective immunity but dissimilar enough so that it did not induce the disease. Unfortunately, the Jennerian approach to making a rotavirus vaccine was, in the end, disappointing.

A Monkey Rotavirus

Rhesus rotavirus (RRV) was isolated from a young monkey with diarrhea in California and tested as a vaccine candidate after 16 passages in cell culture [47]. RRV was similar to human G3 strains but contained a P type distinct from human viruses. Dr. Albert Kapikian at the National Institutes of Health (NIH) directed studies of RRV and subsequently RRV x human rotavirus reassortant vaccines. Kapikian's team consisted of Drs. Harry Greenberg, Yatake Hoshino, Jorge Flores, Richard Wyatt, Karen Midthun, and Roger Glass.

Although the RRV vaccine was safe and immunogenic in children, clinical efficacy varied. The greatest efficacy of RRV was observed in Venezuela. RRV vaccine protected 65% of children from all rotavirus disease and 100% of children from severe disease [48]. Trials in Finland and Sweden showed modest protection (38 and 48%, respectively) against all rotavirus diarrhea but greater efficacy against severe rotavirus diarrhea (67 and 80%, respectively) [49, 50]. Unfortunately, in three trials in the USA, no protection was observed in New York or Arizona [51, 52] and only 29% protection was found in Maryland: [53] the predominant challenge strains in each of these trials were G1.

Trials with RRV showed that protection against the disease could be induced against a challenge strain that was serotypically distinct from the vaccine (heterotypic protection). But heterotypic protection was at best inconsistent and investigators abandoned RRV as a vaccine candidate.

A Cow Rotavirus

At the same time that researchers at the National Institutes of Health (NIH) were evaluating a simian rotavirus as a vaccine candidate, Drs. Fred Clark, Stanley Plotkin, and Paul Offit at The Children's Hospital of Children were evaluating a calf rotavirus strain. This strain was isolated from a calf with diarrhea in Chester County, PA, in 1981 and serially passaged 12 times in African green monkey kidney cells at the Wistar Institute; the strain was called Wistar Calf 3 (WC3) [54]. WC3 did not share either a P or G type with human rotavirus strains. In an initial double-blinded, placebo-controlled, efficacy trial performed in suburban Philadelphia, WC3 vaccine caused a 76% reduction in rotavirus morbidity and 100% protection against moderate-to-severe rotavirus diarrhea [55]. However, in subsequent efficacy trials conducted in Cincinnati and Bangui, Central African Republic, there was little protection against the rotavirus disease [56, 57]. Because (similar to RRV) heterotypic protection afforded by WC3 was inconsistent, it, also was eliminated as a vaccine candidate.

Reassortant Rotaviruses: Combination of Animal and Human Rotavirus Strains

Following disappointing results with RRV and WC3, researchers focused efforts on reassortant rotaviruses – combination viruses that would express rotavirus proteins responsible for inducing protective immune responses but not rotavirus proteins that conferred virulence. The reassortant approach was used for several reasons. First, for the most part, each rotavirus gene segment codes for a single rotavirus protein. Second, rotavirus gene segments can be separated easily on the basis of their molecular weight by polyacrylamide gel electrophoresis. Third, when mixed infections with rotavirus strains occur under experimental conditions, gene segments reassort independently, producing viruses of mixed parentage. Fourth, whereas researchers determined that two rotavirus genes (those coding for vp4 and vp7) each independently evoked neutralizing and protective antibodies, four genes were necessary to confer virulence [58–61]. This meant that reassortant viruses could be made that retained the attenuated virulence characteristics of animal rotavirus strains (RRV and WC3) while at the same time included human rotavirus genes responsible for protective immune responses.

Simian–Human Reassortants: RotaShield

RotaShield, the first reassortant rotavirus vaccine candidate, consisted of simian–human rotavirus reassortants that contained a single human gene (i.e., coding for human vp7) and the remaining ten genes from RRV [62]. As was true for RRV, these RRV–human reassortant viruses were prepared by Kapikian and his team at NIH.

RotaShield, produced by Wyeth-Lederle, contained three simian–human reassortants of G types 1, 2, and 4 on an RRV background. The vaccine also contained RRV alone, which is similar but not identical to human G3 rotaviruses [63, 64]. Field trials of the quadrivalent RRV–human reassortant vaccine were performed using a dose of 1×10^5 plaque-forming units (PFU) per strain (4×10^5 PFU total) [62]. In each of these trials, vaccine was administered orally in three doses at 2, 4, and 6 months of age. Efficacy induced by RotaShield vaccine ranged from 48 to 68% against any rotavirus disease to 64 to 91% against severe disease. Further, the efficacy reported in Venezuela (48% for mild disease and 88% for severe disease) was not significantly different from that reported in the multi-center trial in the USA (57 and 82%), suggesting that Rotashield would likely work well in developing countries.

Prior to licensure, intussusception was found in 5 of 11,000 children who received RotaShield compared with 1 of 4,500 children who received placebo [65]. Intussusception was not found after the first dose in any child but was observed within 7 days after receiving the second or third dose of vaccine in three of the five affected children. Although the incidence of intussusception in vaccine recipients was not greater than estimated background rates, both the CDC and the American Academy of Pediatrics (AAP) warned in their recommendations for use of RotaShield that intussusception might be a consequence of vaccination [65, 66]. In addition, the possible relationship between the vaccine and intussusception was noted in the product insert.

RotaShield was licensed for universal use for infants in the USA in August 1998. In July 1999, after the RRV–human reassortant vaccine had been given to about one million children, 15 cases of intussusception were reported to the vaccine adverse event reporting system (VAERS) [67, 68]. These cases were worrisome because almost all occurred after the first dose, within 1 week of receipt of vaccine, and in very young infants between 2 and 3 months of age (typically, intussusception occurs in older infants 5–9 months of age). For these reasons, the vaccine was temporarily suspended pending results of a case-controlled analysis by the CDC [67].

In August 1999, the CDC found that the relative risk of intussusception within 1 week of receipt of the first or second dose of RotaShield was 37 ($P<0.001$) and 3.8 ($P=0.05$), respectively (Table 3) [69]. Using case–control series and case–control analysis, the attributable risk was estimated to be about 1 case of intussusception per 10,000 immunized children [69, 70]. In October 1999 – after the relationship between the vaccine and intussusception was determined to be causal, not coincidental – the CDC withdrew its recommendation to use RotaShield. Subsequent ecologic studies using hospital discharge diagnoses estimated that the attributable risk of intussusception following administration of RotaShield might have been smaller than the initial estimates [71, 72].

Most unusual about intussusception as a consequence of RotaShield is that intussusception does not appear to be a consequence of natural infection [73]. And wild-type rotaviruses replicate at the intestinal mucosal surface of infants far better than the vaccine virus strains contained in RotaShield. Hence, the etiology of intussusception following RotaShield vaccine remains unclear. The most likely explanation (the "unique strain" hypothesis) is that RRV or RRV–human reassortants might be taken up at an intestinal site or processed by antigen-presenting cells in a manner different from

Table 3 Risk of intussusception following administration of RotaShield

Dose	Risk period (days)	Relative risk (95% CI)	*P* value
All	3–7	14.4 (7.0–29.6)	<0.001
	8–14	5.3 (2.1–13.9)	0.001
	15–21	1.1 (0.3–3.3)	0.91
First	3–7	37.2 (12.6–110.1)	<0.001
	8–14	8.2 (2.4–27.6)	0.001
	15–21	1.1 (0.2–5.4)	0.87
Second	3–7	3.8 (1.0–14.0)	0.05
	8–14	1.8 (0.4–9.5)	0.47
	15–21	0.9 (0.1–8.6)	0.94

Adapted from Murphy et al. [69]
CI confidence interval

that found after natural infection. Antigen-presenting cells involved following vaccine administration may then produce a panel of cytokines different from those induced by natural infection. Indeed, blockage of specific cytokines has been found to be associated with ablation of transient intussusception in experimental animals [74, 75]. The most likely candidate for the "unique strain" contained in the vaccine is RRV. Several biologic features of RRV are unique: (1) RRV is one of the few rotavirus strains that cause diarrhea in a number of species [76]; (2) RRV is the only known rotavirus strain that causes severe and occasionally fatal hepatitis when orally inoculated into immunodeficient and immunocompetent strains of inbred mice [77]; and (3) RRV invades gut-associated lymphoid tissue in mice to a greater extent than RRV×human or WC3×human reassortant viruses [78]. Which, if any, of these unique biologic features are predictive of intussusception in children remains to be determined. In addition, RRV is shed at quantities greater than that found for RRV–human reassortant viruses after the first and second doses of RotaShield; shedding of RRV parallels the increased risk of intussusception observed after the first and second doses [79].

Although RotaShield was withdrawn from use, several investigators argued, reasonably, that given that US children were far more likely to be hospitalized and killed by natural infection than by immunization, RotaShield was still of benefit. More tragic was the loss of RotaShield for use in the developing world. When the CDC withdrew its recommendation, countries in the developing world were also reluctant to use the vaccine, even though the benefit-to-risk ratio was dramatically different than that in the developed world. Fear of using RotaShield vaccine in the developing world was evident at a World Health Organization (WHO) meeting in Geneva in February 2000, 4 months after the CDC withdrew its recommendation for RotaShield for use in American children. It was a heartbreaking moment as representatives from one developing country after another stood up and declared that if RotaShield was unsafe for American children, then it was also unsafe for their children, even though the risk–benefit ratio for the vaccine in the developing world was dramatically different than that in the USA. As a consequence, a technology that had the capacity to save as many as 2,000 lives a day sat on the shelf. Seven years would pass before the next generation of rotavirus vaccines was available to save lives in the developing world.

Bovine–Human Reassortants: RotaTeq

RotaTeq (Merck and Co.) was licensed by the FDA and recommended by the CDC for universal use in infants in February 2006. Like RotaShield, RotaTeq was recommended to be given by mouth to infants at 2, 4, and 6 months of age.

RotaTeq is a live, oral vaccine that contains five reassortant rotaviruses developed from human strains and the bovine strain WC3 [80]. Four of the bovine (WC3)–human reassortant rotaviruses express VP7 (from human serotypes G1, G2, G3, or G4) and VP4 from bovine strain WC3. The fifth reassortant virus contains VP4 (P1A[8]) from a human rotavirus strain and VP7 (G6) from WC3.

The efficacy of RotaTeq against all rotavirus gastroenteritis was also evaluated in a large, placebo-controlled study of more than 70,000 infants [81]. Clark, Offit, and Plotkin had constructed the viruses that comprised RotaTeq, but the key moment in the development of this vaccine came after RotaShield was found to cause intussusception. At this point a line was crossed. Now the FDA wanted companies to prove that the vaccine did not cause a rare adverse reaction (intussusception) *pre-licensure*. Prior to the RotaShield problem such adverse events had been the province of post-licensure studies. Much credit for the decision to perform the kind of study the FDA required (a study that involved more than 70,000 children from 11 countries and cost about $350 million – the largest vaccine safety trial ever performed by a pharmaceutical company) goes to Dr. Edward Scolnick, president of Merck Research Laboratories at the time. Without his support for this massive undertaking, the first licensed rotavirus vaccine following RotaShield would never have been made. Merck's trial showed efficacy against rotavirus disease of any severity of 74.0% and efficacy against severe rotavirus disease of 98%. RotaTeq reduced the rate of rotavirus-associated hospitalizations by 96%, emergency visits by 94%, and office visits by 87%. Efficacy against all gastroenteritis hospitalizations of any etiology was 59%, proving the importance of rotavirus as a cause of severe gastroenteritis in children. The efficacy of RotaTeq in the second rotavirus season following vaccination was 63% against rotavirus gastroenteritis of any severity and 88% against severe rotavirus gastroenteritis.

In the large Phase III efficacy trial, 1 case of intussusception occurred in the vaccine group and 1 in the placebo group within 14 days of inoculation; 6 cases in the vaccine group and 5 in the placebo group within 42 days of inoculation; and 12 cases in the vaccine group and 15 in the placebo group within 1 year of inoculation. No clustering of cases in the RotaTeq group occurred at any time after vaccination [81]. Therefore, RotaTeq did not appear to cause or prevent intussusception, reaffirming the contention that natural rotavirus infection was not a cause of intussusception. Post-licensure monitoring has confirmed that intussusception is not a consequence of RotaTeq vaccination [82].

In February 2006, the ACIP recommended routine immunization of US infants with three doses of RotaTeq to be administered by mouth at 2, 4, and 6 months of age [80]. Since licensure, RotaTeq has caused an approximate 80% decrease in the incidence of moderate-to-severe rotavirus disease in the USA [83]. This degree of protection is disproportionate to the percentage of young children immunized, which is about 50%; this finding is consistent with a significant degree of herd immunity. RotaTeq has been used for approximately 3 years in Nicaragua and is undergoing clinical trial testing in Bangladesh, Vietnam, Ghana, Mali, and Kenya; initial results show substantial protection against severe rotavirus disease [84], but unfortunately lower efficacy than in the U.S.

Attenuated Human Rotavirus: RotaRix

Whereas the first two rotavirus vaccines, RotaShield and RotaTeq, used animal rotavirus strains to attenuate virulence, the next vaccine, RotaRix, used a method first employed by Max Theiler in the development of the yellow fever vaccine: attenuation by serial passage in non-human cells. RotaRix, which consists of a single human rotavirus strain (P1A[8]G1) was licensed by the FDA and recommended for use in all children in the USA in June 2008. The vaccine is now used in more than 100 other countries. The vaccine is given by mouth in two doses at 6–14 weeks and 14–24 weeks of age.

The vaccine virus contained in RotaRix was born of a natural rotavirus outbreak in Cincinnati. Symptomatic or asymptomatic infection of young children with a single circulating wild-type strain in Cincinnati (strain 89-12) was found to provide 100% protection against subsequent rotavirus disease [30]. A vaccine using this strain was developed after serial passage in African Green Monkey Kidney (AGMK) cells. The further-attenuated strain was called RIX4414 and later, RotaRix.

Again, with an interest in ruling out intussusception as a consequence of vaccination, a randomized, placebo-controlled trial of RotaRix was performed in more than 63,000 infants, primarily in 11 countries in Latin America [85]. The efficacy of RotaRix against severe G1 rotavirus disease was 92% while that against non-G1 strains (G3, G4, and G9) that belonged to the same P1A[8] type as the vaccine was 87%. (Although protection was only 41% against G2P[4] strains in this trial, a smaller trial of RotaRix in Europe showed protection against G2 viruses [86].) Like RotaTeq, RotaRix also did not cause intussusception. A total of 13 cases of intussusception were identified during the 31-day window, and seven of these were in placebo recipients. No clustering in the initial 1–2 weeks was identified after either dose. During the entire study period, 16 cases of intussusception occurred in the placebo group and 9 in the vaccine group.

Conclusion

The experience with the first rotavirus vaccine, RotaShield, necessitated Phase III trials of subsequent vaccines that were larger than any other safety trial ever performed by pharmaceutical companies. The interest in ruling out relatively rare adverse events pre-licensure has not been limited to rotavirus vaccines. As a consequence of the rotavirus vaccine experience, much larger safety trials are now performed for all vaccines.

References

1. de Zoysa I, Feachem RG. Interventions for the control of diarrhoeal diseases among young children: rotavirus and cholera immunization. Bull World Health Organ 63:569–583, 1985.
2. Kapikian AZ, Hoshino Y, Chanock RM. Rotaviruses. In Fields BN, Knipe DM, Howley PM (eds). Fields Virology (4th ed). Vol. 2. Philadelphia, Lippincott–Raven, 2001, pp 1787–1833.
3. Brandt CD, Kim HW, Rodriguez JO, et al. Pediatric viral gastroenteritis during eight years of study. J Clin Microbiol 18:71–78, 1983.
4. Glass RI, Kilgore PE, Holman RC, et al. The epidemiology of rotavirus diarrhea in the United States: surveillance and estimates of disease burden. J Infect Dis 174(suppl 1):S5–S11, 1996.
5. Kilgore PE, Holman RC, Clarke MJ, Glass RI. Trends of diarrheal disease–associated mortality in U.S. children, 1968 through 1991. JAMA 274:1143–1148, 1995.
6. Jin S, Kilgore PK, Holman RC, et al. Trends in hospitalizations for diarrhea in United States children from 1979–1992: estimates of the morbidity associated with rotavirus. Pediatr Infect Dis J 15:397–404, 1996.
7. Smith J, Haddix A, Teutsch S, Glass RI. Cost effectiveness analysis of a rotavirus immunization program for the United States. Pediatrics 96:609–615, 1995.
8. Huilan S, Zhen LG, Mathan MM, et al. Etiology of acute diarrhoea among children in developing countries: a multicentre study in five countries. Bull World Health Organ 69:549–555, 1991.
9. Levine MM, Losonsky G, Herrington D, et al. Pediatric diarrhea: the challenge of prevention. Pediatr Infect Dis 5(suppl):S29–S43, 1986.
10. Cook SM, Glass RI, LeBaron CW, Ho M-S. Global seasonality of rotavirus infections. Bull World Health Organ 68:171–177, 1990.
11. Bern C, Martines J, de Zoysa I, Glass RI. The magnitude of the global problem of diarrhoeal disease: a ten-year update. Bull World Health Organ 70:705–714, 1992.
12. Murray CJ, Lopez AD. Global mortality, disability, and the contribution of risk factors: Global Burden of Disease Study. Lancet 349:1436–1442, 1997.
13. Walsh JA, Warren KS. Selective primary health care: an interim strategy for disease control in developing countries. N Engl J Med 301:967–974, 1979.
14. Snyder JD, Merson MH. The magnitude of the global problem of acute diarrhoeal disease: a review of active surveillance data. Bull World Health Organ 60:605–613, 1982.
15. Parashar UD, Hummelman EG, Bresee JS, Miller MA, Glass RI. Global illness and deaths caused by rotavirus disease in children. Emerg Infect Dis 2003;9:565-572.
16. Tallett S, MacKenzie C, Middleton P, et al. Clinical, laboratory, and epidemiologic features of a viral gastroenteritis in infants and children. Pediatrics 60:217–222, 1977.
17. Carr M, McKendrick D, Spyridakis T. The clinical features of infantile gastroenteritis due to rotavirus. Scand J Infect Dis 8:241–243, 1978.
18. Kovacs A, Chan L, Hotrakitya C, et al. Rotavirus gastroenteritis: clinical and laboratory features and use of the Rotazyme test. Am J Dis Child 141:161–166, 1987.
19. Rodriguez W, Kim H, Arrobio J, et al. Clinical features of acute gastroenteritis associated with human reovirus-like agent in infants and young children. J Pediatr 91:188–193, 1977.
20. Estes M. Rotaviruses and their replication. In Fields BN, Knipe DM, Howley PM (eds). Fields Virology (4th ed). Vol. 2. Philadelphia, Lippincott–Raven, 2001, pp 1747–1786.
21. Larralde G, Li B, Kapikian AZ, Gorziglia M. Serotype-specific epitopes present on the VP 8 subunit of rotavirus VP 4 protein. J Virol 65:3213–3218, 1991.
22. Gorziglia MKY, Green K, Nishikawa K, et al. Sequence of the fourth gene of human rotaviruses recovered from asymptomatic or symptomatic infections. J Virol 62:2979–2984, 1988.
23. Hoshino Y, Wyatt RG, Greenberg HB. Serotypic similarity and diversity of rotaviruses of mammalian and avian origin as studied by plaque reduction neutralization. J Infect Dis 149:694–702, 1984.
24. Svensson L, Sheshbaradaran H, Visikari T, et al. Immune response to rotavirus polypeptides after vaccination with heterologous rotavirus vaccines (RIT 4237, RRV-1). J Gen Virol 68:1993–1999, 1987.

25. Ward RL, Knowlton DR, Schiff GM, et al. Relative concentrations of serum neutralizing antibody to VP 3 and VP 7 proteins in adults infected with a human rotavirus. J Virol 62:1543–1549, 1989.
26. Ward RL, Knowlton DR, Greenberg HG, et al. Serum-neutralizing antibody to VP 4 and VP 7 proteins in infants following vaccination with WC3 bovine rotavirus. J Virol 64:2687–2691, 1990.
27. Matsui S, Mackow E, Greenberg H. Molecular determinants of rotavirus neutralization and protection. Adv Virus Res 36:181–214, 1989.
28. Estes MK, Cohen J. Rotavirus gene structure and function. Microbiol Rev 53:410–449, 1989.
29. Bishop R, Barnes G, Cipriani E, Lund J. Clinical immunity after neonatal rotavirus infection: a prospective longitudinal study in young children. N Engl J Med 309:72–76, 1983.
30. Bernstein DI, Sander DS, Smith VE, et al. Protection from rotavirus reinfection: 2-year prospective study. J Infect Dis 164:277–283, 1991.
31. Ward RL, Bernstein D. Protection against rotavirus disease after natural infection. J Infect Dis 169:900–904, 1994.
32. Fisher TK, Valentiner-Branth P, Steinsland H, et al. Protective immunity after natural rotavirus infection: a community cohort study of newborn children in Guinea-Bissau, West Africa. J Infect Dis 186: 593–597, 2002.
33. Yolken R, Wyatt R, Zissis G. Epidemiology of human rotavirus types 1 and 2 as studied by enzyme-linked immunosorbent assay. N Engl J Med 299:1156–1161, 1978.
34. Black R, Greenberg H, Kapikian A, et al. Acquisition of serum antibody to Norwalk virus and rotavirus in relation to diarrhea in a longitudinal study of young children in rural Bangladesh. J Infect Dis 145:483–489, 1982.
35. Mata L, Simhon A, Urratia J, et al. Epidemiology of rotaviruses in a cohort of 45 Guatemalan Mayan Indian children observed from birth to the age of three years. J Infect Dis 148: 452–461, 1983.
36. Chiba S, Nakata S, Urasawa T, et al. Protective effect of naturally acquired homotypic and heterotypic rotavirus antibodies. Lancet 1:417–421, 1986.
37. Linhares A, Gabbay Y, Mascarenhas J, et al. Epidemiology of rotavirus subgroups and serotypes in Belem, Brazil: a three-year study. Ann Inst Pasteur (Virol) 139:89–99, 1988.
38. Georges-Courbot M, Monges J, Beraud-Cassel A, et al. Prospective longitudinal study of rotavirus infections in children from birth to two years of age in Central Africa. Ann Inst Pasteur (Virol) 139:421–428, 1988.
39. Friedman M, Gaul A, Sarov B, et al. Two sequential outbreaks of rotavirus gastroenteritis: evidence for symptomatic and asymptomatic reinfection. J Infect Dis 158:814–822, 1988.
40. Grinstein S, Gomez J, Bercovich J, Biscorn E. Epidemiology of rotavirus infection and gastroenteritis in prospectively monitored Argentine families with young children. Am J Epidemiol 130:300–308, 1989.
41. Reves R, Hossain M, Midthun K, et al. An observational study of naturally acquired immunity in a cohort of 363 Egyptian children. Am J Epidemiol 130:981–988, 1989.
42. O'Ryan M, Matson D, Estes M, et al. Molecular epidemiology of rotavirus in young children attending day care centers in Houston. J Infect Dis 162:810–816, 1990.
43. DeChamps C, Laveran H, Peigue-Lafeville J, et al. Sequential rotavirus infections: characterization of serotypes and electropherotypes. Res Virol 142:39–45, 1991.
44. Matson DO, O'Ryan ML, Herrera I, et al. Fecal antibody responses to symptomatic and asymptomatic rotavirus infections. J Infect Dis 167:577–583, 1993.
45. Coulson B, Grimwood K, Hudson I, et al. Role of coproantibody in clinical protection of children during reinfection with rotavirus. J Clin Microbiol 30:1678–1684, 1992.
46. Moser CA, Coffin SE, Cookinham S, Offit PA. Relative importance of rotavirus-specific effector and memory B cell responses in protection against challenge. J Virol 72:1108–1114, 1998.
47. Kapikian AZ, Midthun K, Hoshino Y, et al. Rhesus rotavirus: a candidate vaccine for prevention of human rotavirus disease. In Lerner RA, Chanock RM, Brown F (eds). Molecular and Chemical Basis of Resistance to Parasitic, Bacterial, and Viral Diseases. Cold Spring Harbor, NY, Cold Spring Harbor Laboratory Press, 1985, pp 357–367.

48. Flores J, Perez-Schael I, Gonzalez M, et al. Protection against severe rotavirus diarrhea by rhesus rotavirus vaccine in Venezuelan children. Lancet 1:882–884, 1987.
49. Vesikari T, Rautanen T, Varis T, et al. Rhesus rotavirus candidate vaccine: clinical trial in children vaccinated between 2 and 5 months of age. Am J Dis Child 144:285–289, 1990.
50. Gothefors L, Wadell G, Juto P, et al. Prolonged efficacy of rhesus rotavirus vaccine in Swedish children. J Infect Dis 159:753–757, 1989.
51. Santosham M, Letson GW, Wolff M, et al. A field study of the safety and efficacy of two candidate rotavirus vaccines in a Native American population. J Infect Dis 163:483–487, 1991.
52. Christy C, Madore HP, Pichichero ME, et al. Field trial of rhesus rotavirus vaccine in infants. Pediatr Infect Dis J 7:645–650, 1988.
53. Rennels MB, Losonsky GA, Young AE, et al. An efficacy trial of the rhesus rotavirus vaccine in Maryland. Am J Dis Child 144:601–604, 1990.
54. Clark HF, Furukawa T, Bell LM, et al. Immune response of infants and children to low-passage bovine rotavirus (strain WC3). Am J Dis Child 140:350–356, 1986.
55. Clark HF, Borian FE, Bell LM, et al. Protective effect of WC3 vaccine against rotavirus diarrhea in infants during a predominantly serotype 1 rotavirus season. J Infect Dis 158:570–587, 1988.
56. Bernstein DI, Smith VE, Sander DS, et al. Evaluation of WC3 rotavirus vaccine and correlates of protection in healthy infants. J Infect Dis 162:1055–1062, 1990.
57. Georges-Courbot MC, Monges J, Siopathis MR, et al. Evaluation of the efficacy of a low-passage bovine rotavirus (strain WC3) vaccine in children in Central Africa. Res Virol 142:405–411, 1991.
58. Hoshino Y, Sereno MM, Midthun K, et al. Independent segregation of two antigenic specificities (VP3 and VP7) involved in neutralization of rotavirus infectivity. Proc Natl Acad Sci U S A 82:8701–8704, 1985.
59. Offit PA, Blavat G. Identification of the two rotavirus genes determining neutralization specificities. J Virol 57:376–378, 1986.
60. Offit PA, Clark HF, Blavat G, Greenberg HB. Reassortant rotaviruses containing structural proteins VP3 and VP7 from different parents induce antibodies protective against each parental serotype. J Virol 60:491–496, 1986.
61. Hoshino Y, Sereno MM, Kapikian AZ, et al. Genetic determinants of rotavirus virulence studied in gnotobiotic piglets. In Vaccines 93. Cold Spring Harbor, NY, Cold Spring Harbor Laboratory Press, 1993, pp 277–282.
62. Kapikian AZ, Hoshino Y, Chanock RM, Perez-Schael I. Efficacy of a quadrivalent rhesus rotavirus–based human rotavirus vaccine aimed at preventing severe rotavirus diarrhea in infants and young children. J Infect Dis 174(suppl 1):S65–S72, 1996.
63. Stuker G, Oshiro LS, Schmidt NH. Antigenic comparisons of two new rotaviruses from rhesus monkeys. J Clin Microbiol 11:202–203, 1980.
64. Nishikawa K, Hoshino Y, Taniguchi K, et al. Rotavirus v.p. 7 neutralization epitopes of serotype 3 strains. Virology 171:503–515, 1989.
65. Centers for Disease Control and Prevention. Rotavirus vaccine for the prevention of rotavirus gastroenteritis among children: recommendations of the Advisory Committee on Immunization Practices. MMWR 48:1–23, 1999.
66. Committee on Infectious Diseases, American Academy of Pediatrics. Prevention of rotavirus disease: guidelines for use of rotavirus vaccine. Pediatrics 102:1483–1491, 1998.
67. Centers for Disease Control and Prevention. Intussusception among recipients of rotavirus vaccine—United States, 1998–1999. MMWR 48:577–581, 1999.
68. Committee on Infectious Diseases, American Academy of Pediatrics. Possible association of intussusception with rotavirus vaccination. Pediatrics 104:575, 1999.
69. Murphy TV, Garguillo PM, Massoudi MS, et al. Intussusception among infants given an oral rotavirus vaccine. N Engl J Med 344:564–572, 2001.
70. Kramarz P, France EK, Destefano F, et al. Population-based study of rotavirus vaccination and intussusception. Pediatr Infect Dis J 20:410–416, 2001.

71. Chang H-G, Smith PF, Ackelsberg J, et al. Intussusception, rotavirus diarrhea, and rotavirus vaccine use among children in New York State. Pediatrics 108:54–60, 2001.
72. Simonsen L, Morens DM, Elixhauser A, et al. Effect of rotavirus vaccination programme on trends in admission of infants to hospital for intussusception. Lancet 358:1224–1229, 2001.
73. Chang E, Zangwill KM, Lee H, Ward JI. Lack of association between rotavirus infection and intussusception: implications for use of attenuated rotavirus vaccines. Pediatr Infect Dis J 21:97–102, 2002.
74. Lin Z, Cohen P, Nissan A, et al. Bacterial wall lipopolysaccharide as a cause of intussusception in mice. J Pediatr Gastroenterol Nutr 27:301–305, 1998.
75. Nissan A, Zhang J, Lin Z, et al. The contribution of inflammatory mediators and nitric oxide to lipopolysaccharide-induced intussusception in mice. J Surg Res 69:205–207, 1997.
76. Ciarlet M, Estes MK, Conner ME. Simian rhesus rotavirus is a unique heterologous (non-lapine) rotavirus strain capable of productive replication and horizontal transmission in rabbits. J Gen Virol 81:1237–1249, 2000.
77. Uhnoo I, Riepenhoff-Talty M, Dharakul T, et al. Extramucosal spread and development of hepatitis with rhesus rotavirus in immunodeficient and normal mice. J Virol 64:361–368, 1990.
78. Moser CA, Dolfi D, DiVietro ML, et al. Hypertrophy, hyperplasia, and infectious virus in gut-associated lymphoid tissue of mice after oral inoculation with simian-human or bovine-human reassortant rotavirus. J Infect Dis 2001;183:1108-1111.
79. Ward RL, Dinsmore AM, Goldbery G, et al. Shedding of rotavirus after administration of the tetravalent rhesus rotavirus vaccine. Pediatr Infect Dis J 17:386–390, 1998.
80. Centers for Disease Control and Prevention: Prevention of rotavirus gastroenteritis among infants and children: recommendations of the Advisory Committee on Immunization Practices. Morbid Mortal Week Rep 58:1-24, 2009.
81. Vesikari T, Matson DO, Dennehy P, et al. Safety and efficacy of a pentavalent human-bovine (WC3) reassortant rotavirus vaccine. N Engl J Med 2006;354:23-33.
82. Centers for Disease Control and Prevention. Postmarketing monitoring of intussusception after RotaTeq vaccination—United States, February 1, 2006-February 15, 2007. MMWR 56: 218-222, 2007.
83. Centers for Disease Control and Prevention. Delayed onset and diminished magnitude of rotavirus activity—United States, November 2007-May 2008. MMWR 57:1-2, 2008.
84. Gouveia, M. Personal communication.
85. Ruiz-Palacios GM, Perez-Schael I, Veláquez FR, et al. Safety and efficacy of an attenuated vaccine against severe rotavirus gastroenteritis. New Eng J Med 354 (1):11-21, 2006.
86. Vesikari T, Prymula R, Schuster V, et al. Efficacy of human rotavirus vaccine against rotavirus gastroenteritis during the first 2 years of life in European infants: randomised, double-blind controlled study. Lancet 370: 1757-1763, 2007.

Veterinary Vaccines in the Development of Vaccination and Vaccinology

Philippe Desmettre

Introduction

In 1796, Jenner, an English physician, performed the very first human immunization using an animal virus. Since he had noticed that dairy farmers infected with cowpox had become resistant to smallpox, he had the idea of inoculating the cowpox (or vaccinia) virus to protect humans against smallpox [1].

As a tribute to Jenner, Pasteur decided in 1881 to extend the meaning of the word vaccine to preventive inoculation with any type of infectious agent [2].

Virus Vaccines

Pasteur's main studies on the etiology and prevention of infectious diseases were devoted to veterinary medicine.

Fowl Cholera Vaccine

Using fowl cholera bacillus, he demonstrated that when the bacillus was passed several times from one culture tube to another, it kept its whole virulence and always killed chickens. However, when the culture was left at room temperature without any passage ("aged culture"), the virulence of the bacillus decreased. When the chickens were inoculated with what Pasteur then called an "attenuated virus," they all became sick but none of them died. Moreover, when they were subsequently challenged with highly infectious virus, they showed resistance [3]. On this

P. Desmettre (✉)
Rhône Mérieux, 254 rue Marcel-Mérieux, 69007 Lyon, France

S.A. Plotkin (ed.), *History of Vaccine Development*,
DOI 10.1007/978-1-4419-1339-5_30, © Springer Science+Business Media, LLC 2011

basis, Pasteur established the principle of "virus vaccines" and stated that in order to create a refractory state, the live virus, of partially attenuated virulence, should be capable of inducing a mild disease and causing fever, just as Jenner's vaccine gave rise to a pustule and caused fever [2].

Anthrax Vaccine

After studying fowl cholera, Pasteur and his co-workers focused their attention on the prevention of anthrax. Because the "ageing" method in the presence of oxygen was found to be inefficient, Pasteur and his co-workers turned to treatment by heat. They had to produce a nonspore-forming culture of anthrax bacillus grown at 42°C in order to obtain the pure mycelial form that could be attenuated due to its sensitivity to oxygen. Once attenuation was achieved, the spores generated from attenuated mycelial forms produced bacilli of the same level of attenuation.

Pasteur's work on anthrax culminated in the experiment carried out at Pouilly-le-Fort in 1881, which greatly contributed to establishing the reputation of his research work [4]. It should, however, be mentioned that since Pasteur's co-workers were most likely concerned with the level of attenuation of the bacterium and anxious to achieve a higher level of safety they surreptitiously added potassium dichromate to the vaccine at a concentration of 1/2,000, about 1 week before use [5].

Swine Erysipelas Vaccine

The knowledge gained in the field of "virus vaccine" then enabled Pasteur and his colleagues to successfully devise an immunization method against swine erysipelas [6].

Rabies Vaccine

Finally, prevention of rabies benefited from the experience gained by Pasteur et al. At that time, in Europe, rabies was mainly street rabies transmitted by dogs. Pasteur stated that all that would be needed to protect mankind against this dreadful scourge would be to develop an appropriate method for controlling rabies in dogs [7].

Pasteur failed to apply the methods he had implemented for the prevention of fowl cholera, anthrax and swine erysipelas to the causative agent of rabies. He conformed to the observation made by Galtier [8], who had suggested using rabbit as an experimental model and established the main characteristics of rabies in dogs and its transmission to rabbits. In addition, Galtier had reported that an intravenous injection of rabies virus in sheep was safe and induced immunity [9]. The process described, although efficient, was difficult to use and possibly dangerous.

Indeed the credit for discovery of antirabies vaccination actually went to Pasteur. He and his colleagues reported results about the preparation of a virus in which virulence had been stabilized by repeated passages in the same animal species (fixed virus). They also reported attenuation of virus virulence by multiple passages in animal species other than dogs. Such processes made it possible to select a range of viruses of variable virulence. Successive inoculations with these viruses, starting with the most highly attenuated one and ending with the most virulent ones, provided dogs with protective immunity and resistance to inoculation of street rabies virus [10].

For the development of an easy, flexible, and reliable method for producing virus of variable virulence, Pasteur and his colleagues discovered the advantages of ageing of spinal cords from rabbits infected with a fixed virus. When stored at 23°C in the presence of potassium, the cords were found to gradually lose virulence, until they no longer caused rabies. Strictly speaking, there was no real attenuation of virulence, but rather a progressive decrease in the quantity of infectious virus [11].

Although the possibility of protecting dogs against virulent challenge, demonstrated by Pasteur as early as 1882, represented a milestone in the development of an antirabies vaccine, the major achievement was the successful postexposure treatment administered to the young Joseph Meister in July 1885 [12].

The Basis of Virus Vaccines

Pasteur performed all of his work on vaccines without knowing the mechanisms involved in the protection of vaccinated animals. For Pasteur, the "virus vaccine" concept involved a mild disease for producing immunity and was based on the idea that resistance was due to the depletion of an element that was crucial for the proliferation of microbes in the organism; this depletion having necessarily to be induced by a mild disease.

In contrast, Chauveau, a teacher at the *École Nationale Vétérinaire de Lyon* (Lyon's National Veterinary School) suggested that "something" appeared in the organism and stopped microbes from proliferating [13]. This hypothesis was then supported by the discovery of antibodies by Behring and Kitasato in 1890 [14].

"Inactivated" Vaccines

As early as 1886, Salmon and Smith prepared a hog cholera vaccine from an inactivated microorganism, thereby demonstrating that the "virus vaccine" concept involving the development of a mild disease was not true [15]. This was then confirmed by the work of Roux and Chamberland who established the possibility of vaccinating in the absence of microorganism with a soluble substance present in the supernatant of *Clostridium septicum* culture. *C. septicum* was subsequently found to be its toxin [16].

The way for inactivated vaccines was opening up and Nocard, a veterinarian, demonstrated in 1892 that it was possible to immunize animals (horses and cows) against tetanus and that substances neutralizing the toxin were found in the serum and milk of vaccinated animals [17]. Inactivated vaccines against infections caused by anaerobic bacteria, in particular vaccines against gangrene, were thus the first inactivated vaccines to be developed in veterinary medicine, as illustrated by the vaccine developed by Leclainche and Vallée in 1923 against blackleg in cattle, a disease caused by *Clostridium chauvoei* [18]. Shortly after, Ramon, another veterinarian working at the Pasteur Institute, made a number of discoveries of major importance. They included (1) the possibility of detoxifying toxins through the combined actions of heat (37°C) and formaldehyde, while preserving their immunogenicity [19, 20]; (2) the possibility of enhancing the immunogenic properties of such toxoids using irritating substances, leading to the discovery of adjuvants [21]; (3) finally, the possibility of combining antidiphtheria and antitetanus toxoids, resulting in the first combined vaccine intended for humans, which paved the way for the successful use of numerous combined vaccines since then in both humans and veterinary medicine [22].

Despite the growing interest in inactivated vaccines, live attenuated vaccines were not abandoned, however.

Bacillus Calmette–Guérin

In 1912, a physician, Calmette, and a veterinarian, Guérin, working at the Pasteur Institute, protected cattle by inoculation with a bovine tuberculosis bacillus attenuated through passages on bile–glycerol medium [23]. As a result of the preliminary studies [24], BCG was developed in 1921 for controlling human tuberculosis and still remains a reference for vaccinal prevention of the disease.

Brucellosis Vaccines

In 1923, Buck isolated a virulent strain of *Brucella abortus* from cattle infected with brucellosis [25]. When left in the laboratory at room temperature for 1 year, this strain became naturally avirulent and stable. Known as B19, it was used to prevent cattle abortion due to brucellosis and is still widely used. Similarly, in 1957, Elberg and Faunce isolated a streptomycin-independent reverse mutant from a streptomycin-dependent strain of *Brucella melitensis*. This strain, called Rev-1, is also used today for the vaccination of sheep against brucellosis [26]. In addition, the need to differentiate vaccinated from infected animals led in 1955 to the development by McEwan of a nonagglutinogenic rough *B. abortus* strain, the 45/20 strain, which resulted in the first veterinary marker vaccine [27].

In Europe, brucellosis served as an example of successful eradication of an animal disease of major economical importance. Systematic vaccination, first used to limit infection and reduce economical losses, was then followed by sanitary measures based on slaughter of infected animals.

Foot-and-Mouth Disease Vaccines

Among the examples of successful eradication of a disease under vaccinal coverage, special emphasis should be given to foot-and-mouth disease, which also marked a major milestone in the development of viral vaccines. Early attempts at developing modified virus vaccines failed. In 1925, the first inactivated vaccine, prepared from ground aphtae and treated with formaldehyde, was developed by Vallée, Carré and Rinjard [28]. In 1939, Schmidt suggested adsorbing the virus onto aluminum hydroxide in order to inactivate it [29]. This inactivation method, although not satisfactory, made it possible for Waldmann et al. to develop in 1941 an inactivated, adsorbed, formolized and heated vaccine prepared from cattle aphtae [30].

This approach, although relatively satisfactory, did not make it possible to produce vaccines in sufficient quantities and reasonable cost. It was Frenkel who developed in 1947 a method for growing foot-and-mouth disease virus in cattle lingual epitheliums [31, 32]. This method is still in use, although virus culture in BHK 21 suspended cells is increasingly used [33].

As was the case for all vaccines mentioned here, the initial foot-and-mouth disease vaccine was gradually improved, in particular by the use of saponin as adjuvant [34], then by oil adjuvants as well as new concentration and purification techniques. Foot-and-mouth disease virus was the first example of industrial production of a viral vaccine.

Combined Vaccine

Due to the increasing number of vaccines required and the complexity of vaccination programs, it appeared essential, both for technical and economical reasons, to develop combined vaccines. The most striking example was the launch, as early as 1972, of a pentavalent combined vaccine protecting dogs against five diseases. This vaccine, which included modified live vaccines against canine distemper (paramyxovirus), Rubarth's hepatitis (adenovirus), and inactivated vaccines against rabies and canine leptospirosis caused by *Leptospiraicter-haemorrhagiae and Leptospira canicola*, was based on freeze-dried live vaccines dissolved in liquid inactivated vaccines at the time of use [35, 36]. Further improvements of this vaccine led to the development of a hexavalent vaccine for dogs, including live modified canine parvovirus vaccine [37]. Such vaccines also served as an example for the development of numerous other combined vaccines for different animal species.

"Subunit" Vaccines

Since veterinary vaccines, just as vaccines for human, must be both safe and efficacious, it makes sense to retain only the protective immunogenic fractions of microorganisms or viruses thereby eliminating the fractions not necessary for protection but are able to cause adverse reactions. Such an approach was adopted in 1981 to develop subunit vaccine against feline rhinotracheitis. The envelope glycoproteins of the feline herpes virus were separated from the viral capsid using polyoxyethylene alcohol as a dissociating agent, then purified, concentrated, and formulated with an oil adjuvant [38]. This resulted in an improved safety without a significant loss in efficacy. Furthermore, this vaccine could then be produced under the economically viable conditions required for veterinary vaccines by multiplying the virus in suspended cell cultures, followed by chromatographic purification and concentration by ultrafiltration. Since then, a similar process has been successfully used to develop other herpesvirus vaccines, such as equine rhinopneumonitis, infectious bovine rhinotracheitis, and pseudorabies vaccines [39].

Rabies Vaccines

If Pasteur and his colleagues demonstrated the possibility of immunizing dogs against rabies, it was only in the 1920s that domestic animal rabies vaccines started to be widely used. The first such vaccines – the so-called Fermi and Semple type inactivated virus – were prepared from brains taken from animals infected with a strain of fixed virus and having shown rabies symptoms. The vaccine consisted of ground and diluted rabid brain tissue suspension with 0.5% phenol [40, 41]. Later, in order to try to eliminate adverse reactions, inactivated vaccines based on brains of newborn animals [42] or duck embryos [43] were used. Live attenuated vaccines using multiple passages of egg-adapted Flury fixed virus strain (the LEP and HEP strains) were also used [44–46], followed by the ERA strain and derivatives [47].

These vaccines were never totally satisfactory, due either to the risk of hypersensitivity linked to the brain material they contained (inactivated vaccines) or to the risk of reversion to virulence (live vaccines). The major improvement came from inactivated vaccines prepared from virus grown in cell cultures together with adjuvants (aluminum hydroxide) [48, 49].

More recently, knowledge gained about glycoprotein, which is the protective immunogenic fraction of rabies virus, made it possible to use it for preparing vaccines [50]. Different expression systems have been successfully employed for producing this glycoprotein by recombinant DNA technology, but only one was successful in leading to the development of a vaccine to be used in practice. Vaccinia virus, a member of the poxvirus family, was used as an expression system [51, 52] of the gene coding for the rabies glycoprotein [53] and resulted in the development of an oral vaccine for immunization of wild animals, which are both reservoirs and vectors of the rabies virus [54].

Present and Future of Veterinary Vaccines

While the first veterinary vaccines relied on Pasteur's work, and the vaccines prepared from the strains of virus or microorganisms of attenuated virulence are still in widespread use, the number of available vaccines and their features has increased, for epidemiological reasons as well as to meet both technical and economical requirements relating to their use.

Except when intended for companion animals, veterinary medicine is an economical medicine and, as a result, a preventative medicine aimed at preserving the health and performances of animals and aiding the protection of human health by protecting animal health is its ultimate goal.

Requirements

In this respect, vaccination plays a key role and vaccines have to meet a number of requirements, which are increasing as animal selection and performances are developing. Such requirements include (1) complete safety, i.e., excellent local tolerance, no postvaccinal systemic reaction, and no decrease in the performances of the animals; (2) outstanding efficacy, including not only protection against clinical signs of the disease but also prevention of the carrier stage and spread of the infectious agent; (3) the possibility of differentiating vaccinated from infected animals, which is a prerequisite for successful eradication of diseases under vaccinal coverage; (4) a reduction in the number and the use of nonparenteral routes of administration in order to secure the well-being of animals, to prevent a possible decrease in their performances, and to lower costs associated with vaccination.

The Impact of New Technologies

Although vaccines developed using conventional technologies (modified live vaccines and inactivated or submit vaccines) meet, in part, these requirements, recent advances in science and technology pave the way to the development of improved vaccines.

In this regard, recombinant DNA technology offers exciting prospects for the development of live vectored vaccines using viral or bacterial vectors, and subunit vaccines prepared from antigens produced in prokaryotic or eukaryotic expression systems and polynucleotidic (or naked DNA) vaccines.

Recombinant DNA vaccines use only the protective immunogenic fractions of pathogens, making them safer than "conventional" vaccines. When combined with immunomodulators, they are also expected to show better efficacy. They allow differentiation between vaccinated and infected animals, and make it easy to develop combined vaccines through coexpression of immunogens. Their production processes are standardized and simplified. Finally, they also make it possible to investigate new routes of administration.

Although recombinant DNA technology has already led to new generation vaccines, its potential remains to be explored and developed, especially regarding polynucleotidic vaccines. Emergence of other technologies, including peptide synthesis, for the development of synthetic vaccines, should not be overlooked, however.

Vaccines, first intended to prevent either viral or bacterial diseases, are also increasingly used to control parasitic diseases. Similarly, immunological products (may we still call them vaccines?) offer new opportunities in the area of animal reproduction and production by immunomodulating their hormonal functions.

Conclusion

One of the main factors that guided Pasteur's work may be found in his statements that great art consisted of designing critical experiments which left the observer no room for imagination.

The opportunity to perform evaluations in target animal species generated the pioneering work carried out in veterinary medicine, which also profited human medicine. However, veterinary vaccinology has benefitted from progress made in human vaccinology. Let us hope the future vaccines using emerging technologies continue to be developed, without borders between human and veterinary medicines and for their mutual benefit.

References

1. Jenner E. *The Origin of the Vaccine Inoculation.* London: Shury, Din Eds. 1801
2. Pasteur L. Des virus vaccins. *Rev Sci* 1881;8:225–8
3. Pasteur L. De l'atténuation du virus du cholera de poules. *CR Acad Sci* 1880;91:673–80
4. Pasteur L, Chamberland C, Roux E. Le vaccin du charbon. *CR Acad Sci* 1881;92:666–8
5. Chamberland C, Roux E. Sur l'atténuation de la virulence de la bactéridie charbonneuse sous l'influence des substances antiseptiques. *CR Acad Sci* 1883;96:1088–91, 1401–12

6. Pasteur L, Thuillier M. La vaccination du rouget des porcs à l'aide du virus mortel atténué de cette maladie. *CR Acad Sci* 1883;97:1163–9
7. Pasteur L, Chamberland C, Roux E, Thuillier M. Nouveaux faits pour sevir à la connaissance de la rage. *CR Acad Sci* 1882;95:1187–92
8. Galtier PV. Études sur la rage. *CR Acad Sci* 1879;89:444–6
9. Galtier PV. Les injections du virus rabique dans le torrent cicurlatoire ne provoquent pas l'éclosion de la rage et semblent conférer l'immunité. *CR Acad Sci* 1881;93:284–5
10. Pasteur L, Roux E, Chamberland C, Thuillier M. Sur la rage. *CR Acad Sci* 1881;92: 1259–60
11. Pasteur L Chamberland C, Roux E. Sure la rage. *CR Acad Sci* 1884;98:1229–31
12. Pasteur L. Méthode pour prévenir la rage après morsure. *CR Acad Sci* 1885;101:765–74
13. Chauveau JB. Des causes qui peuvent faire varier les résultats de l'inoculation charbonneuse sur le moutons algérians. Influence de la quantité des agents infectants. Application à la théorie de l'immunité. *CR Acad Sci* 1880;90:1526–30
14. Behring E, Kitasato S. Über das Zustandekommen der Diphterieimmunität und der Tetanusimmunität bei Tieren. *Dtsch Med Woch* 1890;16:1113–4
15. Salmon D, Smith T. One method of producing immunity from contagious diseases. *Am J Vet Rev* 1886;10:63–9
16. Roux E, Chamberland C. Immunité contre le septicemia conférée par des substances solubles. *Ann Inst Pasteur* 1887;1:561–74
17. Nocard E. Sur la sérothérpie du tetanus en vétérinaire. *Bull Soc Cent Med Vet* 1894; 12 (ns): 723–9
18. Leclainche E, Vallée H. Sur la vaccination contre le charbon symptomatique pas les toxins. *CR Acad Sci* 1923;176:207–10
19. Ramon G. Sur le pouvoir floculant et sur les propriétés immunisantes d'une toxine diphtérique rendue anatoxique (anatoxine). *CR Acad Sci* 1923;177:1338–40
20. Ramon G. Des anatoxines. *CR Acad Sci* 1924;178:1436–9
21. Ramon G. Sur la production des antitoxins. *CR Acad Sci* 1925;181:157–9
22. Ramon G, Zoeller C. Les "vaccins associés" par union d'une anatoxine et d'un vaccin microbien (T.A.B.) ou par mélange d'anatoxines. *CR Soc Biol.* 1926;94:106–9
23. Calmette A, Guérin C. Sur quelques propriétés du bacilli tuberculeux d'origine bovine cultivé sur bile de bœuf glycérinée. *CR Acad Sci* 1909;150:716–8
24. Calmette A, Guérin C. Nouvelles recherches expérimentales sur la vaccination des bovidés contre la tuberculose et sur le sort du bacilli tuberculeux dans l'organisme de vaccinés. *Ann Inst Pasteur* 1931;27:162–9
25. Buck JM. Studies of vaccination during calfhood to prevent bovine infection abortion. *J Agric Res* 1930;43:667–89
26. Elberg S, Faunce K. Immunization against *Brucella* infection. VI. Immunity conferred on goats by non dependent mutat from a streptomycin-dependent mutant strain of *Brucella melitensis*. *J Bacteriol* 1957;73:211–7
27. McEwen A, Samuel J. *Brucella abortus*: heat stable, protective antigen revealed by adjuvant and present in a "rough" variant strain 45/20: immunization experiments on guinea pigs. *Vet Rec* 1955;67:546–8
28. Vallée H. Carré H, Rinjard P. Sur l'immunisation antiphteuse par le vaccin formolé. *Rev Gen Med Vet* 1926;35:129–34
29. Schmidt S. Die Bedeutung der Adsorption für die active Immunisierung gegen Viruskrankheiten. *Arch Ges Virusforsch* 1939;1:215–36
30. Waldmann O, Pyl G, Hobohrn KO, Mohlmann H. Die Entwicklung des Riemse Adsorbatimpfstoffes gegen Maul- und Klauenseuche und seine Herstellung. *Zentralbl Bakeriol I Orig* 1941;148:1
31. Frenkel HS. La culture du virus de la fièrve aphteuse sur l'épithélium de la langue des bovidés. *Bull OIE* 1947;28:155–62
32. Girard HC, Macowiak C. La culture du virus aphteux au stade industriel. *Rev immunol* 1953; 17:224–38

33. Capstick PB, Telling RC, Chapman WG, Stewart DL. Growth of a cloned strain of hamster kidney cells in suspended cultures and their susceptibility to the virus of foot-and-mouth disease. *Nature* 1962;195:1163–4
34. Espinet RG. Nuevo tipo vacuna antiaftosa a complete glucovirico. *Gac Vet* 1951;13:265–77
35. Michel C, Terré J Soulebot JP, Desmettre P, Chappuis G, Stellmann C. Vaccination associée du chien, comparaison entre vaccinations successives et vaccinations simultanées. *Bull Acad Vet* 1971;43:277–84
36. Chappuis G. Terré J, Precausta P, Mougeot H, Moreau Y, Stellmann C. Vaccination associée du chien. Le vaccin pentavalent. *Revue med Vet* 1973;124:877–97
37. Davoust B, Muller G, Chappuis G. Vaccinations associées du chien: response sérologique à un vaccin hexavalent utilisé en rappel. *Revue Med Vet* 1985;136:363–72
38. Chappuis G, Benoit-Jeanin C, Fargeaud D. Rhinotrachéite feline: vaccin inactive, purifié et modèle expérimental. 17th International Congress on Herpesvirus of Man and Animal. Standardization of Immunological Procedures, Lyon, France. In: *Develop Biol Stand* 1981;52:485–91
39. Chappuis G, Fargeaud D, Brun A. Industrial production and control of a subunit vaccine against Aujeszky's disease. In: Van Oirschot JT, ed. *Vaccination and Control of Aujeeszky's Disease.* Dordrecht: Kluwer Academic Publishers, 1989
40. Fermi C. Über die Immunisierung gegen Wutkrankheit. *Z Hyg Infekt Krankh* 1908;58: 233–76
41. Semple D. The preparation of safe and efficient antirabies vaccine. *Bull Inst Pasteur* 1911;9:701
42. Fuenalida E, Palacios R. Rabies vaccine prepared from brains of infected suckling mice. *Biol Inst Bacteriol* 1955;8:3–10
43. Peck FB, Powel HM, Culbertson CG. Duck embryo rabies vaccine: study of fixed virus vaccine grown in embryonated duck eggs and killed with betapropiolactone. *JAMA* 1956;162:1373–6
44. Koprowski H, Cox H. Studies on chick embryo adapted rabies virus. *J Immun* 1948;60:533–54
45. Koprowski H, Black J. Studies on chick embryo adapted rabies virus. II. Pathogenicity for dogs and use of egg adapted stains for vaccination purposes. *J Immun* 1950;64:185–96
46. Koprowski H. Biological modification of rabies virus as a result of its adaptation to chicks and developing chick embryos. *Bull WHO* 1954;10:709–24
47. Abelseth MK. An attenuated rabies vaccine for domestic animals produced in tissue cultures. *Can Vet J* 1964;5:279–86
48. Fenje P. A rabies vaccine from hamster kidney tissue cultures: preparation and evaluation in animals. *Can J Microbiol* 1960;6:605–9
49. Soulebot JP, Precausta P, Brun A et al. Prophlylaxie de la rage animale (ou vétérinaire) par un vaccin inactive monovalent ou associé. *Dev Biol Stand* 1978;41:389–99
50. Wiktor TJ, György E, Schlumberger HD, Sokol F, Koprowski H. Antigenic properties of rabies virus components. *J Immun* 1973;110:269–76
51. Kieny MP, Lathe R, Drillien R et al. Expression of rabies virus glycoprotein from a recombinant vaccinia virus. *Nature* 1984;312:163–6
52. Wiktor TJ, MacFarlan RI, Reagan KJ et al. Protection from rabies by a vaccinia virus recombinant containing the rabies virus glycoprotein gene. *Proc Natl Acad Sci USA* 1984;81:7194–8
53. Anilionis A, Wunner WH, Curtiss PJ. Structure of the glycoprotein gene in rabies virus. *Nature* 1981;294:275–8
54. Blancou J, Kieny MP, Lathe R et al. Oral vaccination of the fox against rabies using recombinant vaccinia virus. *Nature* 1986;322:373–5

Index

S.A. Plotkin (ed.), *History of Vaccine Development*,
DOI 10.1007/978-1-4419-1339-5, © Springer Science+Business Media, LLC 2011

R

S